The Apostles of Apollo

The Journey of the Bible to the Moon and the Untold Stories of America's Race into Space

C. L. Mersch

iUniverse, Inc.
Bloomington

The Apostles of Apollo
The Journey of the Bible to the Moon and the Untold Stories of America's Race into Space

Copyright © 2009, 2011 C. L. Mersch

All rights reserved. No part of this book may be used or reproduced by any means, graphic, electronic, or mechanical, including photocopying, recording, taping or by any information storage retrieval system without the written permission of the publisher except in the case of brief quotations embodied in critical articles and reviews.

As a non-fiction book, this text incorporates information obtained from public and private sources; personal interviews with relevant persons, their families, and those with first-hand knowledge of events; NASA archives; the Nixon Files; and various government sources under the Freedom of Information Act. Where possible, multiple individuals and organizations were contacted to verify facts; however, due to the passage of time, participants may recount differing versions of events. In all cases, information in this book was obtained and researched from what the author determined to be the most accurate sources available. It is up to the reader to research these topics further and determine if what is written is factual. The author disclaims any liability to any party for any loss, damage, or disruption caused by errors or omissions, whether such errors or omissions result from negligence, accident, or any other cause.

iUniverse books may be ordered through booksellers or by contacting:

iUniverse
1663 Liberty Drive
Bloomington, IN 47403
www.iuniverse.com
1-800-Authors (1-800-288-4677)

Because of the dynamic nature of the Internet, any Web addresses or links contained in this book may have changed since publication and may no longer be valid. The views expressed in this work are solely those of the author and do not necessarily reflect the views of the publisher, and the publisher hereby disclaims any responsibility for them.

ISBN: 978-1-4502-6202-6 (pbk)
ISBN: 978-1-4502-6204-0 (cloth)
ISBN: 978-1-4502-6203-3 (ebk)

Printed in the United States of America

iUniverse rev. date: 3/23/2011

Cover Illustration Copyright © 2010 by Crestock Corporation

"I can share my overall verdict right now. A grade of A. Mersch has written a superb book."

- Steve Weinberg, National Book Reviewer and Author

"Well written and captures the human side of space flight."

- Tom Stafford, Commander, Apollo 10
Apollo Commander, Apollo-Soyuz Test Project

"Mersch has produced an amazing book that is written in thoughtful and well-assembled prose, which holds the reader's interest right to the end."

- Colin Burgess, Award-winning Author and Space Historian
"Fallen Astronauts," "Into That Silent Sea,"
& "In the Shadow of the Moon"

"Truly captures the courage and emotion of those in Mission Control as we struggled to find answers and find them fast."

- Jerry Bostick, NASA Chief Flight Dynamics officer
Project Apollo, 1968-1973

"We never thought that we would not save the [Apollo 13] crew, not once. A well researched book."

- Sy Liebergot, EECOM systems Flight Controller, Project Apollo,
Skylab EGIL systems Flight Controller, 1966-1975

"This fascinating book describes an aspect of Project Apollo that I knew very little about."

- Ed Hengeveld, Space Artist and Apollo Historian

In Appreciation

This book would not have been possible without the encouragement and editorial contributions of Apollo 14 astronaut Edgar Mitchell and NASA industrial chaplain Reverend John M. Stout—whose combined efforts succeeded in landing the first Bible on the moon.

"We never stop finding out who we are."

—*Reverend John Stout*
Director, The Apollo Prayer League

Contents

	Introduction	xiii
1.	The "Original 7"	1
2.	First Up	8
3.	A Walk in the Ether	17
4.	The Good Reverend Stout	23
5.	Apollo 1: A Mission in Flames	30
6.	The Aftermath	37
7.	The Apollo Prayer League	44
8.	The Inquest	50
9.	Apollo 4: Up from the Ashes	61
10.	Apollo 7: Resurrection of Apollo	66
11.	Apollo 8: Leap in Faith	72
12.	Shooting the Moon	76
13.	Christmas Eve	82
14.	Apollo 9: Rendezvous in Space	94
15.	Apollo 10: Close Encounter	99
16.	Makings of a Moon Walk	109
17.	The Silver Chalice	115
18.	Apollo 11: Destination Moon	122
19.	One Small Step	127
20.	Return to Earth	137
21.	The Rest of the Story	144
22.	An Atheist Voice	149

23.	40,000 Voices	157
24.	Apollo 12: Space Cowboys	166
25.	The "Icy Commander" is Back	178
26.	Apollo 13: A Space Odyssey	189
27.	The Whole World Prayed	202
28.	Splashdown: The Grace of God	210
29.	Apollo 14: Restoring Faith	223
30.	Lunar Landfall	230
31.	Full Circle	238
32.	A Still Small Voice	246
33.	Apollo 15: One Red Bible	250
34.	Apollo 16: Moonwalker Moment	263
35.	Apollo 17: The Long Last Look	274
36.	Moondust	285
37.	End of an Era	291
	Photos	298
	Bibliography	311
	Index	323

"We choose to go to the moon in this decade and do the other things, not because they are easy, but because they are hard."

- President John F. Kennedy
Rice University, September 12, 1962

INTRODUCTION

It began with a challenge when the Soviet Union foisted a 184-pound basketball-sized alloy sphere called Sputnik into an elliptical near-earth orbit on October 4, 1957. It ended when an astronaut by the name of Eugene Cernan took the last step by man on the surface of the moon. Cernan climbed the ladder and closed the hatch to the lunar module of Apollo 17. Then, on December 14, 1972, the spacecraft launched for rendezvous with the command module silently approaching overhead. Not a single human being has been back since.

What so many thought was a prelude to interplanetary space exploration now reads like an epilogue. For several decades since, the world's gaze has been earthbound, fixed on the chaotic world earthlings inhabit and its many chronic problems. Generations have passed since American astronauts left the moon, so much time that some wonder why the country bothered to go there in the first place.

When President John Kennedy astonished the crowd and his own administration with his proclamation at Rice Stadium in 1962 to land a man on the moon, the public was immediately enthralled by the vision. What is not remembered as well is the veiled geopolitical explanation that immediately followed:

> "…because that goal will serve to organize and measure the best of our energies and skills, because that challenge is one that we are willing to accept, one we are unwilling to postpone, and one which we intend to win."

He did not mention the Soviet Union by name, nor did he mention the Cold War. He did, however, present the mission to the moon as a challenge to "win" with a deadline—which made it a race. This was the second leg of the space race, and the prize was the alignment of the hearts and minds of people and governments around the world with the United States—hearts and minds that would be won by an astonishing display of courage and technological prowess. Were it not for this rivalry between competing political and economic systems, we would surely have stayed home.

But this was the Cold War and the high stakes space race was its high-minded and optimistic manifestation. A brutal proxy war in Southeast Asia was another. As the Vietnam War dragged on with no end in sight, the race to the moon became America's shining hope in the larger geopolitical contest. In the end, America won the race to the moon and abandoned the war in Southeast Asia. The Soviet Union did exactly the opposite. After falling hopelessly behind in the race to the moon, the Soviets simply gave up on manned lunar missions, making the United States the first and last to voyage there.

But in the face of the U.S. "victory" of putting a man on the moon, the American public lost interest in heroic space exploration. As enamored as President Nixon had become with the moon missions, his administration was simply running out of money. The era of epic manned space exploration was over.

As a result, the Apollo missions now occupy a peculiar chapter in the human story. They remain an undiminished achievement towering above all others. To the United States in particular they are the subject of intense national pride and nostalgia. Yet plans for future missions were curiously abandoned and never revived.

This dichotomy is especially apparent in the post-mission lives of the Apollo astronauts themselves. Even today they are enthusiastically received when introduced at public events. People clamor to have pictures taken with them; their autographs are prized and costly. Meeting one of the men who walked on the moon is nothing less than a mystical experience for many because there have been only twelve. As of this writing, only nine are alive. Yet they move anonymously through our world, recognized by only a few when entering a restaurant or boarding an airplane.

This is not how it was for these men some forty years ago. Back then they were the center of the universe. Each had volunteered for a televised mission that was staggeringly dangerous, and in doing so were seen as the most romantic of warriors in the Cold War effort. By the time the Apollo missions reached fruition, the astronauts were seen to be doing something very different. They were altering and elevating human consciousness, and doing it in a way that commanded the attention of the entire world. As they rocketed away, television cameras allowed their fellow earthlings an extended look at themselves and their home planet, a view that no other generation of humans had ever beheld.

As the first Apollo spacecraft entered lunar orbit on Christmas Eve, the world witnessed an "Earthrise" over a pale lunar horizon, a sight that electrified all of humanity. Within our field of view, against the infinite velvety darkness of space, was the composite setting of all human events both past and present. Looking at the world in its totality, we could see the place where the wheel was invented, the place where Christ lived and died, the place where Shakespeare dreamt of Hamlet, the place where the atom was first split, the place where every human being who had ever lived was born. It was a new and mind-bending sight, and it came at the dusk of a traumatic and mind-bending decade.

That moment in time took the world aback. Apollo was no longer the greatest vicarious adventure in history. Suddenly it had become a spiritual odyssey, with a vast population on the planet participating in the experience by live television. Seven months later, on July 20, 1969, a monochrome camera captured the ghostly image of a man walking on the moon and the American flag being planted on the lunar surface. That a particular nation had won the space race was at that moment strangely irrelevant; what mattered to people around the world was that a human being was standing on the moon. In that timeless moment, something larger than geopolitics was at work. What began as a race was ending as a worldwide communion.

We now know that this singular, magical spell would prove surprisingly ephemeral. Within a few months after the first mission to the lunar surface, a consensus began to build in the country that would change everything. The race to land a man on the moon, so the American public's thinking went, had been an exceedingly worthwhile adventure. But the race had been won. It was now time to do in space what we were not prepared to do in Vietnam—declare victory and go home.

However, there was another story to the Apollo missions, one that went largely unreported at the time. It had little to do with geopolitics, technological prowess, or national bragging rights. The story represented what the missions came to mean to many of the men who went into space and to the moon.

These explorers were not only men of science, many were also men of faith. Some would take a moment to pray within the strict confines of their spacecraft, while others brought articles of their belief aboard.

The story is as unique as it is unknown. Overlooked by many historians is the spiritual dimension of the men America sent into space. They were a spiritually diverse cadre and their journeys had a profound effect on how they understood themselves, the world, and their place in the cosmos.

Expressing their faith in space, however, put many of the astronauts and the government agency they worked for in something of a dilemma. The result was an extraordinary drama that unfolded behind the scenes as America rushed headlong to the moon.

From the vast ranks of NASA personnel emerged a chaplain and scientist named John Maxwell Stout with a vision of an organization dedicated to prayerful support of the astronauts and the success of each mission. The organization he founded came to be known as the Apollo Prayer League, and in time he and its membership resolved to ensure that a Bible made landfall on the moon.

This idea did not come out of thin air. The event that caused many within NASA, and the American public in general, to question the true essence of the lunar missions was the fire that took the lives of Gus Grissom, Ed White II, and Roger Chaffee on January 27, 1967, while docked on a launch pad at Cape Kennedy. The Apollo 1 tragedy was a defining moment, not only for the overall Apollo project management, but for the astronauts who would travel to the moon in their wake, and for the American public who underwrote the project.

The deaths of these three astronauts marked the moment that the American people would question the entire purpose of the space program. Up until that point, the price had been paid only in terms of monetary treasure. But at 6:31 p.m. on January 27 on launch pad 34, it was paid in the sacrifice of precious blood.

Before his tragic death, Ed White had said he planned to take a Bible to the moon. The fire that took the young astronaut's life had a profound effect on Reverend Stout. From that moment, one thing became clear to the reverend: we were not leaving earth merely to return with a cargo of knowledge. As mankind reached into the heavens they would be sure to take something with them—something that spoke of the eternal bond between mankind and its Creator. Although Stout had planned to resign his position in the Apollo program at that time, he resolved to stay on and see Ed White's dream realized.

The efforts of the Apollo Prayer League culminated in dramatic fashion four years later when, on February 5, 1971, the spindly legs of Apollo 14 lunar module *Antares* touched down on the powdery surface of the moon carrying the first lunar Bible. By the time the Apollo program ended in 1972, the Apollo Prayer League had grown to over 50,000 members in many foreign countries and remote outposts around the world.

From the retrospective of several decades, it is clear that the deaths of Grissom, White, and Chaffee changed everything, not only NASA protocols and procedures, but the Apollo spacecraft itself. More importantly, the tragedy caused everyone—from congressmen, to NASA flight technicians and contractors, to everyday American citizens—to ponder why we were reaching to the heavens in the first place. Were we going to the moon just to learn what it was made of? Were we going there to plant the stars and stripes on its surface before the Soviets planted the hammer and sickle? Or were we traveling there for altogether different reasons—reasons that, as one astronaut pointed out, nourished the human spirit.

Today Apollo remains as the pinnacle of American space exploration and a testament to the human spirit at a time when the nation was badly in need of hope. America needed heroes and Apollo provided them. The unbridled success of Apollo brought the Russians to their knees and Americans to their feet.

But the spiritual legacy of faith began well before the Apollo program. It began even before President Kennedy's bold proclamation in 1962. The call for the nation's first space heroes came from an unassuming infant space organization that possessed no organized plan, no cohesive infrastructure, and no viable technological platform from which to launch such an audacious endeavor.

It began in 1959 with seven good men.

"In the glory years of space, what the country kept forgetting was that we were people. Each of us was, in fact, four people: adventurer, social lion, would-be business tycoon, and political object. The one thing we were not was heroes."

- *Walter Cunningham, Lunar Module Pilot, Apollo 7*

1
THE "ORIGINAL 7"

There was no doubt that Alan Shepard was unhappy.

"I didn't sign up to be a freaking specimen or test subject!"

Shepard's voice reverberated throughout the halls of the astronaut offices in Hangar S at Cape Canaveral, Florida. Shepard didn't mind the cramped working conditions in the astronaut living quarters. He didn't mind the long hours or the cafeteria food. What he did mind were the *chimpanzees*.

NASA solicited military test pilots to become America's first astronauts because test pilots were trained to think quickly in dangerous situations. What they acquired in the first round of the astronaut selections were seven such pilots, along with six feisty chimps. On April 7, 1959, out of nearly 5,000 applicants, the space agency announced the names of the "Original 7"—America's first Mercury astronauts. Alan Shepard was one. Others were Scott Carpenter, Gordon Cooper, John Glenn, Walter Schirra and Deke Slayton. The seventh man was Gus Grissom.

As a member of the first group of astronauts, Shepard became an icon overnight, and having monkeys throw food at him as he passed their cages was not his idea of proper respect for a new American space hero. The monkeys didn't like the quarters any better than Shepard did, and the encounters escalated into something akin to a turf war. A chimpanzee named Ham was especially adept at hitting his mark with a discarded apple core—or fecal matter if his cage hadn't been cleaned recently.

For two years NASA had been training chimps and monkeys as part of their test pilot program. The plan was to launch "monkeynauts" in the early Redstone rockets before strapping a human in the seat. The chimps were temporarily housed adjacent to the astronaut quarters in a two-story cinderblock building known as Hangar S, subjecting the astronauts to raucous howling and screeching each time they walked down the hall. The astronaut crew quarters were smelly, military, uncomfortable, and too close to the chimpanzee colony. The chimps' reprehensible behavior finally took its toll on Shepard's nerves, and he demanded that NASA move the astronauts to another location. NASA reluctantly obliged.

The seven astronauts were given permission to stay in nearby Cocoa Beach at the Starlight Inn. The motel, run by a man named Henri Landwith, proved to be a story all its own. The foul odors and screaming chimps were immediately replaced with good times and a string of attractive women. Months later when new management assumed control of the Starlight, Landwith moved down the beach to manage a new Holiday Inn and contacted NASA to ask if the astronauts could switch to his new motel. A NASA representative called the next day to say they would accept the offer, but only if the astronauts could be guaranteed a room whenever they were in town. Landwith agreed and even gave them a ridiculously reduced rate of $8 a night. Coincidentally, on the same day the new Holiday Inn opened, the Starlight burst into flames and burned to the ground.

The astronauts had expected to train hard for space, but the storm of publicity came as a surprise. The new space inductees were hailed by the media as red-white-and-blue heroes, fighting for American supremacy in the new frontier of space. The press descended upon the astronauts and their families mercilessly, forcing NASA public affairs officer Walter Bonney to set up an arrangement so the astronauts wouldn't be "pecked to death by ducks" by offering exclusive coverage to the highest media bidder.

Life magazine, a widely-read direct connection to the heartland of America, was the high bidder, offering to buy the astronauts' exclusive stories for a half-million dollars spread over three years, at the time a tidy sum even when split seven ways. The agreement was inked on August 5, 1959, and the September 14 issue of *Life* featured an 18-page cover story lionizing the Mercury Seven. The astronauts were immediately thrust into the public spotlight. Their images continuously embellished the covers of *Life* magazine. The press followed them everywhere, from local restaurants to the sidewalks of their home. Together, they were about to soar irretrievably away from their military peers to create an exclusive seven-man fraternity—and a whole new brand of Cold War celebrity.

NASA groomed the astronauts, told them to wear socks that covered their legs up to the knee so as not to show their hairy legs when they were seated. They learned to stand with their hands in their pockets, thumbs pointed to the rear like fighter jocks. Their attaché cases were to be carried low at their sides, never hugged to the chest.

Public Affairs Officer Paul Haney oversaw public relations at the Manned Spacecraft Center and understood all too well that the basic nature of a jet jockey was not necessarily one of discretion. So at the first scheduled press conference in Washington, D.C., Haney apprised the seven astronauts of the magnitude of the press coverage and the need for diplomacy. But when a reporter asked John Glenn what it was that surprised him most about his indoctrination into the space program, he responded, "They checked orifices I didn't even know I had." Such candid expressions were a shade of things to come for Haney and NASA.

But even as the seven settled into the program, history was unfolding on the other side of the globe. The Soviet Union was operating in stealth mode and rapidly gaining ground in the space race.

Friday, October 4, 1957, is a date not well remembered by most Americans. Newspapers on that day announced that the Milwaukee Braves trounced the New York Yankees 13-5 in the second game of the World Series, the new 1958 Chevrolet Bel-Airs debuted at $2,195, and Sears advertised Jamboree underwear on sale for a penny. *Leave it to Beaver* was premiering on American televisions. But on this otherwise ordinary October day, Russia made a flanking attack in the new and uncharted frontier of space.

The launch of Sputnik, the world's first artificial satellite, put into orbit a visible, blinking reminder of the Soviet Union's domination of space. Chirping in the key of A-flat from outer space, which the press called "deep beep-beep," Sputnik zoomed over America's horizon. The chirp lasted three-tenths of a second, followed by a three-tenths-of-a-second pause. This was repeated over and over again until it passed out of hearing range of the United States—a recurrent painful reminder that the Soviets had taken the lead. The chirp, emitted by a one-watt battery-operated transmitter, could be easily detected by a short-wave radio. Not only could you hear Sputnik, you could see it with the naked eye.

The satellite was silver in color, about the size of a beach ball, and weighed a mere 184 pounds. Yet for all its simplicity, small size, and inability to more than orbit the earth and transmit meaningless radio blips, the impact of Sputnik on the United States was enormous and unprecedented. It was a stunning technical achievement that caught Americans off-guard. In a single weekend, the nation was wrenched out of a mood of social comfort and postwar lethargy.

One month later, the Soviets struck again with Sputnik 2, promptly renamed "Muttnik" by the U.S. press in reference to its canine passenger, a mongrel dog named Laika. Many feared that Russia might already have an operational intercontinental ballistic missile. If the Russians could thrust a dog into orbit, they could certainly do the same with a nuclear warhead. Next on the Russian's space timetable, U.S. officials speculated, would be a Soviet rocket landing on the moon.

"The United States," said U.S. Admiral Felix Stump, "simply cannot afford to *lose* any more."

The advent of Sputnik set teeth to grinding in Washington and captured public imagination throughout the world. Though no one knew at the time, it ushered in another player who would become a key figure behind the scenes of the United States' race into space. That was Professor John Maxwell Stout, a young Texan who was in the right place at the right time—the Brazilian mountains of Minas Gerais in October of 1957.

At the time, John and his wife, Helen, held teaching positions at the University of Lavras in Brazil. Childhood sweethearts, they had walked away from undergraduate studies and gotten married as soon as they learned John would be shipping out for service in World War II. His first taste of missionary work came while serving as an executive officer in an artillery battalion on the Japanese island of Hokkaido. It was there that he saw the unit's chaplain fall wounded. Asked to fill in for the chaplain until a replacement came, he willingly accepted, as his conservative Baptist upbringing in the rural town of Handley, Texas, afforded him a good footing for the call. The experience would have a marked affect on his life.

When a serious war injury while handling an explosive device sent him home with a dismal prognosis of only ten years of active life, Stout was offered retirement as a full colonel if he would serve only three more years of his ten remaining. Stout declined.

"If I only have ten years to live," he said, "I would rather spend it in service to God."

He and Helen returned to the U.S. to complete their degrees at Texas A&M and shipped out together again, this time for a missionary life of academia in the jungles of South America. John accepted a position at the University of Lavras in Brazil where he chaired the Analytical Chemistry and Engineering Design Department and eventually earned a Doctorate in Informologia (information communication) at the University of Pelotas in southern Brazil.

But John's military service had left him with a passion for humanitarian work, and when a yellow fever epidemic broke out at a nearby mission, he and

Helen stepped in to help. Their missionary outreach expanded and before long they were operating four primary schools, a hospital, a clinic, three churches, and an Indian orphanage.

Although his war injury left him with intermittent enduring pain in his leg, he was determined to defy the prognosis. Seeing the desperation around him, he began forays into the great forest of Brazil to provide humanitarian aid to the indigenous tribes. The experience intensified his long-held feeling that he had been called by God to be a minister. So during their first furlough home to Texas, he and Helen completed seminary work at the Austin Presbyterian Theological Seminary, and in 1957 John became an ordained minister. It was an interesting choice. He now had one foot firmly set in scientific academia and the other firmly set as a man of faith. During the furlough, the two stopped at a restaurant in Austin where a chance meeting with George Reedy, public affairs manager for then-Senator Lyndon Johnson, led to an introduction to the senator and his wife at a local church service. Johnson's duties at the time included chairmanship of the Senate Space Committee, and the dialog eventually turned to a discussion of Brazil's ambitious space program and its unique capability for satellite observation, a subject John was intensely interested in.

"Let me put you in touch with the American Society of Professional Photographers," Johnson offered. "They're going to cover the tracking of satellites."

The purpose of the group was to establish surveillance for potential Russian space activity, and Senator Johnson made arrangements for John to receive their orbital mechanics data for tracking objects in space. Armed with his study of celestial mechanics from high school, John could then convert the coordinates to his own azimuth direction in the remote mountains of central Brazil and determine where a satellite would be passing over his missionary location.

In October 1957, unaware of the impending Russian Sputnik launch, John had erected a tripod in the jungle near his home and mounted a home-made camera for use by his students in celestial mechanics at the University of Lavras in photographing Venus. In order to capture moving objects, he devised a series of rotating wooden wheels that would move the film past an Angus lens to automatically track a meteor or satellite through a pre-set arc in the sky for five minutes. It was a technique he had learned in school called "smear" photography. Searching for something to power the wooden wheels, he connected the apparatus to a Westclox alarm clock that would rotate the entire mechanism at an adjustable speed.

A friend of John's from Columbia had loaned him a 30" telescope with a finely ground parabolic lens which he mounted on the side of the camera.

The resulting contraption cost less than $10 and had all the characteristics of a fifth grade science experiment. In truth, it was a structure of sheer genius.

With word of the Russian launch of Sputnik, John hurried through the brush to the tripod and wound the clock connected to the camera. Using the azimuth conversion and coordinates provided him, John and his students had improvised several makeshift "satellite predictors" out of cardboard and celluloid that proved remarkably accurate in class exercises for determining when and where a satellite would appear. These orbital calculations now told Stout that Sputnik's orbit would carry it overhead from the northern horizon of Mato Grasso. It was then that something fortuitous happened. As if to reserve his place in history, the city of Lavras had a complete blackout that night.

And so it was that on the night of October 4, 1957, John Stout stood alone next to his tripod in the dense mountains of Brazil in utter blackness, waiting for the tiny blinking sphere. As Sputnik came into view, the wooden wheels began rolling, and in the ensuing moments Stout snapped one of the first clear photographs of Sputnik ever taken.

His calculations proved more accurate than the Russians' in tracking their own satellite.

"Radio Moscow would sometimes be five minutes off in its predictions of Sputnik times," Stout said. "But our calculations seldom missed by more than 30 seconds."

Local news journalists were incredulous. The *Journal of Brazil*, the largest newspaper in Rio de Janeiro, was downright skeptical and decided to check with the Rio national observatory to confirm if such a thing were even possible. In the meantime, a reporter from a small obscure newspaper in Belo Horizonte, Minas Gerais, found John out, snagged the story, and ran it on October 12 as front-page news. When the director of the *Journal of Brazil* was finally convinced, he called Stout back and offered to buy him his own observatory in exchange for an exclusive on the story. Stout declined. He had already granted permission to the smaller paper. However, a friendship developed between the two, and on October 18 the story hit headlines in Rio de Janeiro newspapers. Eventually, the media mogul offered John control of all his space and religious broadcasting stations on twenty-three networks throughout Brazil. For a man like John, bent toward ministry and celestial science, it was a godsend.

News of John's Sputnik photo circulated rapidly, elevating his name in both Brazilian and U.S. scientific circles; and in 1958, he was asked by the U.S. International Observatories of Satellites to oversee all of Brazil's observatories as a part of a network of satellite tracking stations around the world.

By now John was fluent in the native language of Portuguese and adept at moving on horseback among the oppressive, sometimes hostile, Indian tribes. In spite of advice to the contrary, he never carried a weapon. Although he was captured more than once, his very nature enabled him to be released unharmed.

As he continued his missionary work, he became increasingly fascinated with the emergence of the space race. With the push of an energetic young president, John F. Kennedy, the U.S. was preparing to enter into a new frontier, one that appealed to a man such as John, inclined to scientific endeavors. When John Glenn, the third American astronaut in space, captured the world's attention by successfully orbiting the earth, Stout was inadvertently caught up in the limelight. Local natives, mistaking their own John for Glenn, hoisted him up on their shoulders and carried him around in celebration until he managed to convince them otherwise.

"I had to bathe my ego for awhile," Stout said. "So I didn't stop them too soon."

As an ordained minister and acclaimed satellite authority, Reverend Stout's thinking now began to turn against two strong currents of scientific and religious dogma. As an ardent believer in freedom of thought, he saw both science and religion as mankind's noblest attempts to understand God and His creation. It was a philosophy he put to practice in his personal life as well as in his classroom at the University of Lavras, where he said, "I felt perfectly comfortable with God looking into a test tube alongside my students."

This seeming paradox lay at the very heart of Stout's personal philosophy. He was as well versed in scripture as he was in science, which made him something of an outlier in both communities. Over the years, it was a role he came to relish. And one that would eventually lead him from a small village in Brazil to a position at NASA's Manned Spacecraft Center in Houston—and a place in history he would never have imagined.

But in 1961, John Stout was still in the jungles of Brazil. And the space race in America was only beginning to awaken in the heartbeat of Project Mercury. A handful of astronauts called the Original 7 were the personification of that awakening.

> "I looked at those toggle switches I had to turn on cue. I looked at the dials I had to turn on cue, and I thought to myself, *'My God, just think, this thing was built by the lowest bidder.'* "
>
> *- Alan Shepard, Freedom 7*
> *In a post-flight interview with CBS's Walter Cronkite*

2
First Up

In January 1961, Robert Gilruth, head of NASA's Space Task Group, called the Original 7 astronauts together and announced the verdict they'd all been waiting to hear—the name of the man to take America's first space ride. It came as no surprise to the other Mercury astronauts that Shepard's name emerged at the top of the list. They were the ones who put it there. When the seven were asked in a silent peer-vote, "If you yourself cannot be the one to make the space flight, who do you think it should be?" Shepard edged out Gus Grissom as the first man to fly the Redstone rocket. Grissom would be next. John Glenn would be third.

A hard-nosed Navy test pilot, nicknamed by his peers the "Icy Commander," Shepard was no slouch when it came to leveraging a situation to his own advantage, and he could be merciless in a confrontation. Shepard was considered to be one of the more competent astronauts in the program, and in 1961 he was hell-bent on being the first man to fly in space. When he came home with the news that he had been selected to make the first Project Mercury flight, he told his wife, Louise, "Lady, you can't tell anyone, but you have your arms around the man who'll be first in space!"

"Who let a Russian in here?" she teased.

It was a better joke than she knew.

CIA reports indicated the Soviets were gearing up for another launch. Although NASA had originally planned to launch Shepard's Redstone rocket in late 1960, there were concerns about the safety of the rocket. Twenty-one weeks of delays and unplanned preparation pushed the launch into January 1961. Still uneasy about the spacecraft, a group of NASA flight engineers pressed for a test of the troubled rocket using a chimpanzee as the pilot rather than run the risk of visibly killing America's first man in space. So as a preliminary test, Shepard's dreaded chimp colleague, Ham, was launched on a 435-mile flight over the Atlantic Ocean in the same type of Redstone rocket Shepard was to fly.

Just about everything that could go wrong with a billion dollars' worth of science went wrong—from a leaking capsule to an overshot landing. Temperatures inside the capsule soared to unbearable heights and the electrical system, which was designed to reward the chimp with banana pellets for pulling the correct levers, malfunctioned and delivered electrical shocks instead. But Ham held true to his training and performed the prescribed functions in spite of the miscues. The capsule finally splashed down 422 miles off course and overturned in choppy seas. By the time rescuers arrived thirty minutes later, the capsule had taken on 800 gallons of seawater and was almost completely swamped. When they finally got Ham out of the capsule, he was smiling as if he'd just stepped out of a hot tub.

But Alan Shepard wasn't smiling. Not only was Ham's flight not on the money—it was a disaster. And NASA's chief rocket scientist, Wernher von Braun, was worried. The chimp's troubled flight meant another postponement of Shepard's flight, now scheduled for March. NASA didn't want Shepard's flight to be similarly marred, and they decided to conduct one more unmanned test launch. Despite the life-threatening danger the chimp had withstood—danger Shepard now faced himself—he was furious at the delays. To make matters even worse, politics intervened. A Russian scientist from Massachusetts Institute of Technology (MIT) selected by Kennedy as a technical advisor, warned Kennedy of the damage a dead astronaut could do to his young administration and advised that NASA instead send more chimps into space. At this point, Shepard told John Glenn he was ready to have a "chimp barbecue."

Then, April 12, 1961, less than a month before Shepard's rescheduled liftoff, the Soviets shocked the world once more when 27-year-old Yuri Gagarin became the first human being in space. News that a Russian had beat him into space because NASA had launched a monkey ahead of him sent Shepard ballistic.

"We had 'em by the short hairs," he said. "That extra test flight probably cost me becoming the first man in space."

But NASA had its reasons. One look at the data from Ham's flight revealed that the agency had made the correct decision. Dick Koos, NASA's flight simulation director, had witnessed a series of failures leading up to Shepard's launch. A test rocket carrying a dummy astronaut into space flew off course and had to be destroyed. Three days later, a smaller rocket with a

Mercury capsule on it went off course. It too had to be blown up. Rockets were dancing up and down on the launch pads. "That was a week of 18- to 22-hour days," Koos said, "and we were reconfiguring for Alan Shepard's flight just five days later." As America rushed to match the Russians, the efforts became a public relations disaster.

One day, under secret orders, a group of test pilots were directed to go to Washington, D.C. for a briefing. Apollo 13 astronaut Jim Lovell was in the group.

"They talked to us about the Atlas booster and putting a capsule on top of that with a man in it to try to put a man in space," Lovell said. "And of course Atlas boosters were blowing up every other day down at Cape Canaveral… and it looked like, you know, a very quick way to have a short career."

Getting into space was proving to be a more technically complex and risky enterprise than anyone had foreseen. It was also more frustrating. One day, after a particularly long trial and more of Shepard's incessant complaints, Pad Leader Guenter Wendt quipped, "If you don't like it, I've got someone else who will do it for bananas." Shepard threw an ash tray at him.

Finally, on May 2, just three days before the launch date, NASA announced to the public the name of the astronaut who would man the rocket. All systems were Go for Alan Shepard on *Freedom 7*. NASA had previously refrained from announcing who would be piloting America's first spaceflight in order to avoid media distraction, and Shepard told only his wife and family. Louise Shepard, a devout Christian Scientist, paid prayerful attention to her husband's aspirations in the risky business of space travel; and Alan knew he had the spiritual backing at home to accomplish anything he rightly desired. And being first in space was the shining achievement he cherished.

At 9:15 on the morning of May 5, 1961, three weeks after Yuri Gagarin's feat, all eyes were focused on the Redstone rocket poised on the launch pad at Cape Canaveral. Shepard had readied himself and climbed into the capsule at 7:00 a.m., only to have the countdown repeatedly delayed by flight engineers who were consumed with the fact that there was now a real human being in the capsule. At fifteen minutes before the Go for launch, it was again necessary to halt the countdown. A minor electrical problem resulted in yet another hour-and-twenty-six minute hold, which in turn caused another unexpected problem. Shepard's morning coffee had now worked its way through his system, and no one had anticipated that he might have to relieve himself since it was scheduled to be only a fifteen-minute flight. After another long delay, Shepard finally had to urinate.

"Gordo!" he said, talking to Gordon Cooper, a fellow Mercury Seven astronaut and prelaunch communicator. "Man, I've got to pee."

"You what?"

"You heard me. I've got to pee. I've been up here forever."

Thinking that he could be stuck up there for hours, he told them he was going to do it in his suit. There were no facilities aboard *Freedom 7*, not so much as an empty canister, and the medics were concerned he would short-circuit the leads.

"Tell 'em to turn the power off!" Alan snapped.

"Okay, Alan," Cooper said with a chuckle. "Power's off. Go to it."

Fortunately, his heavy undergarment and the 100 percent oxygen flowing through the suit quickly dried out the problem. Once the power was restored, the countdown resumed. By this time, a nervous Shepard had been strapped in his seat for over three hours and was growing increasingly impatient with the delays. Then, at two minutes and forty seconds to launch, technicians noticed a fuel pressure gauge was running a bit high, and Shepard was told there might be another delay. Each new problem seemed to send NASA ground engineers into near-paranoid frenzy. Shepard finally ran out of patience.

"All right, I'm cooler than you are," he quipped in his laconic voice. "Why don't you fix your little problem and light this candle."

Perhaps the fuel pressure wasn't so high after all, the technicians agreed, and the countdown resumed. As the rocket rumbled to life, 45 million television and radio listeners heard a voice from Cape Canaveral ground control say, "This is it, Alan Shepard. There's no turning back."

Shepard's pulse rate rose from 80 beats per minute to 126 at the signal for liftoff. At 9:34 a.m. the Mercury-Redstone rocket engines ignited 75,000 pounds of thrust.

In Shepard's home in Virginia Beach, his wife, Louise, sat close to the TV holding a transistor radio in her hand and said she felt "the power of good and of God" fill the room. In St. Louis, Shepard's 13-year-old daughter, Laura, stared intently at the black-and-white television with her classmates at Principia Middle School. As the spacecraft rose from the launch pad, the group broke into the hymn, "Shepherd [Shepard] Show Me How to Go."

Eighty-eight seconds into the flight, Shepard's head and helmet were bouncing up and down so hard he couldn't read his panel dials. He was so tightly strapped into his seat that the first time he realized he was weightless was when a loose washer floated by his helmet. Fifteen minutes later, *Freedom 7* splashed down in the Atlantic Ocean and Alan Shepard became a national hero. From beginning to end, the flight had been near-perfect.

Asked later how he became the first American to fly in space, Shepard replied with a smirk, "They ran out of monkeys."

Although discounted by Soviet Premier Nikita Khrushchev as a mere "flea hop" compared to the recent flight of Yuri Gagarin, Shepard's feat was a major morale booster at a time when the American public was in bad need of one. His suborbital flight might have been an up-and-down affair, not Gagarin's free-wheeling excursion in the ether, but it looked pretty good to a country that had just endured the Gagarin triumph, an escalating "proxy war" in Vietnam, and the Bay of Pigs fiasco. John F. Kennedy's surprise at the American public's excitement over Shepard's flight prompted him to consider an epic challenge.

Maybe the United States should set its sights on the moon.

Gus Grissom was standing in the wings. Both Shepard and Grissom shared a passion for fast cars, and in late 1961 local Florida Chevrolet dealer Jim Rathman, winner of the 1960 Indianapolis 500, convinced GM General Manager, Ed Cole, to lease each of the Original 7 astronauts a high-performance Corvette for $1. Shepard and Grissom jumped on the offer. They ordered nearly identical gray 1962 427/435 horsepower Corvette Roadsters and immediately engaged in a friendly rivalry, constantly souping up their roadsters in an attempt to outdo the other as they sped around the Florida countryside.

No one who knew Gus personally took him lightly. Small but powerful, he was known as a fierce competitor. A former Air Force fighter pilot in the Korean War, he brought a hard-driving determination to his spaceflight career. He'd been obsessed with science and flying since he was six in the small town of Mitchell, Indiana, where he was ingrained with the strong fiber of his family's faith in the Church of Christ. Passionate about the quality of his work and dedicated to perfecting the rockets he flew, it wasn't unusual to find Grissom at the McDonnell Aircraft facility at Cape Canaveral at all hours of the night, poring over design work for the new Mercury-Redstone rocket.

On one occasion, while backing up Shepard for his *Freedom 7* flight, Grissom worked late at the space center, then headed back to the Holiday Inn in the early hours of the morning. Not one to give credence to local speed limits, Grissom hopped into his Corvette and jammed the pedal to the floor.

Exiting the security gate at 70 mph, Grissom gathered more speed, throwing the car into fifth gear and swerving onto a deserted U.S. 1 at 100-plus mph—whereupon he was soon pursued by a highway patrolman. This, he knew, could be more than a mere nuisance. Tearing through an intersection, Gus made a high-speed turn onto the 520 Causeway and raced for Cocoa Beach. At that point, a sheriff's deputy joined the patrol car in the chase, raising the entire Space Coast with piercing sirens. Gus sped through another turn onto A1A where the hot pursuit was then joined by the Cocoa Beach police.

Gus jammed the pedal of his turbo-charged V-8 to the floor again, leaving his pursuers hopelessly behind. He made a wide 70-mph turn into the Holiday Inn, where he found his parking space occupied. The space in front of Shepard's room, however, was open. Gus slid his car between the lines and darted to his room two doors away, shedding his clothes in the dark as the flashing lights and screaming sirens pulled up outside. He slipped into his pajamas, peeked outside, and saw the sheriff's deputy and the highway patrolman placing their hands on the hood of his Corvette, feeling the engine heat rising up through the fiberglass.

"This is the room," they announced and pounded on sleeping Alan Shepard's door.

"Is this your car?" the patrolman demanded, pointing to the gray Corvette in front of the door.

Shepard peered out in the dark, not realizing it was Gus' car.

"Sure, that's mine."

With that, Shepard was thrown to the ground and handcuffs locked in place around his wrists. As a sleepy Shepard found himself being hoisted to the cop car, pajama-clad Grissom opened his door and yelled, "Hey guys, can't you keep it down out there? Some of us have to go to work in a couple of hours."

Grissom was the astronaut's engineer—outspoken and tough. Everybody liked him, but nobody messed with him. Grissom and the rest of the Mercury 7 had grown painfully aware of John Glenn's self-perpetuated "Boy Scout" image, so when Grissom saw that Glenn had hired artist Cece Bibby to hand letter "Friendship 7" on his capsule, Grissom decided to suggest a second painting of a different sort.

Cece was an employee of Chrysler Corporation, a subcontractor for NASA, and the only woman artist in the graphics department at Cape Canaveral. One day as she was leaving the Astronaut Office, she passed Grissom on the stairs, and he asked how the Boy Scout's paint job was going. Gus said he thought what she should really do was paint naked ladies on the capsule because that would really shake John up, since Glenn was such a "straight arrow."

"I told Gus that a naked lady on the capsule's exterior was not a good idea," Cece said. As Gus headed up the stairs he made a crack that she was chicken. "I told him it was my job that would be in jeopardy and not his, that they wouldn't fire an astronaut, but an artist was another matter." Gus just made a clucking noise as he went up the stairs.

It was a dare that Cece could not let go.

"I thought about how I could do it and came up with the idea that I could play a joke on John by using the [spacecraft's] periscope view for the naked lady drawing. The periscope had a plug located on the exterior of the capsule and the plug would be removed just prior to the countdown."

When launch day came, what Glenn saw when he looked through the periscope was a naked lady with a message, "It's just you and me against the world, John Baby." Unfortunately, that particular launch was scrubbed, so Glenn and his naked lady never launched together. Glenn did, however, see it—which got around.

Cece's boss didn't dare fire her, but he did try to get her banned from the pad, But that didn't work either. The Mercury 7 guys banded together and told management that they intended to have her design their insignias and paint them on their capsules.

Bolstered by the strong support of the astronauts, Cece put paint to paper again. Just before John Glenn's next launch date, she did another lady for his periscope view. This one wasn't what he expected. It was a rather frumpy old lady in a house dress. She had a mop in one hand and a bucket in the other. The bucket had Friendship on it in the same script as his insignia on the capsule. The caption read: "You were expecting maybe someone else, John Baby?"

Such charades reflected the surging intensity of the competition in the new-found "gotcha" astronaut culture. These were men who thrived on rivalry as much as risk. It was the mindset that got America into space—and got it there in a hurry.

On July 21, 1961, with Shepard's successful lob into space behind him, Grissom piloted the Mercury-Redstone rocket *Liberty Bell 7* into suborbital flight. It nearly ended in disaster, however, when the exploding bolts on the hatch of the small capsule blew off prematurely after splashdown in the Atlantic Ocean. The ocean wave action had been relatively strong that day and *Liberty Bell* was bobbing around like a cork. Grissom quickly exited the capsule, yanked off his helmet, and jumped into the water. Instead of proceeding to extract Grissom, the recovery helicopter hooked onto the capsule, which was quickly filling with water and already too heavy to lift. As Grissom treaded water, his suit slowly began to fill. Exhaustion began to set in as he repeatedly waved at the helicopter to pick him up.

Flight controllers in Mission Control watching Grissom slowly submerging on their television screens, sat frozen in their chairs.

"Somebody get Gus! For God's sake, get Gus!" a controller shouted helplessly at the display.

Then, as the first helicopter grappled with the capsule, a second helicopter appeared out of nowhere and zeroed in on Grissom. Exhausted, Gus draped himself into the copter's rescue collar and was unceremoniously hoisted to safety. The first helicopter was still struggling to lift the capsule. With its gauges telling the pilot the weight of the capsule was putting a dangerous strain on the engine, he finally made the decision to cut the spacecraft loose, sending it 15,000 feet to the bottom of the ocean.

Although disaster had narrowly been averted, the entire spectacle was captured on film and viewed by the entire American public, which made for dramatic contrast to the success that the Soviets were enjoying at the time. Grissom was no sooner on deck aboard the aircraft carrier U.S.S. *Randolph* than a heated debate ensued among NASA engineers as to whether or not Grissom himself had inadvertently triggered the explosive bolts. Grissom insisted the hatch bolts had malfunctioned and blown by themselves and he was outraged when he heard of their speculation. But NASA engineers were indeed skeptical, as were a few of his fellow astronauts. Did Grissom panic and blow the hatch—or was it a system malfunction as he claimed? NASA engineers were certain their spacecraft had performed as designed.

⁓⋏⁓

Meanwhile, as spectacular as Alan Shepard's career began, in early 1964 his life as an astronaut took a devastating turn. He awoke one morning a bit dizzy, then collapsed onto the floor when he tried to stand. In the following months his recurring bouts of nausea and vertigo were diagnosed as Méniére's Syndrome, an inner ear disorder. Not such bad news for an earth-bound person, but for an aviator it was career ending. At a meeting of flight surgeons, the decision was unanimous: the Icy Commander was grounded. Shepard was removed from flight status and relegated to a desk job in the front office. The first man in space, it appeared, would never fly again. .

Grissom, however, did get another shot. Four years later he was selected to fly Gemini 3, the first manned Gemini flight and the spacecraft he personally helped McDonnell Aircraft engineers design. NASA officials, already wary about the practice of letting the astronauts name their own capsules, balked when Grissom proposed naming his Gemini capsule *Molly Brown* after the popular Broadway show *The Unsinkable Molly Brown*—a joking nod to his Mercury capsule *Liberty Bell 7* sitting at the bottom of the ocean. When they ordered Grissom and his Gemini 3 co-pilot, John Young, to come up with a different name, the two submitted *The Titanic*. With this, NASA executives reluctantly gave in to the name *Molly Brown* but refused to use it as an official NASA reference. Nevertheless, *Molly Brown* lived up to her name, achieving a perfect flight and an unsinkable airtight splashdown.

Grissom's experience on Mercury and Gemini put him in line as the astronaut ground communicator for the next mission. Ed White and Jim McDivitt's Gemini 4 mission would be NASA's first attempt at an Extra Vehicular Activity (EVA), as White would actually leave the capsule and float in space. Since astronauts in the spacecraft are more comfortable communicating with one of their own, voice communications between ground control in Houston and the crew in the spacecraft were funneled through a designated astronaut, known as the capsule communicator, or CapCom. Grissom was the designated CapCom for Ed White's space walk on Gemini 4.

Although the Soviet Union was ostensibly outstripping the U.S. in the space race, here was a mission that could finally change America's standing. In June 1965, Gemini 4 promised to be a turning point.

It could, in fact, catapult America into the lead.

> "One thing we do ask, we would like to have the skipper of the recovery carrier move off the landing spot in the Atlantic Ocean so we don't land on his deck."
>
> *--Ed White II, Pilot, Gemini 4*
> *speaking to a reporter June 1965*

3
A Walk in the Ether

"Gemini 4," barked Grissom, "Get back in!"

Grissom in Mission Control tried desperately to get the attention of the Gemini 4 crew circling overhead on their final orbit of the earth. But the Gemini 4 astronauts couldn't hear his calls over their exuberant chatter. The situation was urgent and Grissom wasn't known for patience. There were only four minutes remaining to radio blackout during the transfer of communications to a remote tracking tower and White needed to be back inside the spacecraft before Mission Control lost contact with the capsule.

Ed White, the pilot onboard Gemini 4, dangled in the ether at the end of a life-support line and couldn't hear a word Grissom was saying. Meanwhile, the command pilot, Jim McDivitt, was inside the capsule. Grissom could hear McDivitt but not White, who was enjoying himself during America's first spacewalk. Also listening in were some 190 million people around the world.

By today's standards, the Gemini communications system was primitive. It was set up for one-way communication, which meant that if one person was talking, no one else on the line could get through. During Gemini 4, White could communicate only with McDivitt, who was now singularly focused on White's spacewalk. Had the situation not been serious, it would have been comical. Yet the moment spoke volumes about White and who he was as an astronaut and as a man. In the transcripts, one thing comes through loud and clear: Ed White was utterly happy and at home in space.

White had been NASA's choice to follow Grissom into space on Gemini 4. He was meticulous, clean-cut, and big for an astronaut, standing a shade under six feet tall. College and military records made him a prime candidate for selection in the second round of astronauts for the new space program and NASA considered itself lucky to have him on the roster. When inducted into the Astronaut Corps in September 1962 as a member of the second group of "New 9" astronauts, White, like the new space program itself, was young and hopeful, looking to embark on an epic adventure. Fortunately, there was a president at the helm of the nation who was as excited about the prospects of landing on the moon as Ed White was.

During that same September, President John F. Kennedy made the announcement to the country—and to the world—that would set the tone for science, geopolitics, and national pride for a decade.

> "It is my goal in the next decade to land a man on the moon and return him home safely."

Of those consulted during the space race, no one had been thinking longer about the future in space than Wernher von Braun, a handsome, dashing man with the looks and charisma of a Hollywood leading man. Even when he led development of the V-2 rocket for Nazi Germany during World War II, lobbing errant rockets on London, he dreamed of future manned space journeys. As World War II ended in 1945, he and 126 other German rocket scientists surrendered to Allied forces; and in 1961 von Braun found himself Director of NASA's Redstone Arsenal in Huntsville, Alabama, where his determination to "go where no man has gone before" fostered the development of the NASA rockets and a close friendship with the President himself.

But the Cold War, brought on by the bitter winds of fear and malice, was growing colder, and Moscow was flexing its political muscle by ostensibly outstripping the U.S. in many areas, including space technology.

"In the space race," von Braun cautioned President Kennedy in 1961, "I do not believe that we can win unless we take at least some measures which thus far have been considered acceptable only in times of national emergency."

In Kennedy's eyes, this *was* an emergency.

When Kennedy stood at the podium in the football stadium at Rice University on September 12, 1962, over thirty-five thousand people sat in the sweltering heat, unaware they were witnessing an epic moment in human history. Kennedy announced that America was on its way to the moon. While the audience was still digesting the magnitude of his opening statements, he went on to pose the question he knew was on the minds of the American people.

"But why, some say, the moon? Why choose this as our goal?" Kennedy asked his audience. "And they may well ask, why climb the highest mountain? Why, thirty-five years ago, fly the Atlantic?"

"Why," he added, driving his point home, "does Rice play Texas?"

Then, in one of those evanescent moments in presidential speechmaking history, he answered his own question:

> "We choose to go to the moon in this decade and to do the other things, not because they are easy, but because they are *hard*."

The soaring rhetoric from the young President captured the imagination of the entire nation. President Kennedy's speech was irresistible to the vast majority of the American public. By their very nature, Americans like challenges, especially those with majestic goals. NASA had just lofted Alan Shepard on a fifteen-minute suborbital flight and now Kennedy was giving them less than nine years to put a man on the moon. From the earliest stages of his career, Ed White hoped to be one of those men. He had a thirst for exploration and for pushing the boundaries in his life—an eagerness captured in Kennedy's next statement on that hot summer day:

> "…because that goal will serve to organize and measure the best of our energies and skills, because that challenge is one that we are willing to accept, one we are unwilling to postpone, and one *which we intend to win*."

And Ed White was born to win. A hulk of an athlete with an impressive record of athletic prowess, his entire life had been mapped by strong morals and inner strength. Born and raised in San Antonio, Texas, he could be found with his family on Sundays in the pews of the local Methodist Church. He was reared and grounded in faith, and he took his faith with him wherever he went, as he was soon to demonstrate on Gemini 4.

White's boyish personality and charisma exuded an All-American image reminiscent of astronaut John Glenn, who a few years earlier had traded his career as an astronaut for a career in politics. White loved spaceflight; he was a natural. To NASA administrators, he wore an ineffable halo of success.

The first spacewalk had originally been slated for a later mission, but on March 18, 1965, just three months before White's scheduled Gemini 4 flight,

A Walk in the Ether

Soviet Cosmonaut Alexi Leonov stunned the world when he made a twelve-minute spacewalk.

"[This is] the proudest achievement of the Soviet Union," Leonov told a reporter in broken English. "America is still on the ground and we are in the space. You know it was tremendous psychological lift, a joy."

At that time White's spacesuit and equipment were still on the drawing board and his gear had yet to even be certified for use in space. Lagging behind the Soviets much longer, however, wasn't something NASA was prepared to tolerate; and in a bold move, just seven days prior to the scheduled Gemini 4 liftoff, NASA added a space walk to the docket. They also had another surprise in store.

In a masterful stroke of one-upmanship, and with little time left before the Gemini 4 launch, an enterprising NASA engineer by the name of Harold Johnson devised a hand-held unit equipped with a bottle of compressed nitrogen and a trigger that looked like a gun's that could be used by White to maneuver himself in space. Dubbed the "zip gun," Johnson developed the device as a member of a team of NASA technicians called "model makers." The head of the group was Jack Kinzler, a six-foot-three-inch meticulous man with a mass of silver hair who had been with the space agency since its birth as NACA (National Advisory Committee for Aeronautics) in Pittsburgh in 1941. Kinzler was determined that this would be no ordinary spacewalk and challenged his team to be creative. Johnson's zip gun was the answer. By actuating two triggers, the astronaut could shoot compressed nitrogen through a system of nozzles, giving him complete directional control. White was thrilled at the prospect. While cosmonaut Leonov could merely dangle in space, White could use the zip gun to propel himself in any direction. He spent countless hours in the McDonnell pressure chamber practicing for the flight that, if all went well, would put the U.S. back in the race.

His walk in the ether would be a first for NASA, and like other firsts, it was not without immense risk. When examining the umbilical cord that would be his lifeline during his space walk, White was asked by a reporter, "You certainly have to have a lot of faith in the men who made that, don't you?" Christian that he was, White replied, "No, I have to have faith in something greater than man before I would walk in space using that."

⸻

On June 3, 1965, the Gemini 4 mission became the stuff of storybooks. During their final orbit of earth, White and crewmate Jim McDivitt opened the hatch. White slowly exited and floated into the vacuum. The white, garish rays of unfiltered sunlight picked up White's helmet and shoulder, dramatically catching the red, white and blue of the U.S. Flag sewn on his shoulder, the zip gun clutched in his right hand. The 25-foot-long golden lifeline writhed slowly as it uncoiled from its black storage bag. The excitement in White's voice was a combination of technical observation and childish exuberance as he jetted about with the zip gun, to the point of having to be almost physically reeled back into the spacecraft.

"This is fun," he radioed McDivitt in the capsule. "It's the most natural feeling."

"Yeah," said McDivitt. "You look like you're in your mother's womb."

Meanwhile, in Mission Control, Grissom's frustration was growing. The crew was only four minutes from Bermuda and the loss of signal in a communications relay to another tracking station.

"Gemini 4, Houston. Get back in." Grissom barked.

All the while, the two spacefarers kept chattering over the voice-activated communications system to the point ground control couldn't get a word in edgewise.

"Gemini 4, Houston," Grissom radioed again. "Get back in!"

White, however, still could not hear Grissom's command.

"The sky sure is black."

"Hey, Ed, smile. Let me take a close-up picture of you," McDivitt said as he tried to capture the moment, at which point the free-wheeling White got too close to McDivitt's windshield with his zip propulsion gun. "You smeared up my windshield, you dirty dog. You see how it's all smeared up there!"

Finally McDivitt checked in with Mission Control. "Gus, this is Jim. Got any message for us?"

At that point, an exasperated Flight Operations Director Chris Kraft, who had been listening on his headset to Grissom's repeated efforts, abruptly cut into communications.

"Yeah, *Get back in!*"

Although Kraft was known as a force to be reckoned with, White, who was still unable to hear ground communications, continued to enjoy his space walk.

"I'm coming back down on the spacecraft," he said. "I can sit out here and see the whole California coast …"

Sensing the frustration on the ground, McDivitt began to narrate White's reentry into the ship.

"Yeah, he's standing in the seat now and his legs are down below the instrument panel…"

By the time White finally worked his way back to the top of the spacecraft he had used up all the fuel in the zip gun.

"Aren't you going to hold my hand?" he asked.

"No, come on in …Ed, come on in here."

When McDivitt finally coaxed him into the capsule, White lamented, "It's the saddest moment of my life."

"Well, you're going to find it sadder when we have to come down with this whole thing," said McDivitt.

Once White was inside the capsule, the euphoria quickly faded into the grim reality of a potentially fatal problem—the hatch wouldn't close. The lever White used during training exercises to ratchet the hatch down wouldn't catch, and it was imperative that the hatch be closed and secured for the fiery ride back through the earth's atmosphere. White struggled vigorously with the stubborn hatch. Observing his difficulty, McDivitt jumped in to assist. Finally, with the brute force of both McDivitt and White bearing down together, the latch caught and the hatch firmly sealed. White, utterly exhausted from his spacewalk, fell back onto the couch. Three days later they were speeding home at 17,000 mph, the heat shield of the capsule aglow as it tore through the atmosphere.

Gemini 4 was the first endeavor of its kind, and skeptics predicted the astronauts would suffer horrible side effects from a long-duration flight and that recovery crews would find dead or unconscious astronauts when they opened the hatches. But the recovery helicopter pilot observed an entirely different situation as he hovered overhead.

"They were like a couple of kids playing on a beach, splashing in the salt water," the pilot reported. White was doing exercises that, according to the pilot, resembled deep knee bends. Later, discussing unsubstantiated reports of the astronauts' "distinct aroma" after the flight, White quipped, "I thought we smelled fine, it was all those people on the carrier that smelled strange."

"We look at life through a frosted glass. All are looking for understanding. We try to explain God in the terms of man. We try to explain His love in the words of man. And we can't."

- Reverend John Stout
Director, The Apollo Prayer League

4
THE GOOD REVEREND STOUT

At the time the world was captivated by Ed White's Gemini 4 walk in space, Reverend John Stout had left the jungles of Brazil and was working as an information manager at the Air Force Eastern Test Range in Cape Canaveral, tracking the parts used in construction of the very Gemini launch vehicle that catapulted White into orbit.

After serving as missionaries in South America for eleven years, he and Helen left Brazil and returned to the U.S., where in 1964 John's reputation as an authority in satellite tracking earned him a position with Pan American Airways at their Guided Missile Ranch Division at the Cape. His responsibilities included missile flight data processing, tracking instrumentation, and communication links between launch sites and downrange tracking stations. By this time, he had six degrees, including a double degree in chemical and petroleum engineering, two masters in chemical engineering and theology, a master's equivalency in education, and a doctorate in linguistics.

Given his intellect, education, and scientific notoriety, in 1965 Stout was asked by International Telephone & Telegraph (ITT), an Apollo project subcontractor under the Federal Electric space division, to transfer to the burgeoning new NASA space organization in Houston. The Stouts had planned to retire in Florida, and although John wasn't especially interested in the position, he accepted an invitation to fly to Houston for an interview.

At that time, the Johnson Space Center had yet to be built and NASA facilities were scattered among different buildings across Houston. NASA officials realized they needed someone to consolidate the disparate components to support the next leg of the space race designed to put a man on the moon. This meant bringing together a hodgepodge of infinitely complex information networks. NASA had over 20,000 suppliers scattered around the world, managing over two million components used in the assembly of the Apollo command module, and every part had to be serialized, tracked, and merged into a single network of data on an IBM 1401 computer at nearby Ellington Air Force Base. The task, if he accepted, would fall to Stout.

NASA gave him the red carpet treatment. The job they offered interested Stout, but he hesitated to accept.

"I will do it so long as I can serve as the industry chaplain," he told them. As a missionary and ordained pastor, Stout required approval from both NASA and the Presbyterian Church to undertake the position as a "tent making ministry." This meant that, following the example of the apostle Paul, he would not be paid anything by the church for his ministerial services.

Given the uncertain nature of the rocket business, NASA recognized the need for someone with Stout's unconventional mix of skills and welcomed him as an ordained Presbyterian minister and volunteer chaplain. Stout accepted the offer as a relatively short-term assignment, his intent being to develop a method of networking the computer needs of the Apollo project and then move on. He told himself he would stay for three years, no more.

The newly constructed Manned Spacecraft Center in Houston was the hub of activity within NASA, and Stout found that his combined role as both chaplain and information scientist gave him a unique perspective on the enormous challenge of the Apollo undertaking—something that engendered an uncommon alliance between him and many of the astronauts. Here was someone they could talk to who understood not only the enormity of the challenge, but to whom they could also confide the personal problems that such a challenge imposed on them and their families.

"In trying to understand the astronauts," Stout said, "I found their own pastors didn't know them. They kept things from their pastors that they didn't want them to know because NASA was trying to keep them as idols. They were our heroes—and we needed heroes."

As an ordained minister, Reverend Stout mixed easily with the astronauts. He listened intently to their problems, their hopes, their frustrations, and sometimes shared their fears. Some he came to know personally. One was a young astronaut by the name of Edward White, II.

On occasion, Stout and White spent time together contemplating the risks and enormity of what lay ahead. Through prayer they hoped to advance a mission that, from the knowing perspective of the information scientist and the astronaut alike, seemed impossible to fulfill by the will of man alone. Both knew firsthand that the spacecraft being developed, assembled, and tested was the most complex machine ever wrought by the hand of man, far outstripping the atomic or hydrogen bombs, nuclear-powered aircraft carriers, and supersonic aircraft—indeed, any device or any assembly of devices. From the point of view of the chaplain and the astronaut, the machine was unfathomable.

With Project Apollo looming on the horizon as America's dark horse in the race to the moon, Ed White was emerging as one of the first men to make the ride. Few were more mindful of the risk and complexity of the Apollo spacecraft than his friend Reverend Stout.

From the moment he was inducted into the space program in September 1962, it was clear that Ed White wanted to take a Bible to the moon. And there wasn't any doubt among those who knew him that he would accomplish his objective. The young astronaut knew what he wanted to accomplish in life and possessed the wherewithal to make it happen.

The son of a retired Air Force general, White graduated from West Point in 1952. He was a golden boy and a fierce advocate of all the basic virtues—God, country, and family. West Point was tradition in the family. His father, Edward White Sr., had been born into a prominent Fort Wayne family whose forebears included Midwest pioneers and veterans of colonial wars. At West Point, the saying goes, "Most of the history we teach was made by people we taught."

The young astronaut was also a gifted athlete who barely missed qualifying for the 1952 Olympics as a 400-meter hurdler. He competed in the high hurdles and the 440, 600-yard, and half-mile runs. He was on the half-mile relay team that held the Academy record for eight years. He was on the starting lineup of the Army varsity soccer team for two years, and could knock off fifty sit-ups and fifty pushups without breaking a sweat.

On top of his athletic prowess, he was an affable guy, well liked among the NASA engineers, technicians, and astronauts, and equally friendly to all regardless of rank or position. He was a highly focused but fun-loving astronaut whose life was sprinkled with a healthy dose of humor. One of White's West Point classmates delighted in telling the story that on graduation day Ed departed the ceremony dressed in his dashing Air Force officer's uniform and headed out to meet his fiancée, Pat Finegan, for the graduation ball. The next morning, the young officer and his date were discovered in the dormitory shower room grinning, Ed wearing Pat's gown and Pat dressed in Ed's dress white cadet uniform.

Ed had met Pat, a petite blonde, at a West Point football game, and in 1953 they were married. Shortly after his induction into the astronaut program, White moved his wife and two children to Houston near the NASA training facilities, where they joined Seabrook Methodist Church, and Ed became close friends with the church pastor, Reverend Conrad Winborn. Over time, their friendship grew beyond that of parishioner and pastor. Reverend Winborn came to view Ed almost as a son.

Like many of the men at NASA, White retained a special admiration for President John F. Kennedy. In a very personal sense, he was *their* President. Kennedy not only pointed the way to the moon, he embodied the youthful spirit of the U.S. space program. In turn, the young President was enamored with the bravery of the astronauts and the imagination and technical ingenuity of the people who were taking America into space.

In late November 1963, President Kennedy and First Lady Jacqueline traveled to Texas to manage minor political in-fighting within the Democratic Party. The President's second stop was Houston, where he addressed a Latin American citizen organization before ending the day in Fort Worth.

The next morning, a light drizzle fell outside the Hotel Texas as a breakfast crowd gathered to hear him speak. The President took the opportunity to talk of America's need to be "second to none" in defense and in space and of the willingness of Americans "to assume the burdens of leadership." The speech was warmly received by the thousand-plus crowd that had waited over an hour outside the hotel in the rain. After the speech, the President and Mrs. Kennedy boarded a limousine. Since it was no longer raining, the plastic bubble top was removed.

The next stop on their tour was Dallas.

Reverend Stout, his wife, Helen, and 10-year-old son, Jonathan, were having lunch at Kumbak's restaurant in Austin, Texas when the news came in. A waitress rushed over to the restaurant television and switched the channel to view an uncut pre-broadcast showing Kennedy's assassination. Young Jonathan had received a personal note from Kennedy just days before thanking him for his replica of Kennedy's PT-109 torpedo boat that was rammed and sunk during World War II. Now, in a flash, the boy's hero was gone.

The news from Parkland Hospital that day would have a lasting effect on the people at NASA, especially the men who would venture into space. In a speech at Brooks Air Force Base in 1965 to honor his Gemini 4 space walk, Ed White spoke with halting humility of his immense respect for the slain President. Two years earlier, Kennedy had spoken at the dedication of the Aerospace Medical Complex at the same podium where White now stood.

> I believe that President Kennedy epitomized not only the things we were trying to do in the space world… I believe he hit upon the hearts of all of us in a lot of different ways. I believe he epitomized the things that we would like to be ourselves.

After his spacewalk, White wrote a note to Mrs. John F. Kennedy saying his walk in space made him feel "red, white, and blue all over." The first American flag to appear on a NASA spacesuit was on his sleeve and he had unofficially named his Gemini craft *American Eagle*. His patriotism was plain to all, and despite suggestions that he had disobeyed an order by not returning to the spacecraft when told to do so, Ed White was a West Point man, and as such, would never have disobeyed an order. Duty, honor, and country not only described him—they *were* him.

In memory of the President who launched America's dream of landing a man on the moon, in 1963 Kennedy's successor, President Lyndon Johnson, changed the name of Cape Canaveral to Cape Kennedy. By then, however, the term Cape Canaveral had become a metonym for the launch area; and despite their respect for the late President, many of those who worked and lived there had difficulty dropping the original name.

By 1965, White was one of America's premiere astronauts, and the broader exchange during White's Gemini 4 space walk testified to his complete sense of comfort during the spaceflight. Where others struggled with the complexities of space travel, disorientation, and motion sickness, White flourished. He seemed to possess an intuitive feel for and understanding of the many systems and how they worked in concert with each other.

In addition to achieving America's first space walk, he could also claim a lesser-known first: as a man of faith, he was the first person to carry a religious symbol into space. It came as no surprise to those who knew him that he had taken with him in his left leg pocket a gold crucifix, a St. Christopher's medal, and a Star of David during his historic spacewalk.

"I had great faith in myself," he told a reporter, "and also I think that I had a great faith in my God. So the reason I took these symbols was that I think that this was the most important thing I had going for me, and I felt that while I couldn't take one for every religion in the country, I could take the three most familiar to me."

For White, the space program wasn't only about science, technology, or world opinion. For him, his time in space brought him closer to God. His natural ease in space allowed him the presence of mind to grasp the enormous spiritual dimension of the endeavor. In an interview with a Texas newspaper, he was not ashamed to speak of his faith, openly expressing his intention to take a small condensed Bible to the moon.

"The best all-a-round volume of literature you could take would be the Bible. I do plan to take one."

To White, the Bible was much more than a religious document or a book of great historical significance. It was a symbol of all that humanity hoped to become and could possibly achieve. It embodied the faith and courage required to venture to the moon in the first place. For these reasons, he felt it was important on a spiritual level for mankind to take the Scripture into space when men eventually left earth. And if landing on the moon happened to be his lot, Ed White would see to it that a Bible landed there with him.

By January 1967, America was in a dead heat with Russia in the race to the moon, and White's name had once again cycled to the top of the selection roster to fly an Apollo mission. He could not have been more excited. If all went well, AS (Apollo-Saturn)-204 would become Apollo 1, the first manned Apollo mission—the final link in an ambitious program to put a man on the moon. Commanding the mission would be Grissom, who had nothing but respect for White.

When asked by a reporter to give a commander's assessment of White, Grissom remarked, "Ed's a real hard driver. I don't care what kind of job you give Ed, he's going to get it done; he's going to get it finished. Just like on his space walk. That hatch that wasn't working right just didn't bother him a bit. By God, he goes ahead and gets the job done."

The relationship among the crew was natural, which allowed a frankness with one another even in front of reporters. The result was a projection of true camaraderie between America's most accomplished astronauts. And there was plenty to rib about. Although NASA doctors failed to find one ounce of fat on White's 170-pound frame, White could put away two full course dinners and then ask for dessert with a straight face. So awesome was White's appetite that Grissom told reporters that on their upcoming fourteen-day Apollo mission, "I'm going to keep my own food supply under lock and key."

Filling out the crew was Roger Chaffee, a thirty-one-year-old space rookie and an astronaut since 1963. Chaffee had never been in space, but was a solid Navy aviator. Sources in astronaut circles said he had piloted one of the RA-3B jet reconnaissance missions over Cuba that took photos of the Russian missiles cited by President Kennedy in his October 1962 television address which served to initiate the Cuban Missile Crisis. Chaffee was confident and exceedingly disciplined. He was also dedicated to the idea of space exploration.

A few days before a scheduled systems ground test of the first Apollo rocket, a reporter from the *Washington Star* questioned him on the need to do space exploration beyond the moon. Chaffee replied, "We have to go there."

"Well, Mars and Jupiter are there—and so are the stars—do we have to go to them too?" the reporter queried.

"Of course we do," was all Chaffee had to say to that.

"If we die we want people to accept it. We are in a risky business and we hope that if anything happens to us it will not delay the program. The conquest of space is worth the risk of life."

- *Gus Grissom, Commander, Apollo 1*

5
APOLLO 1: A MISSION IN FLAMES

There was something about that day—something palpably *wrong*. One glitch, it seemed, was followed by another. But there was also an undeniable sense of great promise. Two years after Ed White's spacewalk, an altogether newly designed capsule stood on Launch Pad 34 atop NASA's new Saturn 1B rocket. The stack of the first Apollo spacecraft rose skyward nearly forty stories from her truncated bottom to her needle-tipped nose. Earlier that day NASA officials announced the designation of the flight as AS-204. If the test that day proved successful, they could be on track to launch in less than a month, and AS-204 would become Apollo 1, the first manned flight of the monstrous new Saturn rocket.

On Friday, January 27, 1967, the three-man crew emerged from the transfer van at Pad 34, donned in their new A-1C space suits. The purpose of the test that afternoon was a "plugs out" test, the first of three major tests of the spacecraft required prior to launch. During the test, all electrical, environmental, and ground checkout cables would be disconnected to verify that the spacecraft could function solely on internal power alone. Since this was only a routine test, there was no fuel in the rocket.

Arriving an hour after noon, the crew slid into the spacecraft couch. The capsule would be pressurized with pure oxygen, which eliminated unnecessary weight inside the capsule, since a natural air mixture was considerably heavier, and every ounce of weight had to be carefully calculated.

The Apollo 1 also had an improvement in the hatch design. As a result of Grissom's problem with the exploding bolts and White's difficulty in getting it secured during his Gemini flight two years earlier, Apollo 1 was designed with two hatches: the first, an inner hatch that opened inward towards the astronauts and utilized internal pressure to force the seal; the second, an outer hatch that opened away from the spacecraft. The two-piece design was also considerably lighter, yet still remained problematic, requiring the use of a ratchet and removal of several bolts to open. On top of the assembly was a third booster cover cap.

After securing themselves in their couch positions, the crew began preliminary checks, listened to the countdown, and complained about persistent communications problems that caused intermittent delays. Near sunset on the winter evening, communications problems again caused a delay, this time for ten minutes. To everyone's aggravation, it occurred even before the plugs could be pulled. Thus, the test that should have long since been finished had not yet even begun.

Time passed. The shadow of the Apollo spacecraft grew over the Indian River and eventually vanished altogether as the winter sun settled into the palms to the west.

"How are we going to get to the moon if we can't even talk between two buildings?"

It was now 6:29:41 p.m., and Gus Grissom's voice betrayed the strain and frustration of the lengthening line of miscues that had defined the new Apollo program up to that moment. It wasn't that things had been going extremely wrong; it was just that they weren't going extremely right either.

Command Pilot Grissom sat strapped in the capsule of the Apollo "Block 1" AS-204 rocket in the Merritt Island Launch Area of Cape Kennedy in Florida. Strapped in next to him listening to his aggravation were fellow crewmembers, Senior Pilot Edward White and Pilot Roger Chaffee.

In spite of the recurring spacecraft problems, NASA had come down with a malady known as "go fever"—a mixture of raw optimism and determined imperative to beat the Soviets to the moon. The astronauts were determined to fly regardless of the consequences, even if their own safety was on the line. Things needed to be fixed, but fixes meant delays, and delays had to be justified. Things had to keep moving forward as no one knew exactly where the Russians were in their program. It was a heady program and the pervasive mindset in the Apollo boot camp was that if NASA could manage to launch even a wash tub into space, a good astronaut would be able to fly it.

On this January evening, everything was readied for a simulated launch, a vital step in determining whether the spacecraft would be ready to launch the following month. NASA and North American Aviation (NAA) engineers and technicians were taking constant notes on future corrections to the next capsule design because there wasn't time to make changes to the current design. As a result, specifications and safety protocols in the current design were reworked and compromised to meet increasingly demanding deadlines.

Up to this point, there had been no disasters during any previous manned mission, and there was no reason to expect otherwise this time.

Delays popped up almost from the moment the astronauts entered the capsule. The first occurred when Gus Grissom hooked up his space suit to the oxygen supply. He described a strange odor in the spacesuit umbilical loop as a "sour smell." The tech crew halted the test to take a sample of the suit loop, and after discussion with Grissom, decided to continue the test.

The next was a high oxygen flow indication which periodically triggered the master alarm. The men discussed the problem with environmental control system personnel, who theorized the high reading resulted from movement of the crew. The matter was never fully resolved. Finally, at 2:45 p.m. the cumbersome three-layered hatch was secured in place. Once sealed shut, the interior of the cabin was pressurized with pure oxygen to 16 psi to approximate atmospheric pressures in space—an environment that could turn the flame resistant Velcro covering the inner walls of the capsule into tinder at the slightest electrical spark. Even the capsule's aluminum fittings would combust in such an oxygen rich environment. The original specifications required anything in the capsule to be nonflammable up to temperatures of four hundred degrees. But everyone seemed to forget that the center of a tiny electrical spark could reach six thousand degrees Fahrenheit. Although a fire could be rather easily contained at 5 psi pressure in space, no one had contemplated a fire on the ground.

A third serious problem then arose in communications. Initially, the problem seemed to exist only between Grissom and the control room. Later, the difficulty extended to include intermittent loss of communications between the operations and checkout building and the blockhouse at the launch complex. The men could speak over four channels by radio or telephone line, but the tie-in with the test supervisors and technicians was troublesome. Somewhere there was an unattended live microphone that could not be tracked down and turned off. To further complicate matters, no single person controlled the troubleshooting.

Meanwhile, the communications problems dragged on for hours and the countdown was stalled at T-minus ten minutes—the time remaining before the "plugs" would actually be pulled from the rocket and the test would finally begin. Near sunset, yet another communication problem caused a delay, this time for ten minutes. An emergency egress practice had yet to be done. The crew had now been sitting on the pad for over six hours and Grissom's patience was wearing thin.

"Hold on, Gus," came a satirical voice from Mission Control. "We're going to get a couple of tin cans and some string so we can hear each other better."

As far as Grissom was concerned, Apollo had learned very little from the previous space programs, Mercury and Gemini. When Grissom pointed out inconsistencies and potential flaws in the Apollo design, the NAA engineers working on the prototype spacecraft didn't want to hear it, a cavalier attitude that had begun pervading the space program after several years of glitch-free operations. Alan Shepard's famous fifteen-minute venture into space had set the space race off and running at a blistering pace, and NAA wasn't about to slow it down.

In earlier missions, Grissom had the ear of the engineers at the McDonnell Aircraft Company, the firm that designed the Mercury and Gemini spacecrafts, through its director, James "Mr. Mac" McDonnell. Unfortunately, Gus didn't carry the same weight with the new NAA engineers. The firm was just too big, and if changes didn't fit into some overall scheme, they were instantly lost in the shuffle. And as things were developing, there were plenty of changes to be made to the new 224-foot Saturn 1B before it would be ready for space.

But Grissom was no rookie to be cast aside so easily. As one of the Original 7, he already had two space flights under his belt. He also had a degree in mechanical engineering which he put to regular use in assessing the readiness of any spacecraft he was asked to fly. And things about this spacecraft bothered him. The spacecraft's cooling system had leaks, spilling a nasty water-glycol mixture onto the floor. After the water evaporated, a flammable glycol residue persisted. The new complex hatch design was another Apollo monstrosity. Backup pilot John Young sarcastically labeled the entire ensemble as nothing but "a bucket of bolts."

Betty Grissom said her husband "became such a gadfly in trying to enforce his standard of excellence that some contractor personnel tried to work paths around him." The Apollo program, she said, "was several orders of magnitude more complex than either of the two preceding manned space programs."

Everything in Apollo was run by computers, while everything in Mercury and Gemini had been stick and rudder. In Mercury, switches and dials flew the ship. In Apollo, all the controls plugged into computers which ran the ship. Astronauts weren't just pilots anymore, they also had to be computer experts. Bundles of wires were everywhere. The spacecraft had nearly thirty miles of wiring.

It wasn't that Grissom felt the Apollo 1 capsule posed any immediate danger. They were test pilots, after all, and accustomed to working with potentially dangerous experimental aircraft. But the continual malfunctions caused him an overall sense of uneasiness about the design and safety parameters of the first Apollo mission. Everything seemed awkward. A service module, the engine and tankage part of the Apollo spacecraft, had blown up in the course of a test at the factory in California. Gus was there that day.

As a symbol of his frustration, just three days before the pre-flight test he yanked the largest lemon he could find off of the citrus tree in his backyard and hung it on the command module simulator.

Apollo 1 was looking to be the Edsel of space. Sharing his uneasiness was Wally Schirra, a close friend and member of the Apollo 1 backup crew. "If anything seems the slightest bit wrong," he advised Grissom, "just get the hell out."

NASA had experienced pre-flight problems with earlier missions and this one appeared to be no exception. Everyone was still confident it would fly. Granted, Mercury and Gemini had been experimental programs, bristling with incredible risk. But the complexity of the Apollo spacecraft magnified those risks several fold.

No one understood this better than Apollo 1 Command Pilot Gus Grissom.

On the first floor of the launch complex, Gary Propst, an RCA employee, watched a television monitor with its transmitting camera trained on the window of the command module. Deke Slayton, Flight Operations Manager, sat half a kilometer away at a console in the blockhouse next to Stuart Roosa, the astronaut serving as the capsule communicator. North American employee Donald Babbitt, the pad leader, and several system technicians were standing by in the enclosed room adjacent to the capsule waiting to pull the plugs on signal.

After five-plus hours of being locked inside the capsule, the crew was growing edgy. No one on the circuit had responded to Grissom's earlier complaint.

"I said, how are we going to get to the moon if we can't even talk between two buildings?"

"They can't hear a thing you are saying," White quipped.

Then, at 6:30:55 p.m., something happened inside the capsule. Somewhere in those thirty miles of wiring, one segment arced across another creating a tiny spark of flame, and with that the greatest catastrophe the U.S. space program had ever known was underway.

One of the crew, presumably Grissom, moved slightly. Then came a two-and-a-half second interruption of power. The monitors indicated an inexplicable increase of oxygen flow into the crew's space suits. White's heart and respiration rates suddenly and noticeably increased. The airways fell silent for roughly a minute until 6:31:05 p.m., at which point a cry was heard over the headset.

"Fire!" shouted Chaffee.

"Did he say 'fire'?" Babbitt asked, turning to the man next to him. The technician on the headset next to him nodded silently, listening intently to what came next.

Two seconds later White yelled, "Fire! We've got a fire in the cockpit!"

"We have a bad fire!" Chaffee shouted. "We're burning up!"

In the blockhouse located near the pad and at the various control centers, surprised engineers and technicians looked up from their consoles to television monitors showing the interior of the command module. To their horror they were witness to an unbelievable scene—intense flames were raging inside the spacecraft and dense smoke was rapidly blurring the picture.

Stunned, Babbitt looked up from his desk and shouted, "Get them out of there!" As Babbitt spun and reached for a squawk box to notify the blockhouse, a sheet of flame flashed from the spacecraft and he was hurled to the door by the explosion.

Spacecraft technicians ran toward the sealed Apollo capsule, but before they could reach it, the command module ruptured. Flames and thick black clouds of smoke billowed out, filling the room. The intense heat and dense smoke drove one after another back, choking and gasping as they ran in and out of the capsule enclosure, attempting to remove the spacecraft's hatches. From time to time, one or another would leave to gasp for air.

Meanwhile, Propst's television picture showed a bright glow inside the spacecraft, followed by flames flaring around the window. For nearly three minutes the flames increased steadily. As the room surrounding the spacecraft filled with smoke, Propst watched with horror as silver-clad arms behind the capsule window fumbled for the hatch. Ed White could be seen grasping to open the hatch above him. A second pair of gloved hands—presumably Grissom's—reached across to help.

"Blow the hatch, why don't they blow the hatch?" yelled Propst.

What he didn't know was that the hatch could not be opened explosively. The two-piece hatch that opened inwardly required the crew to undo several bolts in order to remove the inner section. In short, it was impossible to open quickly. The hot gasses produced by the fire held the hatch shut, and within a few seconds the air pressure raised enough to literally seal the crew inside. A split second later it rose so high it ruptured the capsule. Seventeen seconds after their first report of fire, Mission Control and the horrified controllers received the last transmission from the crew. It was a brief scream.

Beyond 6:31:23 p.m. on January 27, 1967, there were no more transmissions from the crew of Apollo 1. Slayton and Roosa watched a television monitor, aghast, as smoke and fire billowed up. For the following few minutes NASA tapes recorded urgent voices from various test stations. Roosa tried and tried to break the communications barrier with the spacecraft.

"Can you egress at this time? Confirm it."

"Pad leader, can you get in there and help 'em?" another voice injected urgently.

"Alright crew, did we get verification? Can you egress at this time?"

"Gus, can you read us? ..."

"Pad leader? ..."

The static on the line rasped monotonously for several minutes—then lapsed abruptly into a sobering silence.

All telemetry information was lost shortly after that.

"The measure of his life, any man will tell you, is to be found not in its length, but in the quality of its commitment to God and man."

- Reverend Conrad Winborn
Pastor, Seabrook Methodist Church

6
THE AFTERMATH

Walt Cunningham had been at the launch complex at Cape Kennedy most of the day. When it was announced that the "plugs-out" test would be delayed past 5:00 p.m., he and the rest of the backup crew climbed aboard T-38s and headed for Houston. After the two-hour flight, the crew landed and taxied to the NASA hangar. He immediately noticed something was different. It wasn't the usual lineman who greeted him. Instead, it was Bud Ream, the number two man in flight operations. Bud was a NASA test pilot. He was short, solid, and had a crisp manner. There was no reason for Bud to be there. He stood on the concrete apron and waited for the backup crew to climb down. His face was grim. "The prime crew has had an accident," he said.

At first no one spoke. Shocked at the news, they just looked at each other.

"The crew is dead," Bud said. "They're dead. All of them."

As anguished officials gathered at the scene of the disaster, the pad was cleared of unnecessary personnel and guards were posted. At 7:15 that evening the preliminary investigation was already underway.

The causes of the tragedy appeared to have been hiding in plain sight all along. Initial evidence suggested that the crew had struggled to open the hatch, a complex task requiring more time than the fire allowed. Technicians calculated that even with a ratchet it would have been impossible to open the hatch sealed shut by the intense pressure inside the cabin. Although rescuers were only a few feet from the spacecraft, they were blocked by the dense smoke pouring from the cockpit. They had no firefighting equipment or gas masks. Their clean room nylon gloves melted in seconds, as did their shop

The Aftermath

coats. Speculations put the top heat inside the capsule at 2,500 degrees, a temperature reached within seconds of the first report of fire. One by one, the support crew removed the booster cover cap and the outer and inner hatches, prying out the last one five and a half minutes after the alarm sounded. By the time they finally opened the hatch, the crew members had perished.

In the aftermath of the tragedy, doctors treated twenty-six men for smoke inhalation. Two were hospitalized. The outer surface of the capsule had blistered and blackened in places, evidence that the blaze had erupted through the light skin of the airtight craft. The heat had been so intense that molten metal was reported running down the outside of the capsule. The lifeless bodies of Grissom and White were half out of their seats. Chaffee, as the exit drill required, was still in his. The three astronauts were left undisturbed in their couches for more than seven hours as officials tried to piece together the cause of the disaster. All data was impounded pending an official investigation.

The wives and families of the astronauts were steeled for death, but not prepared for it. How do you prepare five-year-old Steve and eight-year-old Sheryl Chaffee, who would never again chase frogs and lightning bugs as their father worked meticulously in his rock garden? How could you prepare thirteen-year-old Mark and sixteen-year-old Scott Grissom, who would never again play slot cars with their father? AIA Slot Speedway owner Bob DeMoss said, "Gus turned into a 'pit marshal' when he came in with the boys. Grissom stood inside the slot car tracks, bent over all night long, picking up the miniature cars for the boys as they spun off the tracks."

How could you prepare thirteen-year-old Ed White, Jr. and ten-year-old Bonnie White, who would never again go bicycling with their dad? At their home near Houston, White had set up a 40-foot climbing rope and a horizontal bar where he put the youngsters through their paces. During their workout, Eddie Jr. and Bonnie Lynn would compete against their father. Eddie Jr. had recently received a new shotgun and experienced the thrill of hunting with his dad. But Ed's faith was central to his family life. "Ed's Sundays with his children were sacred," said a close friend who had known the astronaut for more than ten years. "There was no hunting on Sundays or anything else."

Pat White had reluctantly shared her husband with the nation. Hers had been an all-American marriage, tighter than most of those buffeted by the winds of NASA fame. Her husband wasn't the kind of man who parked on a bar stool, drinking and flirting—a popular fighter pilot recreation. The 36-year-old Texan drew his wife, Pat, and their children close to him as he shared with them his experiences in space. Before Ed set out on his Gemini 4 flight, Pat White told a reporter that she and her husband lived one day at a time. "You can't live your life in advance. You make plans and you have faith, but you take each day as it comes."

All three crew families were home that Friday evening. Each was notified of the tragedy by either an astronaut or the wife of an astronaut. One fact became exceedingly clear to all: the families of the astronauts were a closely knit group. They would close ranks and see one another through this harrowing crisis.

The next morning, a warm sun broke over Houston's sprawling Manned Spacecraft Center and its wooded suburbs. But for a Saturday in the city, January 28 was startlingly quiet. Martha Chaffee left early from her yellow-brick home. Where she was going, no one knew. Her two children came out of the house with a neighbor lady, blinking at the television cameras.

"We soon realized that you don't keep them from the TV set," one of the wives related, "and you don't cover up, because the children not only do well themselves, but at a time like this they're a big support for their mothers."

Houston's Mayor Louie Welch ordered the police to stand guard at the Grissom home in the small suburb of Timber Cove. Gus had cherished his privacy. He and Wally Schirra had built ranch-style homes next door to each other on Pine Shadows Drive. The houses appeared almost identical, both built without windows in the front so their families would be shielded from onlookers. Even before NASA officials arrived, Betty Grissom already knew she had lost her husband. She appeared calm and restrained.

"If you're going to get killed," she said, "you expect it to be in the war. Not in an automobile accident on the way there."

The Armstrong family lived next door, and Neil Armstrong's wife, Jan, was the first to the Whites' house. Seeing the look on her face, Pat White knew instantly something had gone deathly wrong. Upon hearing the news, the White children were hurried to a neighbor's house. Ed White's father and mother arrived the next day from Florida. Ed's brother, James, flew in from the Air Force Academy.

Ironically, the evening Grissom, White, and Chaffee died was the same evening they and their wives had planned to attend a party in honor of the fifty astronauts, a celebration for having made it through another year. The dinner and desserts were ready and hairdos were done. When the "plugs-out" tests ran late, the party was cancelled.

"They're bearing up mighty well," a friend said of the astronauts' wives. "We know, though, that the next few weeks and months are going to be pretty horrible."

In the Houston suburb of El Lago, neighbors gathered in quiet groups

The Aftermath

along the sidewalks on which Ed White had taken his early morning jogs around the block.

"Of all the astronauts we knew," one boy said, "Mr. White was the nicest."

The mood in Houston's NASA Space Center was somber. Astronaut Tom Stafford and several other astronauts left to console Pat White. "I have no more idea than you what to say about it," Stafford told reporters." About the only thing I can say is that this was a tragic loss that will really set the program back."

When he and others arrived at the White family house, they found a note neatly affixed to the front door of the contemporary-style house. It read: "Please knock. Do not ring the door bell."

At the White House, diplomats from sixty nations were gathered to sign a new "Peace in Space" treaty which President Lyndon Johnson described as "an inspiring moment in the human race." Soviet Ambassador Anotaly F. Dobrynin told the East Room capital audience, "Let us hope we will not have to wait long for the solution of earthly problems." The treaty was aimed at preventing territorial and military rivalries in outer space and included blocking the use of nuclear warheads. Astronauts Neil Armstrong, Jim Lovell, Dick Gordon, Gordon Cooper, and Scott Carpenter were on hand for the signing, along with Wernher von Braun. The two-thousand-word treaty was also aimed at preventing territorial claims in space, such as asserting national title to real estate on the moon. The benevolent atmosphere was shattered by news of the disaster from the Cape.

"This reminds one again," Bill Allen, the president of Boeing, said to Wernher von Braun in the immediate aftermath of the accident, "that we are not in the business of making shoes."

What alarmed NASA, North American Aviation, the astronauts, and the American public, was that Gus Grissom, Ed White and Roger Chaffee had not died on their way to the moon. They had not burned up during reentry. Their rocket had not exploded at launch, nor had they crashed into the moon. These three men had not died in a blaze of glory, but rather in the midst of a routine test conducted during a simulated countdown on the launch pad just three feet from safety.

As the sun rose over the Atlantic Ocean on January 28, 1967, the moon seemed a long way off indeed. In Houston, a city of 1.5 million souls, and throughout the nation, flags were hung at half-mast. Pope Paul VI and world leaders offered their condolences. Tributes to the courage of the dead astronauts poured in from leaders of the world community and from the man on the street. But behind the messages of condolences, nearly everyone was struck by the irony that these three brave men whose hopes and dreams had soared so high had died on the ground. Britain's morning papers hurriedly made over their front pages to tell the story of the tragedy. Some Britons said the United States should reconsider its goal of putting a man on the moon by 1970 and predicted a hot debate over the new ground crew, new spaceship, and new contractor. On a local level, a twelve-year-old Tennessee boy spent all his savings at a Houston florist. His three $1 bills arrived with a note: "This is my savings. Please send each of the astronauts' widows one red rose."

Vice President Hubert H. Humphrey, President Johnson's special counsel in the field of space, often visited the Cape Kennedy Space Center and knew the astronauts well. The news hit him especially hard.

"The death of these three brilliant young men," he said, "true pioneers and wonderfully brave, is a profound and personal loss to me. I have such close relationships with them that my sorrow is very deep. My heart goes out to their families and loved ones."

The memorial service for Ed White at Seabrook Methodist Church on January 30 was closed to all but family, friends, and a few dignitaries. White's friend and pastor, Reverend Conrad Winborn, stood at the front of the sanctuary. Behind him was a large wreath of yellow flowers in the shape of an astronaut pin on a background of white flowers placed there by Ed's fellow parishioner and astronaut, Tom Stafford. No one put the meaning of White's death into clearer perspective than Reverend Winborn:

> "Ed's life speaks for itself and makes a eulogy unnecessary. He wrote his own eulogy in the fabric of day to day living. We know what Ed stood for, and it was right and good. His life is best described through the use of the word commitment.
>
> Commitment to nation and space exploration.
>
> Commitment to family.
>
> Commitment to Church and community.
>
> God reveals in Christ that commitment is costly."

Mourners gathered the next morning in the brisk, winter air at Arlington National Cemetery for Gus Grissom's funeral. The Reverend Roy Van Tassel, Grissom's Church of Christ minister from Mitchell, Indiana, spoke briefly:

> "Forty-one years ago through one of the mysteries of life, Gus Grissom came into this world. Last Friday, in another of life's mysteries, he left this world. He knew this was going to happen. But, like us, he did not know when."

President Lyndon Johnson was in attendance, along with the six remaining Mercury astronauts, all of whom wore full-dress uniforms in honor of their fallen friend and comrade. Four hours later, with similar honors, Roger Chaffee was laid to rest alongside Grissom. Chaffee's pastor, Reverend Ernest Dimaline, delivered the eulogy:

> "Oh Lord, support us all the day long until the shadows lengthen and the evening comes, and the busy world is hushed, and the fever of life is over, and our work is done. And in thy mercy grant us a safe lodging, and a holy rest, and peace at the last."

All of the fallen astronauts were in their uniforms complete with tiers of chest medals. The official cause of death was listed as asphyxiation; they had suffered only minor burns.

The tragedy occurred amidst a particularly chaotic moment in a particularly chaotic decade. Two years earlier, President Johnson had signed the Voting Rights Act of 1965 giving blacks the right to vote—a full century after the signing of the Emancipation Proclamation. In August of that year some of the most violent riots the country had ever endured broke out in outlying communities of Los Angeles.

"What the hell do we care whether we have the right to vote if we don't have the right to live?" a black man yelled at a city council meeting. To them, Lyndon Johnson was just another white man. Riots erupted in Newark, New Jersey, and other major American cities. The social unrest, the war in Vietnam, and the tragedy of Apollo 1 cast a pall over the country.

On January 31, the day of the funerals, Lady Bird Johnson wrote in her diary, "A miasma of trouble lies over everything." As President Johnson departed for Grissom and Chaffee's funeral in Arlington, Lady Bird left for Ed White's funeral at West Point.

The flag hung limp at half-mast in the brisk winter air as the last astronaut was laid to rest in the hallowed grounds at West Point. Reverend Winborn delivered White's final eulogy at the West Point chapel:

"Our Father God calls us to be thankful for the hope which is ours in Christ Jesus. 'God so loved the world that He gave His only begotten Son that whosoever believes in Him should not perish, but have everlasting life.' Death is swallowed up in victory through the resurrection of Jesus.

Grave, where is thy victory?

Death, where is thy sting?

Life with God begins in the eternal now. God seeks us in love and draws us into eternal companionship with Him through Faith in His Christ. Let us give thanks for this lively hope.

We now trust our loved one, Edward White, into the gracious care and keeping of our Father. Let there be for us the lingering memory of a thankful life, well lived."

Astronaut Ed White returned to West Point Academy and the hill country he loved to rest among the hero dead. His grave lies just across a white gravel roadway from a stone wall that tops the high cliff overlooking the Hudson River and the oval track of Shea Stadium. On that track, Ed White set the West Point record in the 400-meter hurdles. The last time he had walked on those same grounds, he presented the Corps of Cadets with an American flag and the West Point Sesquicentennial Medallion he had carried on his Gemini 4 space walk. Now it was the Corps' turn to honor him. At graveside, Vice President Hubert H. Humphrey and Mrs. Lyndon Johnson bowed their heads and joined the small crowd in prayer.

"Our Father which art in heaven …"

―⁂―

At the time of his death, Ed White's record in the 400-meter hurdles had not been equaled.

"In the design and flight of the current Apollo spacecraft, we cannot speak in terms of blind faith, however sincere that may be. We wish we could, but we cannot. It would be tempting God if we did a little and expected him to do the rest. We must do everything that is in our power to do—and doing this with His help—and then leave the rest to God."

*- Reverend John Stout
Director, The Apollo Prayer League*

7
THE APOLLO PRAYER LEAGUE

Reverend John Stout was in Oklahoma City giving a talk on the Apollo program on what would become known as "The Day of the Fire." At first the tragedy was described to him as a "minor mishap." But then the truth came out. He hung up the phone in stunned disbelief. Ed White, the devout young astronaut he had come to know and admire, and who had frequently confided in him, was dead. So too were Grissom and Chaffee. For the moment, however, there was little time for him to grieve. NASA had made an aircraft available for the trip, and he returned to the Manned Spacecraft Center immediately. It was late that Friday evening when the plane landed.

As NASA's chief information scientist charged with collecting and indexing spacecraft information, Stout had access to data that would help determine what had gone wrong on Pad 34. He went directly to his office and worked through the weekend, retrieving and analyzing information from the myriad of data and communications systems.

The following Monday morning, a grim-faced Frank Borman appeared in his office. Borman was the astronaut representative appointed to head the internal committee charged with investigating the tragedy. In a few words Borman explained what he wanted from Stout and his staff. "I already have that information for you," Stout replied. He then handed Borman a six-inch stack of documents containing the results of work he had compiled over the weekend. "You'll find your answer here," he said. "It took me all weekend to get it ready for you, but you'll find your answer here."

Borman took the documents and left. It was clear the stack of paper was not the work of a typical government scientist, but the obsession of a man desperate to know the cause of his friend's death. After Borman and his team had gotten through the materials, he returned to Stout's office and thanked him. Stout's work had cut by almost half the time Borman's committee required to investigate the tragedy.

In the aftermath of the fire, Reverend Stout was shocked but not completely surprised. The potential for disaster had weighed on his mind for some time, a possibility that he both feared and anticipated. The program had been going too well for too long—and had proceeded too fast. NASA staff and engineers, he felt, had rushed headlong into a world in which they felt utterly invincible. But the truth was, the relative simplicity of the Mercury and Gemini missions paled in comparison to the staggering complexity of Apollo. This stark reality seemed to have escaped nearly everyone at the Manned Spacecraft Center.

"At the time of the Apollo 1 fire at Cape [Kennedy] in which Grissom, White, and Chaffee perished," Stout said, "we felt that we could do anything because of our previous successes. We had just completed successfully flying all of our Gemini missions, a new landmark in the space program had been passed, and we were proud and happy. Champagne flowed freely and party girls were available. No one was concerned too much about the scientific information that verified Gus Grissom's remarks that Apollo 1 was an unsafe spacecraft."

Reverend Stout and Ed White had been friends for several years and often discussed the risks. Now White was gone. Not so long ago, Stout's own younger brother, Joe, had been shot and killed while trying to prevent a robbery in South Fork, Colorado. For Stout, the sudden death of Ed White was a painful reminder of a promising life cut short. The news was also a crushing confirmation of the overwhelming nature of the mission.

The computer system Stout managed coordinated the workings of some 1,200 scientists and engineers on a computer capable of running searches for roughly one-million technical reports per week. The sheer scale of the Apollo project was one that boggled the mind. The more one understood, the more puzzling it became. For Stout, this massive project was akin to a miracle unfolding in slow motion before his eyes. Although it was a project carried out by man, there was something unmistakably divine about the reality of celestial exploration. Stout was far from alone in his awe.

The magnitude of the program amazed everyone associated with it. Apollo 16 astronaut T.K. (Ken) Mattingly claimed that the individual astronaut had "no idea how to make the whole thing work." The same was true for the hundreds of scientists, engineers, designers, and controllers. As a result, each man became master of his own independent domain. No single earthly entity, it appeared, could fathom the entire massive collaboration.

Stout and White's scientific appreciation of the Apollo spacecraft's complexity fit hand-in-glove with something they shared on a spiritual level—the sense that a successful lunar landing would require the hand of God. There were hundreds of thousands of men and women scattered across America and around the globe, working for thousands of contractors and subcontractors, all working to bring the more than five million Apollo components together. All this scientific and engineering expertise coming from every corner of the country would be gathered, organized, and concentrated into a spacecraft on a single launch pad at Cape Kennedy. The fragile assembly of spacecraft and its crew would be sitting atop an engine that produced nearly one million pounds of thrust for six minutes and then hurled into the most unforgiving environment into which man had ever ventured.

And there was another matter few outside the realm of the Apollo Project were aware of. Program managers were in the midst of testing a rather radical concept called the "all-up" doctrine. The doctrine stipulated that none of the million-plus individual components of the Apollo spacecraft be tested individually before launch to see whether they worked properly. The first combined use would be on the intended mission itself. This meant that each component had to work *perfectly and synchronously with every other component the first time—beginning with launch.* Risky as this strategy was, it dramatically reduced the number of tests each mission required, making President Kennedy's before-the-decade-is-out deadline feasible.

It wasn't until after the fire that the germ of two related ideas grew in the mind of Reverend Stout. Both ideas were bold, but one would prove historic. The first was to form a prayer group, which in time became known as the Apollo Prayer League (APL). The second, which he initially had only the vaguest idea of how to implement, was to land a Bible on the moon.

Stout realized that an historic opportunity was at hand. Mankind was either going to attempt to take the Holy Scripture to its first extra-terrestrial destination or the opportunity would likely evaporate into the ether of unmade history. Of course, there was also a more personal reason. He was aware of White's aspiration to take a Bible to the moon and was determined to see his friend's goal achieved.

Although an ordained minister, Stout was also a scientist and, as such, saw an underlying connection between science and religion, advising that "Man's search for truth in science and technology is not incompatible with his search for spiritual truth." While some in the religious community were suspicious and even fearful of science and the idea of manned space flight, Stout was not among them. Nor was he among those who saw it as an exclusively "Christian" project. He saw the Apollo missions as a collective human spiritual odyssey that spanned all religions in every corner of the world. He viewed science as a noble human endeavor when accompanied by faith and one of mankind's greatest hopes for future generations. From his perspective, science held great promise.

Stout's versatility as a scientist and religious leader was to become an invaluable resource to the space agency. His secondary duties included overseeing the work of thirteen volunteer ministers in NASA through a group he organized called Aerospace Ministries, an organization similar to one he had formed at Cape Canaveral to coordinate the diverse ministries throughout the agency. One of his principle tasks was to see that the cloister of chaplains kept a few degrees of separation between their work and their religious activities.

"Some of them wanted to evangelize people," he said. "And you can't go telling people they've got to be saved. The whole thing at NASA was a stressful situation and you just can't do things that way."

By now the immense spiritual significance of the space program was revealing itself to the public, and wary NASA officials wanted to nip any and all public relations problems in the bud. Moreover, it was becoming clear to all that there was a need to maintain a respectful distance between the Apollo Prayer League and NASA activities. Stout took his concerns to bestselling author and pastor Norman Vincent Peale, who was experienced and influential in religious circles. Peale had authored *The Power of Positive Thinking*, a widely-read book, and founded *Guidepost* magazine, an immensely popular Christian publication with a widespread and growing circulation. He was an enthusiastic supporter of the space program and urged Stout to work toward fulfilling Ed White's dream of landing a Bible on the moon. He also offered any assistance he could to the APL. At the time, it was feared the Soviets would get to the moon first, taking with them their own "bible," so to speak—Marx's *Communist Manifesto*. Peale was an ardent anti-Communist and wanted to see an American carrying a Bible to the moon on what some were now calling the "chariots of Apollo." The effort would take a prodigious amount of planning and work, all of which would have to be handled with political aplomb.

"We need people who believe in the power of prayer to be on the team," Stout told his co-workers at the Manned Spacecraft Center in the days following the fire. And that's exactly what he got.

The concept of the Apollo Prayer League was first hatched by Stout's wife, Helen, during a conversation in their tan road-worn Jeep one afternoon when she suggested to John that they take the model of her small prayer group of fourteen NASA wives and apply it to the whole of NASA's 300,000 employees around the world. She would supply administrative support for the League, including publication and distribution of an APL newsletter and APL News Wire Service. The first official organizational meeting was held with a handful of NASA volunteers around a small table at the Manned Spacecraft Center.

The bylaws of the Prayer League were drawn up specifying that the governing committee be comprised of one person to represent NASA contractors, one person to represent the Church, and one to represent NASA. In addition, one person would be designated to represent the astronaut involved for each mission, usually his pastor, a relative, or friend. Stout was elected as director.

Together the group devised a plan to drive membership, which would proliferate by way of a strategy that called for each to contact five friends, who in turn would be asked to contact five of their friends, and so on.

It was a uniquely American enterprise. What began as a grassroots initiative within NASA quickly became an international movement and spread to NASA sites outside the U.S. Anyone could join regardless of their faith. Members could be anyone who believed in the power of prayer, united together to pray for success of the mission, the astronauts' safety, and for God's guidance over the skill and foresight of NASA ground personnel involved in Apollo. Reverend Stout described the germ of the group's formation in an APL newsletter:

> When the Apollo Prayer League was first formed, we had just experienced a heartbreaking setback, for three of our finest astronauts had been killed during the testing of our first manned Apollo spacecraft. Some of the employees at the Manned Spacecraft Center suggested that people of good faith from all over the world join them in taking their petition to God. They referred to this new fellowship as the Apollo Prayer League. We feel that God has answered our prayer and has guided in the development of our space program. We feel He has given us special guidance in the development of the Apollo Prayer League as His first interplanetary church.

As an industrial chaplain within NASA, John was able to leverage the cause of the prayer group. At a time when the space program was struggling to get back on its feet, the movement quickly gained momentum. Within a month after its formation, the League had nearly 5,000 members. Soon prayer groups were popping up around the world.

Although the Apollo Prayer League received no funds or grants from NASA or any other government agency, it required that all members of its board of directors be employed at the Manned Spacecraft Center. As a part of Aerospace Ministries, the APL would serve to complement NASA, which could not become involved in religious matters. As a nonprofit organization, its sole support would come from contributions of individuals and churches.

Nevertheless, Reverend Stout, Aerospace Ministries, and the Apollo Prayer League were in for a long and tumultuous journey. As the secular NASA embarked on the greatest scientific and spiritual journey of its time, Reverend Stout was about to become acquainted with a cantankerous atheist from Austin by the name of Madalyn Murray O'Hair.

O'Hair, the antagonistic and foul-mouthed atheist renowned for her success in persuading the Supreme Court to ban prayer in public schools, was now poised to pounce on any religious activities in space as a clear violation of the U.S. Constitution requiring separation of church and state. Following her 1963 *Murray v. Curlett* Supreme Court victory to oust prayer from schools, she founded the organization of American Atheists. It was controversies like this that Madalyn O'Hair lived for.

"Religious people are throw-away people," she snarled.

At the National Annual Convention of the American Atheists, a sampling of her many targets included the exemption from taxation enjoyed by religious bodies, prayers at breakfast at the White House, and crosses as part of Christmas decorations on Federal Buildings. O'Hair reportedly toppled bingo tables in churches. Watching a female orangutan on television, she snipped, "The Virgin just made another appearance." People would hear her speak live and their mouths would hang open.

Stout knew the woman and her tactics well. O'Hair, he surmised, was circling overhead watching APL activities like a hungry hawk. But for now, he would need to see a young agency through its darkest hour.

> "If you could ask any of the men and women in the space program who have died in the line of duty, or those who will yet surely do so, I believe they would offer a simple answer: We should go forward for reasons that nourish the human spirit, simply because we are capable of doing so."
>
> *- Edgar Mitchell, Lunar Module Pilot, Apollo 14*

8

The Inquest

The day after the fire, Ed White's father, retired Major General Edward White Sr., made a statement to the press from the front porch of his home in St. Petersburg, Florida.

"My son died doing a job for his country, and I'm sure he would have wanted it this way... We realize that these things are inevitable," he added, "but the world must go on."

This remarkably selfless statement, made at the depths of a grieving father's anguish, offered hope to the entire U.S. space program. The lifeblood of the Apollo program was dependent upon the goodwill of the American taxpayer, and NASA officials were quietly but deeply concerned that the deaths of these astronauts would prove corrosive to their already fragile political sentiment. Although an investigation into its cause had yet to get underway, certain facts were inescapable.

As details concerning the cause slowly leaked out, the American public initially seemed prepared to accept the dangers and difficulties involved in space exploration and continued to be resolute in accomplishing President Kennedy's goal of putting a man on the moon by the end of the decade. Nevertheless, it would take time for Congress, who held NASA's purse strings, to take the temperature of their individual constituencies. Three of America's best and brightest had just died in a horrific accident in the pursuit of a distant goal that was still, literally, a long way off. In the aftermath of the tragedy, nothing of a political nature could be taken for granted.

There was also another concern that loomed over all others: putting a man on the moon was a tough sell at a time when the country was simultaneously waging an expensive war on the other side of the globe, a war in which the country was deeply divided.

Projects Mercury and Gemini set the stage for Apollo, where it was all supposed to come together. This was the final leg of a race to put Americans on the lunar surface before the Soviets. But it was now 1967. The end of the decade was less than three years away, and spacecraft contractors, NASA personnel, and the astronauts themselves were putting in grueling hours. The grind of work and endless training, coupled with the memories of departed friends, were beginning to take their toll on the entire project.

Marshall and Mildred Wilkes both worked for NASA and were members of the Apollo Prayer League. Marshall was a financial analyst and Mildred was executive secretary for Dr. Rolf Lanzkron, the leading scientific engineer charged with the examination and investigation of the fire. Lanzkron was responsible for all project engineering activities for the spacecraft launched at Cape Canaveral. A tough orthodox Jew, Lanzkron had given priority attention to the Apollo 1 spacecraft and coordinated directly with North American's Chief Project Engineer, Ray Pyle. Mildred handled all of Lanzkron's calls and correspondence and took the tragedy personally.

"The fire was terrible," she said. "The strange thing was that Dr. Lanzkron had dictated a letter to [Deputy Center Director] George Low for me to type up changing the wiring in the spacecraft that would have prevented the fire from happening. But George Low only got it the day of the fire. There wasn't anything they could do about it then. It was a sad time."

So far nineteen Americans had flown in space. All had survived and none had been seriously injured while ascending to or descending from the heavens. It was a remarkable record and the result of procedures painstakingly designed to prevent tragedy in space. Airplane crashes had already claimed the lives of Astronauts Elliot See, Charlie Bassett, and Ted Freeman. Then, less than five months after the Apollo 1 fire, another astronaut, Ed Givens, from the fifth group of astronauts, was killed in a freak car accident near Houston. But as traumatic as these deaths had been, they had not been the result of faulty design and engineering of a spacecraft; their deaths had not been the fault of their colleagues in the space program. On the other hand, the loss of three men during a ground test for the first manned Apollo flight had an utterly demoralizing effect on everyone associated with the Apollo project. "The guilt people felt was just tremendous," observed Flight Director Glynn Lunney.

Many within and outside of NASA contended that the accident was completely avoidable. Internally, NASA had feared that danger to future crews would increase as spacecraft became more sophisticated and the missions more ambitious. Ambition added complexity, which in turn heightened the danger. NASA officials had warned against undue optimism, pointing out that any program as large and ambitious as Apollo would inevitably take its toll in lives. Man is fallible. Nowhere was this more abundantly clear than in a program made up of several hundred thousand people scattered across several continents.

NASA executives had pushed the entire Apollo team to strive for perfection, and they had been nothing if not imaginative. One NASA program featured the "Lunar Roll of Honor." The first lunar landing party would carry this Roll of Honor, which was a microfilm list of 300,000 names of those in NASA who had aided significantly in this epic endeavor. The program was intended to inspire extraordinary devotion to excellence. After the fire, the idea was abruptly dropped. NASA officials realized that it was just as difficult to fix blame for failure as it was to assign credit for success.

The day following the accident, NASA Administrator James Webb asked Floyd Thompson, director of the Langley Research Center, to head up an "AS-204 Review Board." To most Americans, the muffled drums and the somber eulogies were the only suitable form of tribute left to honor Grissom, White and Chaffee. But to the stunned Apollo technicians there could be no more fitting service to the memory of these astronauts than an exhaustive inquest into the burned-out spacecraft. Thompson knew the scope of the challenge he was up against. This was without question the biggest news story of the day and the panel would be working amid a media circus. He expected it to be hard, thankless work—and it was.

On April 10, a House of Representatives Subcommittee charged with NASA Oversight began a "full and complete review of the Apollo accident, its causes and effects." From the beginning, misleading accusations were leveled that the investigation was a sham, that it would essentially be a case of NASA investigating itself. Some of the more sensational headlines accused the space agency of "murder." Gus Grissom's father asserted that if McDonnell Douglas had built the spacecraft instead of NAA, his son would still be alive.

In fact, any and all skeletons in NASA's closet were about to be brought to light. Two years earlier a report compiled by NASA Apollo Program Director Major General Samuel Phillips for NASA Assistant Administrator George Mueller to review NASA's contract with North American Aviation listed a plethora of problems. Quality control troubles were among the most critical problems cited in Phillips' report. Ironically, the report was shelved at the time because it was thought to be irrelevant in light of the program's unblemished record of success. But after the catastrophe, ABC news correspondent Jules Bergman notified Senator Walter Mondale, a member of the Congressional oversight committee, that he had seen a copy of the Phillips Report during a recent visit to the Manned Spacecraft Center. NASA Administrator James E. Webb, however, claimed to be unaware of its existence.

Upon hearing of the contents of the Phillips Report, Senator Mondale lit into George Mueller during a public hearing. The questioning turned into an embarrassment for NASA officials. It became apparent that there was indeed a Phillips Report issued two years before the fire at the request of George Mueller himself, and that it had outlined specific failings, project delays, and budget overruns. Revelations in the report intimated that NASA officials had been playing fast and loose with the lives of the astronauts.

James Webb immediately became a key figure in the hearings. On a personal visit with Mondale, he criticized the senator for bringing up the Phillips Report during an open hearing, arguing that Mondale should have met with him privately on the matter rather than in a public forum. Mondale disagreed.

"I think it's rotten and I'm going to blow it out of the water," Mondale replied. "I'm a United States senator and this is public business." For the first time in his entire tenure at NASA, Webb felt himself losing control of the politics of the situation. From that moment on Senator Mondale would be calling the shots.

In the face of endless scrutiny, Apollo managers were made to take a long, hard look at the details of their track record. Was there a grain of truth to the accusations? Had they really done all they could in the name of crew safety? Procedures for emergency escape called for the hatch to be opened in a minimum of ninety seconds, but in practice the crew had never accomplished it in less than two minutes. In truth, Grissom had to take time to lower White's headrest so White could reach above and behind his left shoulder to access a ratchet-type device to release the first in a series of six latches. According to one source, White made only part of a full turn with the ratchet before he was overcome by smoke. It was also revealed that in March 1965, the "crew systems" personnel in Houston had wrestled with the question of pure oxygen for space travel and the inherent fire danger it presented. Most of the studies were based on the possibility of fire while in space, however, and concluded that a pure oxygen system was safer, less complicated, and lighter than the 79% nitrogen, 21% oxygen system used by the Russians. Thus, it made for a more fuel-efficient spacecraft. The best way to guard against fire, they concluded, was to keep flammable materials out of the cabin. But this didn't happen. Velcro, for instance, was everywhere. There was also highly flammable netting and nylon on their suits, and the crew had brought in a large block of Styrofoam to help relieve pressure on their backs during long hours of testing. The Styrofoam exploded like a bomb.

As investigators delved ever deeper into the causes of the tragedy, NASA officials were confronted with even more skeletons. NASA's relationship with North American had been an odd one from the very beginning. There was speculation as to why the giant aerospace manufacturer had been chosen as the prime contractor. McDonnell Aircraft Company had built the very successful Gemini and Mercury spacecrafts and had essentially written the book on spacecraft technology. Why, then, was North American awarded the enormous Apollo contract even though they had little, if any, experience in the design and manufacture of the spacecraft? Their mandate was to build the most advanced spacecraft ever assembled, utilizing an entirely new generation of digital technology.

There was also the question of morale among NAA employees, as cited in a 50-page report compiled by Robert Baron, a missile preflight inspector for NAA. In a lengthy hearing, U.S. Congressman Edward Gurney, a member of the oversight committee, quizzed Baron on the report:

Mr. Gurney: Mr. Baron, you mentioned something about the morale... Do you think that there was a really serious morale factor with people generally dissatisfied all over the place with their jobs and what they were doing?

Mr. Baron: I would say for the most part, yes.

Mr. Gurney: Assuming what you say about morale is true, do you think this affected the work on this job?

Mr. Baron: Yes, sir, I do.

Mr. Gurney: In what way?

Mr. Baron: Well, especially in reference to safety, lackadaisical in some job operations, sleeping on the job, people, just a lot of them, didn't care one way or the other, and I am not talking about isolated instances, many times of book reading and sleeping and things of this nature... Another technician witnessed one evening when he was working, three technicians who were supposed to flush out [the capsule] by purging the environmental control unit with an alcohol solution to clean it and get it ready for proper use. He disclosed to me that a 55-gallon drum of 190-proof alcohol had been delivered to the site...

>The three men who were assigned to flushing this unit out were – well, one of them took a 5-gallon jug of this stuff home and one other, or perhaps all three of them, I don't really recall, had mixed this stuff and cut it with water and were drinking it right there at the site.... .

[Edward Gurney yields to John Wydler]

Mr. Wydler: Well, that doesn't have anything to do with this particular wire or this particular door?

Mr. Baron: *Possibly so, because they were working on that unit and the spacecraft...*

But a surprise lay in store for many of the critics of NAA. As it turned out, not all problems with the Apollo 1 spacecraft could be placed at NAA's doorstep. NASA management—and even the astronauts themselves—were responsible for some of the problems. North American, for instance, had initially proposed that the hatch open outward and be outfitted with explosive bolts for rapid exit in the event of an emergency. They had also wanted an environmental system that would provide the cabin with a non-flammable atmosphere of oxygen-nitrogen mixture comparable to that used by Russian spacecrafts, as the danger of fire in a pure oxygen environment was no small matter. NASA, and Grissom in particular, disagreed on both counts. The hatch, they argued, could be accidentally opened, resulting in an incident similar to Grissom's misadventure on *Liberty Bell 7*.

Stephen Clemmons, a 36-year-old NAA Senior Spacecraft Checkout Mechanic, remembered the spacecraft condition vividly.

"The lower equipment and floor directly under the seats was wide open with no protective covers, exposing miles and miles of wire, aluminum tubing, valves and electrical devices. NAA wanted to install specially designed covers on the lower equipment bay floors to protect these items, but NASA engineers said they didn't have time to design the covers for ground operations and used four-inch-thick rubber mats instead."

"There was an attitude of 'moon fever,'" Clemmons said. "During the manufacture and testing, there were over 1800 critical discrepancies written, which took time to investigate and find fixes, but NASA was adamant that the schedule would not slip, so many of them fell by the wayside."

Finally, there was the issue of Velcro. The astronauts themselves had been its biggest advocates. So fond were they of the material they had literally wallpapered the cabins of the Gemini spacecrafts with it. North American, however, objected to its use due to its highly combustible nature. With its fuzzy skin, Velcro was like kindling. They cited that the rule of having no more than 500 square inches of Velcro inside the spacecraft cabin had been ignored and the capsule had over 5000 square inches of Velcro in it.

Incredibly, in each case NAA's concerns had been voted down by either NASA or the astronauts. Previous missions had succeeded because of the virtues and in spite of the flaws of the status quo, and it was hard to argue with success. As the hearings progressed, it became increasingly difficult to assign blame. Prior to the fire, there had been little reason to question the tried and true. North American Aviation had pointed out the risks of each issue and then yielded to its client's wishes.

In an appearance before the Congressional committee, Webb made a moving plea for sanity in the hearings.

> If any man wants to ask for whom the Apollo bell tolls, I can tell him. It tolls for him and for me as well as for Grissom, White, and Chaffee. It tolls for every astronaut or test plot who will lose his life in a space-simulated vacuum of a test chamber or the real vacuum of space. It tolls for every astronaut-scientist who will lose his life on the moon or Mars...We have a grave responsibility to work together and to purge what is bad in the system that we together have created and supported. We have perhaps an even graver responsibility to preserve what is good and what represents still at this hour a high point in all mankind's vision.

Unmoved by poetic appeals, the Congressional committee expressed its disgust regarding the capsule's pure oxygen environment, the frayed wires, and fatally clumsy hatch.

"The level of incompetence and carelessness we have seen here is just unimaginable," they reported.

On the other hand, it was pointed out by many that there was manifest unfairness in the scrutiny being focused on the failings of NASA at a time when the entire U.S. government was failing Americans in its military objectives abroad. No one seemed able or willing to contrast the unusually high standards expected of NASA with regard to losing three astronauts in the same year that 56,000 American soldiers were wounded in action and 9,400 were killed in Southeast Asia in a war whose outcome was uncertain, whose purpose was ill-defined, and whose execution was flawed. After all, NASA astronauts knew theirs was a risky adventure and climbed willingly into their capsule. The soldiers in Vietnam had little or no choice about their exposure to risk.

As the Congressional hearings lingered on with no end in sight, NASA developed stringent new safety guidelines on its own. They understood what had gone wrong. They understood the mindset that had allowed the tragedy to occur in the first place. Regardless, NASA still had a job to do—and it came with a deadline. It was clear the hearings were not going to end any time soon.

Eventually astronauts Frank Borman, Wally Schirra, Deke Slayton, Alan Shepard, and Jim McDivitt appeared before Congress with a plea to end the witch hunt. After repeated questioning by the committee as to whether they maintained confidence in the program, Borman answered a congressman's doubts:

> You are asking us do we have confidence in the spacecraft…I am almost embarrassed because our answers appear to be a party line…The response we have given is the same because it is the truth. We are trying to tell you that we are confident in our management, and in our engineering, and in ourselves. I think the question is really, *are you confident in us?*

By the time Borman vented his exasperation to Congress, both NASA and North American Aviation had effectively responded to each point of criticism brought to light by the investigation. A new "unified" single hatch would be a dramatic redesign of the complicated two-hatch system. The new system was heavier than the old, but it could be opened outwardly in five seconds. NASA would stick with the pure oxygen system and the use of Velcro, but extensive new measures would be taken to mitigate the dangers.

What Congress would slowly come to realize was that space exploration was an inherently dangerous enterprise. And America, it turned out, was not the only country dealing with this reality. Tragic news from Russia arrived in Washington, D.C. on April 24, 1967, squarely in the middle of the Apollo 1 Congressional hearings. One of the great Soviet space-faring heroes, Cosmonaut Vladimir Komarov, had died when the parachute of his Soyuz I spacecraft failed to deploy properly after reentry. The spacecraft then crashed into the grassy steppes of Kazakhstan. But there was more to the story.

It was clear the Russians knew they had problems for over twenty-four hours before Komarov died and had been fighting to correct them. A U.S. national security analyst monitoring Komarov's radio signal from a NATO facility later recounted the story. According to unsubstantiated reports, the last events leading up to the disaster were heartbreaking.

"The Kremlin called Komarov personally. They were crying and told Komarov he was a hero. The guy's wife got on the phone too, and they talked for a while. He told her how to handle their affairs and what to do with the kids. Towards the last few minutes he was falling apart. In a lot of ways, having the sort of job we did humanized the Russians. You study them so much and listen to them for so many hours that pretty soon you come to know them more than your own people."

In the midst of the Cold War, the hearts of the American people went out to the Russians in a way they had never done before. The news could not have arrived at a more poignant time and prompted a heartfelt question from NASA administrator James Webb.

"Could the lives already lost have been saved if we had known each other's hopes, aspirations, and plans?" he posited. "Or could they have been saved if full cooperation had been the order of the day?"

But the day of international cooperation had not yet arrived. In April 1967, both countries still saw themselves as strategic competitors in a race to the moon. And both countries would endure the deaths of their best and brightest as they continued the struggle to win that epic race.

Shortly after Borman's Congressional testimony, he was asked to oversee the recovery efforts at North American Aviation. It was a time he came to look back upon as the most productive episode of his career. Addressing the many problems of the Apollo spacecraft was taking longer than anyone had expected, but Borman knew the Saturn V spacecraft was destined to become an extraordinary ship. He also understood the profound psychological significance of getting it to the launch pad as soon as possible. A successful mission would lift NASA out of the shadow of the fire and restore its shaken confidence.

But there were lingering problems within the space agency. NASA wasn't merely an organization of machines, it was also an entity made up of individuals. In the tortured emotional aftermath of the accident, Assistant Flight Director Gene Kranz, along with several others in leadership of NASA, became acutely aware of a new problem: Mission Control personnel and backroom technicians had just lived through their worst nightmare, and the psychological toll had all but immobilized them. Many felt indirectly complicit in the deaths of their friends. They were also very young and had never before been so closely associated with disaster and death. Many of the controllers were in their twenties. Kranz himself was only thirty-three years old.

Dutch Von Ehrenfriend, one of the controllers who monitored the guidance console during the fire, was deeply traumatized by the event. He was, as Kranz depicted him, "a vulnerable young man who had witnessed his friends' death." John Aaron, a bright 27-year-old flight controller, nearly drove himself crazy playing back the data over and over and studying the electrical current spike believed to have set in motion the fire.

"[Aaron] pushed himself beyond exhaustion and finally had to be driven home," Kranz said.

Kranz understood that the men who were taking America to the moon—the astronauts who put their lives on the line—as well as NASA technicians, would somehow need to rise above the trauma of this experience before they could fully focus again on their jobs. They were unusually resilient young men, but the collective sense of guilt threatened to cripple the entire program.

On the Monday morning following the fire, Kranz called a meeting of his branch and flight control teams, who still remained in a state of emotional turmoil. In an auditorium at Mission Control he delivered what was destined to become a legendary discourse on integrity and leadership. It came to be known as the "Tough but Competent" speech, and none of the people in the auditorium that day would ever forget it.

The talk opened with a review of the accident and an announcement of the formation of an investigative review board led by Thompson. Kranz then did something that became the stuff of legend in the halls of NASA, something that would eventually allow his people to harness their crippling grief and move beyond it. He concluded his meeting by identifying what he believed to be the problem: The people in that room and throughout the agency were "not tough enough," he said. They were, in Kranz's mind, avoiding their responsibilities. "We had the opportunity to call it all off, to say, 'This isn't right. Let's shut it down,' and none of us did." From that moment on, he demanded, they would all be hard-headed.

Somehow Kranz intuitively understood that they did not need or want sympathy. What these young men needed before they could get back to their jobs was a dose of cold, hard reality. And that's exactly what he gave them:

> Spaceflight will never tolerate carelessness, incapacity, and neglect. Somewhere, somehow, we screwed up. It could have been in design, build, or test. Whatever it was, we should have caught it. We were too gung ho about the schedule and we locked out all of the problems we saw each day in our work.

Every element of the program was in trouble and so were we. The simulators were not working, Mission Control was behind in virtually every area, and the flight and test procedures changed daily. Nothing we did had any shelf life... I don't know what Thompson's committee will find as the cause, but I know what I find. We are the cause! We were not ready! We did not do our job. We were rolling the dice, hoping that things would come together by launch day, when in our hearts we knew it would take a miracle. From this day forward, Flight Control will be known by two words: "Tough and Competent"... We will never again compromise our responsibilities... Each day when you enter the room these words will remind you of the price paid by Grissom, White, and Chaffee. These words are the price of admission to the ranks of Mission Control.

North American Aviation made changes to its top management and brought in a new team to run its Space Division. NASA replaced the Apollo Program Manager with a new man from their own ranks, George Low, who became a pillar in the program during the pivotal years that followed. At the time of the accident, the flight schedule had projected a possible lunar landing before the end of 1968. This goal now appeared to be no more feasible than a pie in the sky, and Low and everyone else knew it. NASA was already nearly five years into the program and had yet to even get an Apollo Saturn V rocket off the launch pad.

In June of 1967, Kennedy's vision of landing a man on the moon before the end of the decade appeared to be fading fast.

"Your father had tremendous persistence... He believed that the exploration of the universe must and will go on, that it is our destiny as children of God to keep seeking new challenges, asking new questions, finding new answers. He knew the farther we go, the more mysteries we encounter. But this only proved to him the infinite power and majesty of God."

- Ed White Sr.
in a note to his grandson, Edward White III.

9
APOLLO 4: UP FROM THE ASHES

If Apollo had seemed complicated before the fire, it became even more so afterward. If NASA gave the impression of being hurried in late 1966, it gathered even more momentum in the months that followed.

By the fall of 1967 the Apollo program had undergone a revival that by all accounts would have been unimaginable nine months earlier. The mammoth Saturn V rocket was the crowning achievement of rocket scientist Wernher von Braun and the heaviest mechanical device ever heaved into the atmosphere. Launching one, NASA officials knew, would make for stunning live television.

Apollo 4 would be the first unmanned flight of the Saturn V launch vehicle. It would also be the first liftoff from Launch Complex 39, specifically built for the Saturn, and the first time a spacecraft would reenter the earth's atmosphere at speeds approaching those of a lunar return trajectory. On its nose rested an unmanned spacecraft, along with newly designed features, including an improved heat shield and a new hatch. This list of firsts came with 4,098 measuring and monitoring instruments aboard the rocket and the spacecraft.

The stakes could not have been higher. The Saturn V rocket itself had yet to be fully tested. It was entirely possible that it would explode in an epic fireball on live television and become an unprecedented public relations nightmare. On the other hand, if the mission succeeded, it would be just the push forward the space agency needed. There was no middle ground and the people at the Flight Control Office—many of whom found themselves working around the clock—understood the consequences.

Just prior to the Gemini 4 mission, Gene Kranz moved from the role of Assistant Flight Director to that of a full-fledged Flight Director, and he was now being inundated with a barrage of data and tasks for Apollo flights he had never before encountered. Concerned about assimilating so much data in his head, Kranz developed a scheme of indexed handbooks, one for each mission, in which he color coded and highlighted areas of critical constraints. He spent seemingly endless days and hours pouring over the handbooks, even working evenings after his children had gone to bed. His greatest fear as launch day approached was that he would lose one of the handbooks. To assure they were easy to spot, he tore pages of bikini-clad women out of the *Sports Illustrated* swimsuit edition and glued them to the covers.

The hours were long and the pace grueling for NASA personnel clear up to launch. Sy Liebergot, a lead flight controller, observed certain personnel exhibiting a "lack of clothing changes, weight loss, and late nights at the office." Often people simply spent the night where they worked as they obsessed over "the tiniest detail of every aspect of the booster design." It was the kind of hours and obsessive behavior that eventually led to a rash of divorces among many of the astronauts and NASA employees. The driving force behind their obsession was the epic dimension of the project itself. It was an historic opportunity, something they never dreamed they would be a part of.

"It was like we were chug-a-lugging a fine wine," John Aaron told Glynn Lunney.

The Apollo 4 spacecraft would carry a mock-up of the lunar module (LM) which, for the purposes of this mission, would essentially serve as ballast. Once the launch vehicles and spacecraft were stacked at the launch pad, the immense fueling process began. Three days later, on November 9, 1967, at 7:00 a.m., the birds, reptiles, and animals of higher and lower order that gathered at the Florida Wildlife Game Refuge near the launch site received a tremendous jolt. When the five engines in the first stage of the Saturn V ignited, a man-made earthquake and shockwave erupted.

Before a worldwide television audience of tens of millions, Apollo 4 slowly lifted off the pad, taking a full twelve seconds just to clear the tower. The shockwaves radiating from the massive engines were far more powerful than anticipated, shaking tiles and insulation from the ceiling of the press building five miles away where CBS news anchor Walter Cronkite and reporters from around the world gathered. As the rocket pitched over on its back and reached skyward, the Saturn V created the second loudest man-made explosion ever. Only a nuclear blast was louder.

As the service module engine was fired, the spacecraft rocketed off at some 24,000 miles per hour. After several laps around the planet lasting four and a half hours, the engine was fired once again, sending the capsule back into the earth's atmosphere. The heat shield performed as planned and the capsule splashed down on target in the Pacific Ocean, just north of Midway Island. Everything worked so well and with so little trouble, NASA was confident Apollo was once again on the way to the moon.

The entire cast of NASA celebrated. "With this one mission," Liebergot said, "the lunar program took a giant step forward."

―――

Just a few months earlier, in April 1967, Deke Slayton, Director of Flight Crew Operations, had called a select group of astronauts together for an unforgettable meeting at the Manned Spacecraft Center in Houston. "The eighteen of you," he said in his characteristically direct tone, "are going to fly the missions that get us to the moon, including the lunar landing."

It was a stunningly bold statement. Slayton wasn't mincing words. He was putting the astronauts, and therefore the entire U.S. space program, on notice. In his mind, a *manned* spaceflight was the only catalyst that would restore the program's previous momentum. And yet the Apollo spacecraft was far from being ready for manned space flight. The lunar landing module, the "ugly, buggy, leggy lander" the astronauts would use to navigate their descent from the command module to the surface of the moon, was revealing itself to be riddled with enormous design and quality issues.

The problems with the lunar lander were as unique as the spacecraft itself. Since it was designed to fly only in a weightless atmosphere, there was no way to test it on the ground, as its spindly spider legs would collapse under its own weight. Strict weight limits could not be exceeded, since the module had to be light enough to be pushed out of earth's strong gravitational pull and then transported a quarter of a million miles into lunar orbit. Once undocked from the command module, it had to land on and launch from an airless world, making it the first "flying" machine without a rotor, a wing, or any other kind of aerodynamic surface. Since there was no conceivable way to test it, you wouldn't know if it worked on the moon until it did—or didn't. Many of the problems with the lunar module were old-fashioned quality control issues. It was an extraordinarily complicated machine that was being "invented" at a pace that was virtually outstripping the capacity of the engineers, designers, and contractors to build it.

Delays with the lunar lander caused a bottleneck in the entire flight schedule. After four months of painstaking tests, the lunar lander was finally shipped from the Grumman development lab to Cape Kennedy. But once again the sense of triumph was short-lived. The ascent engine on an identical lunar module being tested at the Grumman plant failed, so the entire assembly was sent back to the drawing board. Everyone at NASA understood they still had a long way to go.

Of all the problems Grumman was experiencing, the most forbidding was the persistent failure of the lunar module's ascent engine—the rockets needed to launch the vehicle from the lunar surface to the command module orbiting overhead, which was their only ride home. This was the type of problem that played games with the mind. If the ascent engine failed, two astronauts would be left stranded on the moon for what remained of their very short lives, a terrifying possibility of which NASA, Grumman, and every astronaut who aspired to walk on the moon were painfully aware.

Nevertheless, in late January 1968, NASA prepared for yet another unmanned launch—Apollo 5—which would test the lunar module in earth orbit. Since the lunar spacecraft's "legs" were not yet ready, it would fly without them. The lander would also be windowless, as the windows too had failed during testing. But at least the retooled lunar descent and ascent engines could be tested in earth orbit and the lunar module control, guidance and environmental systems could be run through their paces, all in a fraction of a second with the lunar module still firmly affixed. The launch vehicle for the Apollo 5 flight would be the very Saturn 1B rocket mounted under Apollo 1 at the time of the fire.

"Naturally, those of us who were training for a lunar landing flight were very interested in the Apollo 5 mission," said Buzz Aldrin.

The challenge was understandable. NASA was trying to launch a vehicle the size and mass of a Navy destroyer, with eleven new engines powered by new fuels, new pumps, and using new guidance.

Finally, on January 22, after six hours of delays, Apollo 5 was flung into lunar orbit with the lunar module. Although the lunar module's ascent and descent engines fired and shut down on command, it experienced a slew of software problems. As one problem was resolved, another sprung up in its place; and by mission's end NASA officials had come to the inescapable conclusion that the lunar lander would not be ready for manned testing for the balance of the year.

But steady progress was being made on other fronts, and by the spring of 1968 there was a gathering sense that, however slowly, the embryonic Apollo spacecraft was finally coming to full term. As developments on the lunar module inched forward, the command and service modules were taking giant strides. By winter the latter two were scheduled for an in-space mission, along with a dummy lunar module, all of which would be stacked and launched on the nose of a Saturn V rocket.

On April 4, 1968, another unmanned rocket, Apollo 6, launched in a cataract of flame and thunderous noise. This time, however, the television audience was virtually absent as news from Memphis, Tennessee, flooded the networks with reports that Dr. Martin Luther King Jr. had been shot and killed. Urban riots erupted throughout the country later that day. For the moment, Americans could not be bothered with the space race. The contrast between the ambitious race to another celestial world and the immediate tragic events on earth could not have been more stark. Project Apollo, the great technological, spiritual, and geopolitical endeavor, suddenly appeared strangely superfluous and out of step with reality.

As though echoing the chaos on the streets of American cities, problems arose as Apollo 6 roared into the heavens. Shortly after take-off, the unthinkable appeared to be happening. The massive Saturn V booster rocket began contracting and expanding like an "angry alligator" during its furious ascent. It looked like a runaway circus ride. After a long, white-knuckle moment, the problem subsided just as the spacecraft reached into space.

Wernher von Braun and Chuck Casey, head of the Booster Flight Control Office, suspected that tons of liquid fuel rushing through the engines was burning unevenly, causing violent fore-and-aft contraction and expansion. The problem was so acute that the spacecraft headed for an elliptical orbit rather than its planned circular orbit. But the rocket held together and managed to foist its payload into space. Later, the service module engine fired as planned and thrust the spacecraft into a higher orbit and it eventually ran through its paces.

In spite of the many technical failures, Apollo 6 was viewed as a success. Ground control had averted disaster, and the mission itself had revealed a laundry list of critical flaws, all of which would be addressed before risking a human life in space. What mattered most was that another rung on the ladder to the moon had been cleared. By the summer of 1968, NASA was at long last ready to return to the dangerous business of sending men, rather than machines, into space.

> "If you have been a fighter pilot for any length of time, you have seen your friends get killed—often. You build up a certain immunity. I flew with such men and knew them well, men whose faces I can no longer remember. They are frozen in time now as shadows in old group photos. My young brother Ken will always be the twenty-nine-year-old fighter pilot."
>
> *- Walt Cunningham, Lunar Module Pilot, Apollo 7*

10
APOLLO 7: RESURRECTION OF APOLLO

Even twenty months after the fire, the shadow of Apollo 1 still hung heavily in the air. The press was everywhere, and it didn't go without notice that the first Apollo crew to launch, Apollo 7 Commander Walter (Wally) Schirra, Command Module Pilot Donn Eisele, and Lunar Module Pilot Walt Cunningham, had been the backup crew for Grissom, White, and Chaffee. It was well known that Schirra was a neighbor and close friend of Grissom's. He was executor of Grissom's will and felt the tragic loss deeper than most. He also knew that Gus would have wanted the Apollo program to move on. "If you're sad, you mourn the loss," Schirra said, "but you don't wear the black arm band forever."

Schirra's reputation as a jokester was well known within NASA, so his actions on an earlier Gemini 6 flight had come as no surprise when he and Tom Stafford commenced to perform the first ever space rendezvous. Frank Borman and Jim Lovell had launched seven days earlier and were already orbiting the earth in Gemini 7 in preparation for the rendezvous. Once in formation, the two Gemini capsules flew in close proximity for five hours, coming within a foot of each other but never touching. During the rendezvous, Schirra, a commissioned officer in the U.S. Navy, grinned and held up a sign to his window: "Beat Army," for the benefit of Gemini 7 commander and West Point graduate Borman.

While it was all good fun and games, the competitiveness among them was nothing short of ferocious. "Like the guys sinking Wally Schirra's mast," fellow astronaut Edgar Mitchell said. According to Mitchell, Wally had a cruiser anchored in Clear Lake, and the Original 7 astronauts, with Shepard as the ringleader, arranged for the harbor master to haul Wally's boat away and sink a mast right where his boat had its anchor. Shepard then had the harbor master call Schirra with the bad news: "Wally, I think you'd better come over here, there's something wrong with your boat." The seven were watching from behind the bushes when Schirra's jaw dropped in disbelief at the sight of a sunken mast where his boat used to be.

"I observed—but I wasn't in on it," Mitchell quickly added.

Schirra was fearless and known to attempt any task with unbridled exuberance. It was expected that Apollo 7 would be no exception.

As part of the crew called upon to resurrect the program, Walt Cunningham sensed the growing anticipation. "As the public saw it," he said, "the prior crew had been killed, Apollo was still trying to get off the pad, and we—the crew of Apollo 7—were the men entrusted with doing it, with getting the U.S. back into the game." He noticed a renewed sense of purpose infusing the ranks of NASA. "My whole mood changed," Cunningham said, "the confidence in my approach to problems, my daily rhythm. I slept better. Food tasted better. My wife looked better. For the first time around the office our crew was being envied instead of being envious."

Two days before launch, the media was given a preview of coming attractions: NASA rolled out the mammoth Saturn 1B onto Pad 34—the same pad Apollo 1 had burned on. "U.S. Prepares Moon Shot in December," the *New York Times* announced in bold headlines. It was brilliant stagecraft. For the past several months rumors had circulated that there had been a change in plans. Apollo 8, the rumors held, was going to the moon, and NASA officials would neither accept nor deny them. A future mission to circle the moon, they said, was "a possibility."

On October 11, 1968, Schirra, Eisele, and Cunningham tore into the heavens. The launch was flawless. But the crew of Apollo 7, as well as the troops on the ground, were in for the most frustrating mission to date. While the press celebrated the first manned flight of the Apollo spacecraft, the command module system was in the midst of enduring some fifty "malfunctions," which included a nine-minute communications blackout. Fortunately, the extended mission called for nearly eleven days in space, which allowed NASA the time to painstakingly iron out each problem. And that's exactly what they did. What didn't make it into the newspapers or TV were the difficulties and frustrations experienced by the crew.

Schirra was one of NASA's most experienced astronauts and the only one to have flown both Mercury and Gemini missions. He clearly felt the full weight of the program riding on a successful mission and as a result became more openly critical. Two weeks before the scheduled launch, Schirra announced he was retiring from the astronaut corps, NASA, and the Navy. The announcement, he felt, freed him to say his piece. And that's exactly what he did. Schirra's third mission proved to be by far his most challenging. After a few hours in space he developed a bad head cold, which instantly took a toll on his energy and mood. While on the ground, a cold is merely an inconvenience. In space, it could become a mission-threatening ailment.

By the third day, he began sneezing and blowing his nose. Within a short time the crew was stuffing used tissues in every unused nook and cranny they could find. Wally didn't seem interested in conserving them, and Donn and Walt began to worry whether a tissue crisis would end the mission.

As the flight continued, the crew's mood quickly began to sour. Mission Control inadvertently exacerbated the problem with a growing list of tests for the crew to conduct. The three astronauts were confined in an enclosed space of some 250 cubic feet for eleven days, carrying out a to-do list that kept getting longer and longer. Throughout this seemingly never-ending ordeal, the crew was further aggravated by an inability to contain the commander's nasal discharge in zero gravity.

The endless number of tests was the result of a tightened schedule. NASA hadn't attempted a manned mission in more than twenty months, and the next mission was tentatively slated just two months after the splashdown of the current one. This meant NASA had to cram a huge backlog of undone tests into this one flight. All the while, episodes of the mission were scheduled for live television, which only heightened tensions between Mission Control and the crew. A random dose of humor provided much needed relief from the tension. Donn Eisele reported an errant radio signal he was picking up:

Eisele: We have some music in the background. Is that you?

CapCom: You must be picking up the twilight zone there.

Eisele: Is someone trying to plug in a radio program to us, or are we just picking that up spiritually?

CapCom: That must be a spurious signal. No, we don't have anything like that.

Eisele: OK, I'm getting a hot tip on some hospital insurance plan from some guy.

CapCom: Maybe they are trying to tell you something.

Eisele: Yeah. Maybe they know something I don't.

One of the unique contrivances of manned space travel was the manner in which the astronauts disposed of waste matter. A urine bag with a series of valves permitted the astronauts to dump the contents of the bag overboard.

"Now that was something worth taking a picture of," noted Walt Cunningham. If one off-loaded at sunset, he noted, the flecks of ice coming off the urine dump nozzle would look like a million stars and it would be impossible to take star sightings for about five minutes. "Of course," he said, "it's a real experience to see your own urine take on a cosmic quality in space."

And so it went, peaks and valleys, a little mischief, and occasionally a growing sense of the magnitude of the endeavor. But a crew adopts the mood of its commander, which in this case dampened the mood aloft and on the ground. Schirra was being particularly difficult, which surprised many, as this was a test flight and they were, in fact, the test pilots. Here was the Apollo 7 mission commander railing on about this or that system and challenging any requests from the ground to deviate from the flight plan. Then Cunningham dismissed a test of the backup cooling system as a "Mickey Mouse" procedure. Mission Control was taken aback. They'd never experienced this kind of pointed rebellion and couldn't quite fathom what the fuss was all about. All they could do was manage the situation from afar through diplomacy.

Eventually the exchanges and insubordination became embarrassing. Glynn Lunney, the lead flight director, became so enraged by the astronauts' conduct that he directed the flight controllers to target splashdown in the midst of a hurricane bearing down on Hawaii. Of course, that didn't happen. According to Ed Buckbee, a senior public affairs official and Schirra's close friend, Wally was "utterly terrified of that capsule. It *really* bothered him." Schirra had been the backup for his friend, Gus, and knew all too well it could have easily been him in the Apollo 1 spacecraft on that fateful day.

Such exchanges spoke directly to the intense pressure that NASA in general, and the astronauts in particular, were under. Each mission was wildly more ambitious than the last and each packed with tests that had to be successfully completed in order to make possible the next mission. As if that were not enough, the missions themselves were to be broadcast on television. And everyone in the space program understood the profound importance of the media in the overall scheme of things.

Given Schirra's reputation for light-hearted antics, his behavior on board Apollo 7 was somewhat unexpected. On his earlier Gemini mission, Schirra attracted notoriety for playing "Jingle Bells" on a four-hole Hohner harmonica he smuggled on board. Regardless, in spite of his messy illness, the camera was often rolling live as the commander managed to put forth a brave face. The crew of Apollo 7 not only broadcast the first televised moments from space, they broadcast nearly every day; and the effort proved immensely popular with TV viewers throughout the country. The broadcasts were dubbed "The Wally, Walt, and Donn Show." On a placard floating in weightlessness the crew wrote: "Hello From The Lovely Apollo Room, High Above Everything." Another read, "Keep those cards and letters coming, folks." The public couldn't get enough. This was the kind of rapport that was going to take NASA all the way to the moon.

But these light-hearted public relations moments would not be the hallmark of Apollo 7. A few of the intemperate remarks even reached the media before the mission's end. During a press conference a reporter asked Glynn Lunney: "I've covered sixteen flights...and I don't recall ever finding a bunch of people up there growling the way these guys are. Now you're either doing a bad job down here, or they're a bunch of malcontents. Which is it?"

"I would be a little hard pressed to answer that one," was all Lunney would say.

Meanwhile, each "morning" Schirra was marking off the days until their return to earth. The mission had become a physically and mentally agonizing ordeal, one he found himself staggering through day by day until the end of the flight. Gone was the sense of wonder and joy experienced during Ed White's Gemini 4 spacewalk. Instead of communing with God, these men were just trying to get stuff done. "Our universe," Cunningham explained, "had contracted to the spacecraft and the mission in which we were immersed."

To make matters even more aggravating, their sleep schedules were at odds with their scheduled meals, causing them to awake to a meal of dinner instead of breakfast. In addition, they had been shorted one meal entirely. Fortunately, the bacon bites they stuffed in their space suits prior to liftoff made do for their final breakfast.

As the end of the mission neared, Wally's mood failed to improve. Due to his head cold, he didn't want to wear a helmet during reentry for fear of damaging his ear drums, even though this was contrary to mission rules. Flight Operations Director Chris Kraft issued a direct order to Schirra: "Wear the helmet."

Wally's response was equally direct and simple: "Go away." Kraft was mortified. He'd never experienced anything like this before; there was nothing he believed in more strongly than order and discipline within the Astronaut Corps.

Cunningham later described the mission as "a pressure cooker," especially for Schirra. Even during less ambitious missions, he said, "NASA liked to get its money's worth from every crew—and did." Halfway through the Apollo 7 flight the crew had completed seventy-five percent of the mission's objectives and by flight's end had completed more tests than any other.

Schirra had arranged a special celebration at splashdown to commemorate the mission's success. Before liftoff he approached pad leader Guenter Wendt with the idea of smuggling onboard a small bottle of scotch. This was certainly something that management would not condone, so they proceeded with the utmost secrecy. Schirra secretly tested it in the altitude chamber during routine tests, wrapped it in Beta cloth, and decided where to stow it.

Now, at splashdown, as the Apollo 7 capsule bobbed in the Pacific Ocean awaiting the recovery team, Wally broke out his tiny bottle of scotch and made a toast. During his post-flight physical, the medics would have no explanation for the traces of alcohol that showed up in his bloodstream.

The first manned flight of the Apollo spacecraft had been, in Schirra's words, a "101 percent success." He was correct. Awaiting their arrival onboard the aircraft carrier U.S.S. *Essex* were hundreds of letters, cards, and telegrams from complete strangers, telling the crew of Apollo 7 they had shared in their flight and how much it meant to them. For Schirra personally, Apollo 7 brought to a successful completion essentially the same mission previously assigned to his friend Gus Grissom—a fitting tribute that assuaged the Apollo 7 crew in every sense.

Jerry Bostick, working behind a console in Mission Control, knew better than most the demands placed on the crew.

"It would remain an unsung mission," he said. "They never have gotten the credit they deserve, because it was the most jam-packed flight plan we had ever flown."

But NASA still had a long way to go. By October 22, 1968, thirty-eight human beings had been in space, yet none of them had been "across the abyss" to the moon. That was about to change. All that remained was what President Kennedy referred to as "the most hazardous and dangerous and greatest adventure on which mankind has ever embarked"—the half-million mile voyage to the moon and back.

The drudgery of Apollo 7 had cleared the way for the grandeur of the ultimate goal. The next step up the celestial ladder would dwarf by far all previous missions in terms of sheer danger and brazenness. It would also dwarf all previous missions in terms of spiritual import.

Three thousand miles away in Houston, Texas, three men now learned that they would be making the most ambitious spaceflight ever conceived. They would also see the world in a way no one ever had before—and they would share the view in a way that would stun all of humanity.

"When you see that puny little blob we live on from space, see its black nothingness, you think there isn't room for all those people, some of whom even hate one another. A look at earth from space will change your scale of thought. It's a very warm and inviting place. We thank God we are a part of it."

- *William Anders, Lunar Module Pilot, Apollo 8*

11
APOLLO 8: LEAP IN FAITH

In the spring of 1968, George Low, the man brought in to help heal and resurrect the space agency in the wake of the fire, quietly made what some initially interpreted as a preposterous proposal. He wanted to send Apollo 8 around the moon. The problem, of course, was that the lunar lander still wasn't ready, and without it Apollo 8 was a mission without a purpose. Low's idea was nothing if not bold. *Why not send Apollo 8 around the moon without the lunar lander?* Why would Apollo 8 need to haul a lunar landing module if they weren't even going to test it or land it on the moon? The beauty of Low's plan was that it would grant NASA and Grumman engineers much-needed time to complete development of the lunar module without delaying the flight schedule. Meanwhile, it would allow the U.S. space program an entire mission devoted to developing the many critical procedures necessary for a successful circumlunar journey.

"When I first heard of the plan," Glynn Lunney said, "I thought it was crazy. Then I thought that maybe it was a workable idea, and then I thought it was brilliant. This process didn't take me a week, a day, or even a few hours. It took something like a couple of minutes."

The issue remained unsettled until Flight Operations Director Chris Kraft called a surprise staff meeting in his office. Gene Kranz and team leaders from flight control gathered around the table as Kraft presented a synopsis of the missions up to that time.

"The lunar schedule is in trouble," he said. "We must understand and fix the problems with the Saturn... The [lunar module] is overweight and the software for its computer is not ready." Then Kraft wanted to hear an airing on the Apollo 8 issue from his own people.

What happened next exemplified the Apollo team at its finest: a brilliant staff collectively focusing on an issue, followed by open speculation and argument that concluded with realistic alternatives. This brought Kraft back to Low's plan as "a possible alternate sequence."

Within the month Low's plan had been revised so that it was even more audacious. John Mayer, the chief of mission planning, had a conceptual plan drawn out that called for ten lunar orbits instead of a mere one-time figure-8, out-and-back flight.

Many within NASA were skeptical. They worried that any circumlunar mission at this point crossed the fine line between the bold and the reckless, as it would require one "first" after another. The biggest first was in attempting trans-lunar injection (TLI) and leaving earth orbit. Another was navigating a distance of nearly a quarter-million miles. This would be the first time humans had ever escaped earth's gravitational field, the first time humans communicated over such vast distances, the first time they ever entered the gravitational field of another celestial body, and the first time they ever navigated on the far side of the moon. Each of the firsts made for challenges that could prove beyond the ability of the crew or Mission Control to overcome from over two hundred thousand miles. Gene Kranz, for one, was incredulous. "The decision to go to the moon with Apollo 8 was being made before we had ever flown a manned Apollo spacecraft."

By April 1968, however, the decade had but twenty months of life remaining. Thus, many began to see Low's plan as a prerequisite for making the deadline. The audacity of Low's plan could not have been more jarring to NASA executives. When NASA Administrator James Webb was briefed on the plan, he reportedly said, "Are you out of your *mind*?" Webb was a political animal, not an engineer. Ever since the beginning of the Kennedy Administration, Webb had adeptly cultivated the agency's rapport with Congress and kept the money flowing. After the fire he actually managed to get the NASA budget *increased* by $500 million. Webb knew exactly the type of space disaster that could result from a hastily rearranged flight schedule. Moreover, the operational challenges were enormous and very real. Although Webb initially considered Low's idea reckless, he promised to keep an open mind. Meanwhile, Low quietly lobbied NASA personnel.

Gene Kranz had yet to be convinced. "I recognized Low's plan as a bold move that would let us get to the moon by the most direct path and buy us some badly needed schedule time… provided it worked." Kranz claimed to have "mixed emotions" after learning of the proposal. "Personally I believed the best track to reach the moon was the current sequence. Low's plan would heighten the risks." But even the admittedly conservative Kranz understood its brilliance. "Apollo succeeded at critical moments like this," he said, "because the bosses had no hesitation about assigning crucial tasks to one individual, trusting his judgment, and then getting out of his way." Hard-line skeptics, however, echoed Webb's concerns.

In August 1968, Deke Slayton informed his men of the change in plans for Apollo 8. Jim McDivitt, Dave Scott and Rusty Schweickart were scheduled to be next in line. But Slayton had other plans. He wanted Frank Borman, Jim Lovell and Bill Anders. Lovell was excited at the prospect. Borman, a fighter pilot renowned for possessing a broad cautious streak, was to be the commander and was far more measured in his response. Bill Anders, the newcomer, accepted the assignment with mixed feelings. He was to be the lunar module pilot, but there would be no lunar module on the flight. His job, Lovell teased him, "was to sit there and look intelligent." Lovell, as the command module pilot, would have his hands full navigating across the immense vacuum of deep space.

By September, support for Low's plan received a boost when a Russian spacecraft carrying a life-size mannequin made the journey around the moon and back on a free-return trajectory. Once again, NASA was caught off guard. Soon thereafter, the CIA received word that the Soviet Union was moving a spacecraft in place capable of carrying two men to the moon. Many in NAA questioned whether or not the Soviets were truly ready to chance a manned lunar mission so soon. No one knew for certain, but by now they had grown weary of being up-staged once again by the Soviets. NASA watched and waited. Nothing happened.

Shortly after learning of his Apollo 8 assignment, Jim Lovell went to his wife, Marilyn. The family had planned a vacation in Acapulco over the Christmas holiday, a vacation that had been repeatedly postponed over the past few years due to the enormous time demands of Lovell's work. Once again he told her they wouldn't be going to Acapulco or anywhere else over Christmas.

"Why not?" she asked, deeply disappointed, as they had already told their four children and made reservations at the Las Brisas resort.

"I know, I know," Lovell said. "But I thought Frank and Bill and I might go somewhere else instead."

"Like where?" she asked.

"Oh, I don't know," Lovell replied. "Maybe the moon."

Marilyn Lovell, a woman of deep faith and a veteran of her husband's two Gemini spaceflights, was accustomed to the anxiety that comes with being an astronaut's wife. He had, after all, spent more time in space than any other human being. But with this latest news, she felt herself tremble. She'd been waiting for years to hear her husband say that he was going "to the moon." Now that he actually was, she felt a deep sense of dread.

This voyage, she understood, would be different from all the rest.

The changes in the mission plans for Apollo 8 necessitated a change in the timing and location of splashdown. The capsule was now slated to splash into the Pacific just after Christmas day and a recovery vessel would need to be spotted nearby.

American military forces were engaged in a bloody, unpopular war in Vietnam, and the Navy was planning to grant much-needed Christmas leave to as many sailors as possible. In order to arrange for the necessary naval resources and manpower needed to recover the capsule, NASA would need the cooperation of the commander in charge of the Pacific Fleet.

The task of arranging this fell to Chris Kraft. In Hawaii, he addressed an amphitheater teeming with captains, admirals, and generals from every branch of the military. The man whose assistance he truly needed was Admiral John McCain, father of a fighter pilot being held as a prisoner of war in Hanoi. Without Admiral McCain's cooperation, all manner of glitches would spring up. In light of the ongoing war, Kraft sensed that it was going to be a tough sell. Men who had earned and were in desperate need of shore leave were not going to get it, at least not this Christmas.

"Neither before nor since have I stood in front of that kind of high-powered audience," Kraft said. "McCain's eyes only left my face to look at my charts and graphs. I don't remember many of my exact words. I simply ran through the mission and told him what we wanted to do. I know that I stressed the importance of the flight and its risks, and that the greatness of the United States of America was about to be tested in space.

"Then I got to the real point: 'Admiral, I realize that the Navy has made its Christmas plans and I'm asking you to change them. I'm here to request that the Navy support us and have ships out there before we launch and through Christmas. We need you.' "

There was complete silence in the room for maybe five seconds.

"McCain was smoking this big long cigar, and all of a sudden he stood up and threw it down on the table."

Kraft braced himself for an angry response.

"Best damn briefing I've ever had," McCain announced. "Give this young man anything he wants."

> "Say something appropriate."
>
> - *NASA instructions for the Apollo 8 Christmas Eve broadcast*

> "In the beginning, God created the heaven and the earth, and the earth was without form, and void."
>
> - *Frank Borman, Commander, Apollo 8*
> *Reading from Genesis 1*

12
SHOOTING THE MOON

Chris Kraft set the crew's odds of survival at fifty-fifty—a coin toss. And he wasn't being unduly pessimistic. "It took a lot of nerve to do Apollo 8," he said. "The biggest thing about Apollo 8 that I was impressed with was that the country let us do it."

As Susan Borman, wife of Apollo 8 Commander Frank Borman, put it, "We knew that the Russians were hell-bent to do the same thing, and by golly we were going to get there first. But I didn't really think they would get back. I just didn't see how they could. Everything was for the first time. Everything."

Borman's wife had known all along what lay ahead, but she hadn't expected it to happen on such short notice. When her husband came home and announced that he had just signed up for the moon shot, she was aghast. "And this is August, you haven't tested the capsule yet," she said. "December, that's what, three some odd months? But usually you train for a year…To the *moon*?"

During the run-up to launch, excitement continued to build. For the first time in history, humans were poised to depart from earth and venture out to another world. The crew was cast into the media spotlight and celebrated by a fascinated public. "These three brave men are the Columbuses of space," the *New York Times* wrote.

For four months the flight crew of Apollo 8 and their ground crew trained relentlessly. "It compressed everything that we were doing," Commander

Borman said. "It compressed building the software in the computers to go to the moon. It compressed measuring the instrumentation that we looked at in the spacecraft. And we convinced ourselves… Okay, we're now ready to go, we think we can do the flight control, we think we can do the computers, we're satisfied the spacecraft is going to work, so let's get ready and go."

On December 21, 1968, the countdown to one of the greatest moments in exploration began. The first manned lunar flight, Apollo 8, was scheduled to travel over 230,000 miles into space, orbit the moon ten times, and safely return home. As the Apollo 8 crew embarked upon their six-day mission, they left behind a deeply troubled planet. Through 1968, America had suffered through one shock after another. On top of the escalating war in Vietnam and virulent anti-war protests, came the assassination of Dr. Martin Luther King Jr., and on its heels, the assassination of presidential candidate Robert F. Kennedy.

The mission would provide a welcome reprieve from the social chaos—provided everything went well. Within minutes of ignition, the rocket would fling them out of the grip of earth's gravity and the crew of Apollo 8 would be farther from their home planet than anyone had ever been. The only thing that lay between the crew and the infinity of space was the moon. Not only was the moon their destination, it was also their ticket home, since the moon's gravity would provide the celestial pull that would keep the spacecraft from flying straight past the moon into oblivion. On the downside, it could also pull the spacecraft into its rocky surface.

Immediately after launch the spacecraft would orbit the earth twice, then fire the rocket that would send it moonward. Some three days later it would arrive. If the crew of Apollo 8 could enter the moon's gravitational field slowly enough so as not to fly past the moon, yet fast enough not to slam into it, they would enter that blessed state called lunar orbit. The technological challenges for pulling this off were formidable. Many things had to go right—and absolutely nothing wrong. It would be impossible for Mission Control to determine the exact position of the spacecraft at any particular moment, and they would lose contact with the spacecraft altogether as it passed behind the moon. They would have no idea as to its success or failure until it re-emerged on the front side. So Chris Kraft was hardly exaggerating the dangers of the mission at fifty percent odds. The wives and families of the three men understood well the risk of the upcoming six-day mission.

"We'd say how proud we were," Susan Borman later confided, "how confident we were, and then I'd go back in the house and kick a door in. I thought, 'They're rushing it, they're leap-frogging… This time it's not just another test flight.'"

The other astronauts' wives tried to camouflage the same jagged emotions.

"As much as I tried to hide my fears," Marilyn Lovell said, "it wasn't easy on any of us—especially the older children, who understood what was happening."

"We were very aware," said Valerie Anders, "that when we said 'goodbye,' it could be *goodbye*."

But in the business of space exploration, peril walked hand-in-hand with progress. As 1968 came to an end, NASA was finally prepared to take one giant stride after another on its way to the moon. If all went well on Apollo 8, NASA flight planners believed they would be able to put two men on the lunar surface within seven months.

Flying to the moon had been a dream as old as mankind itself. That the country was now poised to actually fulfill that dream seemed miraculous to people around the world. This sense dovetailed with certain poignant aspects of the Apollo 8 mission.

"We found out that we were going to burn into lunar orbit Christmas Eve," Jim Lovell said. "All three of us decided that this would be a significant time to say something. But what can we say? We tried new verses to 'The Night before Christmas' and 'Jingle Bells,' but nothing really seemed appropriate."

"We wanted to do something not so much religious," Bill Anders said, "as something to give them sort of a shock in the psychological solar plexus, to help them remember Apollo 8 and humankind's first venture from the earth."

It was a riddle all three astronauts became intent on solving as they prepared for the mission. On the morning of December 21, 1968, the Apollo 8 spacecraft stood within a matrix of lights in the ambient pre-dawn darkness. Tens of thousands of spectators from around the world descended upon the Cape to witness the launch. Around 5:30 a.m. the crew boarded the command module. NASA personnel closed the hatch, gave a thumbs-up, and sealed the three astronauts inside. All three men wondered if they were actually going to launch or if a bit of ratty telemetry data might appear on someone's console in Mission Control that would scrub the mission. In the astronaut corps, nothing was taken for granted. Nothing was ever assumed to happen until it had actually happened.

For the next two hours all systems were checked. In the final five minutes the access arm to the tower swung away. At T-minus sixty seconds, thousands of tons of super-cooled fuel was pressurized and the rocket stood poised for the heavens. With ten seconds remaining, the astronauts sensed a jolt as the booster rocket came to life. Only then did they realize they really were headed for space. Intent on the console monitor in front of him, Dick Profitt, the Cape launch controller, began the final countdown, *"Ten, nine, eight, seven, six..."* Then came the final confirmation: *"...ignition."*

The massive Apollo 8 spacecraft slowly—almost tentatively—rose, lurching in a perfectly vertical column, slowly lifting skyward as it strained to clear the tower; and the Saturn V rocket became a fire-snorting dragon. The intense shaking took them by surprise, as it was something the crew hadn't experienced in the static confines of the ground simulator. Only seconds after Profitt announced liftoff, the massive Apollo 8 spacecraft was climbing through tiers of the atmosphere. One minute after launch, the missile went hypersonic—five times the speed of sound.

Over the course of the next three hours, Apollo 8 circled the earth twice while the crew checked and rechecked the status of the systems. Beyond the window loomed the blue and white immensity of the earth, a mesmerizing sight to Anders, who had never before been in space. To Borman and Lovell, however, the view and sensation of weightlessness were familiar from the fourteen-day marathon flight they endured together three years earlier on Gemini 7. When Lovell caught a glance of Anders looking out the window, he abruptly suggested that he get back to business. Anders returned to the never-ending mission checklist, but his eyes were constantly tempted by the stunning beauty just outside the window.

For the next two hours the systems performed flawlessly as the spacecraft silently wheeled 115 miles above the earth. Then, as thousands in NASA listened intently, CapCom Mike Collins transmitted the official go ahead for translunar injection.

"Apollo 8, Houston... You are Go for TLI."

The engines came to life as the spacecraft directed for a trajectory that would send it into the void of space. Mission Control tracked their progress as they headed into deep space, accelerating to an incredible speed and suddenly morphing into a star as they bulleted out of earth's atmosphere.

"You're looking good here," Collins told the crew of Apollo 8, "right down the old center line."

Americans came home from church on Sunday, December 22, 1968, to watch the NFL playoffs between the Baltimore Colts and the Minnesota Vikings. The game was being broadcast on CBS, and the entire country, it seemed, was riveted. On any other Sunday, Commander Frank Borman onboard Apollo 8 would have been in the congregation of St. Christopher Episcopal Church in League City, Texas. This day, the fourth Sunday in

Advent, however, he was on his way to the moon. On board was something Borman initially argued against bringing—a television camera. It weighed twelve pounds, and as Borman explained, "We were cutting out everything," including extra meals, each of which weighed a mere sixteen ounces, and now NASA wanted them to take a twelve-pound camera. He also didn't want the distraction of live television during such a dangerous mission. But NASA officials were insistent. The pictures, everyone knew, would be spectacular. In the end Borman relented.

During day two in space, the crew did a bit of housekeeping and prepared the camera for their first broadcast. A moment later, back on earth, CBS interrupted the playoff game for what they hoped would be the first-ever live broadcast from deep space. A picture of the home planet, greatly diminished in size from such a distance, had been planned, but problems with the lens prevented it. Instead, audiences saw Borman at the controls of the spacecraft, Jim Lovell conducting star sightings and wishing his mother happy birthday, and a few antics by Anders playing with his toothbrush in weightlessness. When coverage of the playoff game resumed, CBS headquarters was inundated with complaints about the interruption.

The outbound journey to the moon was off to an uncertain start in more ominous ways. Borman, the first crew member scheduled for sleep, had to resort to taking a sleeping pill, which brought about a vague, unsatisfying rest. He "woke up" feeling sick to his stomach and announced that he was about to lose his lunch. His crewmate Anders described the incident in painful detail: "The one nice thing about being on earth," Anders said, "if someone gets sick, it lands on the floor. In zero-gravity it floats around." Which is what it did.

Borman was reluctant to inform Houston of his illness because he didn't want it becoming worldwide news. Anders urged him to do so anyway, which led to a disconcertingly long discussion among NASA physicians. Incredibly, the doctors raised the possibility of aborting the mission. If it were a flu virus, they argued, it was only a matter of time before the rest of the crew also became ill. But the spacecraft couldn't simply stop, turn around, and head back to earth. The only way home was around the moon, and trimming twenty hours off of the total flight time wouldn't solve much of the problem. The discussion was finally dropped when Borman recovered.

Twenty-four hours after the first broadcast, the crew of Apollo 8 passed the invisible gravitational divide between the earth and the moon. At this point the spacecraft was no longer pushing away from the earth, but rather falling toward the moon. In another twelve hours they would fire the spacecraft's engine while flying "backward," slowing it down just enough to enter lunar orbit. They were aiming for a narrow and invisible seam that lay just sixty-nine miles above the jagged mountains and valleys of the moon.

On the morning of Christmas Eve, Apollo 8 was given a Go for lunar orbit and passed into radio silence behind the moon. The next several minutes of the mission would prove to be emotionally agonizing for the families of the Apollo 8 crew. If the mission were to fail, now, it seemed, was when it was most likely to happen. NASA had never before placed a manned spacecraft into lunar orbit, and the margin of error was exceedingly thin.

Just before the spacecraft curved behind the moon and lost radio contact with earth, Susan Borman asked the CapCom, astronaut Jerry Carr, to send her husband a special message. Carr agreed.

"The custard is in the oven at 350," Carr reported to Borman.

"No comprendo," was Frank's reply to Carr. However, after a moment of silence, he added: "Roger."

Susan Borman knew that her husband understood the message. It was an old line from his days as a test pilot at Edwards Air Force Base. It had to do with the division of labor that made their family life unique and had come to define their respective roles. Frank told Susan, "You worry about the custard and I'll worry about the flying." Each was doing what had to be done, and now Borman's job was to get home from the other side of the moon. At 3:59 a.m. Houston time, Apollo 8 slipped behind the moon. If all went well, they would emerge again in exactly thirty-six minutes.

"Thanks a lot, troops," Anders said in farewell to the people at Mission Control.

"We'll see you on the other side," Lovell added.

Shortly after losing communications with Mission Control, the Apollo 8 retro-fired its engine for four full minutes, slowing the spacecraft's speed to a mere 3,700 miles per hour and lowering their altitude above the lunar surface. They were now alone, completely unable to communicate with Mission Control. Houston wouldn't know of their fate until the spacecraft emerged from behind the moon.

―⁂―

Meanwhile, in the sleepy Houston suburb of Timber Cove, a black Rolls Royce limousine wound through the row of small ranch houses on Lazy Wood Lane. It slowed to a stop in front of the Lovell's' home. The driver emerged carrying a large box in royal blue wrapping paper decorated with two Styrofoam balls—one representing the earth, the other the moon. Prior to the launch, Jim Lovell had become acutely aware of the anxiety the mission had placed on his wife and family. Inside the box was a mink coat with a note: "To Marilyn, from the Man on the Moon."

Marilyn Lovell's own personal Santa was 230,000 miles away on a sleigh ride.

> "For all the people on earth, the crew of Apollo 8 has a message we would like to send you…"
>
> *- William Anders, Commander, Apollo 8*
> *1968 Christmas Eve broadcast from the Apollo 8 spacecraft*

13
CHRISTMAS EVE

The moon lay on its back in a perfect crescent when people in the unclouded regions of the world looked up at the night sky on Christmas Eve, 1968. For the crew of Apollo 8, the view was very different.

As the spacecraft slipped behind the moon, the eyes of the crew fell upon a vast region of a world that has never been visible from the earth. It was a strangely clear sight no other human beings had ever directly beheld. From the tiny window the crew could easily see the details of impact craters that were hundreds of miles away. There was nothing to diminish the resolution of the perfectly etched scars on the landscape of this desolate world, nothing to impair the vision of the human eye, as the moon has no atmosphere. Distant objects were as clearly visible to the astronauts as those nearer objects directly beneath them. "We had our noses pressed to the glass," Jim Lovell said. "It was really an amazing sight."

As the spacecraft emerged from the far side of the moon, communications with Mission Control resumed, eliciting raucous cheers from the crowds gathered around the consoles. After all these many years, after the fire that claimed three comrades, after this Herculean national effort, an American spacecraft was finally orbiting the moon. For the crew, the joy came in the form of relief that another burn had gone off without a problem, that they were still alive.

But the emotional relief was soon replaced with complete awe. It was then, just as they emerged from the back side of the moon, that they were struck by the most stunning sight any of them had ever witnessed. Suspended just above the gray lunar horizon was a beautiful blue and white sphere. The contrast with the bleakness of the moon was astonishing. When Lovell saw it, he directed Anders, "Go get the cameras." They were not scheduled for photography at that point, but this was a site that had to be captured. Anders, whose utter enchantment with the beauty of earth from space had glued him to the capsule window, hurriedly snapped a photograph of the small jeweled marble earth as it rose into view over the crusty edge of the moon, capturing "Earthrise," a photograph that would forever alter mankind's view of our planet and our place in the universe.

As breathtaking as the view was, it also underscored the exceedingly precarious nature of their personal safety; these three men were farther from their home than earthlings had ever been. For the next twenty hours and nine orbits, the crew surveyed the lunar surface, searching for potential landing sites for future missions. Then they began naming various features of the moon after the people who had been instrumental in taking man to the moon. They named features after Gus Grissom, Roger Chaffee and Ed White. They named one after Christopher Kraft.

Lovell took the mike and narrated a description of the lunar surface, reporting, "The view at this altitude is tremendous. There is no trouble picking out features we learned on the map." Lovell's job while in lunar orbit was to study the lunar surface from the perspective of navigation, to verify that their lunar charts were accurate and could be used to locate future lunar landing locations, such as the Sea of Tranquility, the crater Taruntius, and the mountains that skirted both.

"The mountain range has got more contrast, because of the sun angle," he noted. Then he inexplicably added, "I can see the initial point right now—Mount Marilyn."

At Mission Control, Mike Collins responded, "Roger," though he had no idea what Lovell was referring to, as his knowledge of the area around the Sea of Tranquility included no such mountain. And until Lovell arrived there, that had been true. Ancient astronomers had never seen Mount Marilyn with their earthbound telescopes. NASA hadn't given it a name when their unmanned probes had photographed it. And neither Borman nor Anders had noticed it when they studied the maps prior to launch. But just as he promised her, Lovell had brought his wife, Marilyn, with him to the moon. Even if some scientific institution might not consider the name Mount Marilyn official, it was now that mountain's name.

Lovell had chosen shrewdly. Located on the edge of the Sea of Tranquility just east of the Apollo 11 lunar landing site, Mount Marilyn would be a reference point on maps used by Armstrong and Aldrin during their historic approach six months later. In fact, they would fly directly over it as they came in for a landing.

In Houston, all eyes and ears were focused on every word, every maneuver.

"Oh, I was fortunate," Gene Kranz said. "I was probably the most fortunate person in Mission Control, because I wasn't working the mission. And I was absolutely mesmerized by what was going on… And then the thing that really came down and grabbed me was the crew would describe the surface as they saw it. They would describe the back side, and then they started naming surface features for the people they thought got them there. The gurus and the Krafts, and you know, then their names, their portions, astronauts and the pioneers who had died on the way to the moon. And you sit there and you say, my God, I'm glad I'm a spectator at this thing as opposed to having to do something because I got so involved in what the crew was saying."

After another engine was fired to alter Apollo 8's orbit at the start of the third orbit, Borman asked Mission Control if a man by the name of Rod Rose was there.

Rose was expecting the call. He and Borman had been quietly working on a special task. The men were neighbors in El Lago and both were members of the vestry of the local St. Christopher Episcopal Church. Two Sundays before Apollo 8 blasted off, Borman learned that he was on the duty list as lay reader for the Christmas Eve communion, and he wanted to do something special to mark the occasion.

Knowing that he would be circling the moon at that time, Borman arranged to deliver his scheduled reading from lunar orbit, which would be recorded and later played for the congregation by Reverend Jim Buckner during the St. Christopher Episcopal Church service on Christmas Eve. The selection for Borman's reading was the *Prayer for Vision, Faith and Works* by G. F. Weld. Rose had given a copy of the prayer to Borman.

"We decided to call it 'Experiment P1,'" Rose said, "and Frank agreed to give me one lunar orbit notice before he read the prayer so the recording could be finalized."

Now, a few minutes after Borman's call, CapCom Mike Collins replied, "Rod Rose is sitting up in the viewing room. He can hear what you say."

"I wonder if he is ready for Experiment P1?" Borman asked.

"He says thumbs up on P1," Collins replied.

When Collins gave the go-ahead, Borman continued: "Okay. This is to Rod Rose and to the people at St. Christopher's, actually to people everywhere.

> Give us, O God, the vision which can see Thy love in the world in spite of human failure. Give us the faith, the trust, the goodness

in spite of all our ignorance and weakness. Give us the knowledge that we may continue to pray with understanding hearts, and show us what each of us can do to set forth the day of the universal peace. Amen."

"Amen," Collins replied.

"I was supposed to lay read tonight and I couldn't quite make it," Borman explained.

"Roger, I think they understand," Collins said.

So concluded the first prayer broadcast from space. But it was their second broadcast on Christmas Eve during their final orbit around the moon that would amaze the entire world.

Prior to liftoff, the crew of Apollo 8 knew they wanted to do something special for the scheduled Christmas broadcast. When queried, NASA officials had merely told them: "You'll have a TV appearance on Christmas Eve. You're going to be seen by more people than anybody, witnessed by more people than anyone has ever seen before, and you've got to be prepared."

"Well, what do you want us to do?" Borman asked.

The response he received from NASA Public Affairs Chief Julian Scheer, was elegant in its simplicity: "Say something appropriate."

Borman was amazed at the latitude he and his crew were being given by NASA. So in the midst of preparing for their mission, the crew attempted to come up with just that. But what, exactly, would be appropriate? Here was an utterly unprecedented occasion—and it came with its own unique set of complications.

"Almost the whole world would be listening to us on Christmas Eve," Lovell said. "But the whole world does not consist of Christians." They wanted to say something that would be meaningful to a majority of the people on earth. Borman and Anders concurred.

As they hashed over various ideas, Anders proposed that it shouldn't be "so much a religious reading, but more of a significant statement that not just Christians and Jews would understand, but that all people of all religions would react to in a deep and moving way to help them remember this event of exploration." The answer finally came to the crew in a most circuitous fashion. During an intensely busy run-up to the mission, even though the mission was unprecedented in complexity, the crew had been given only a third of the time typically allotted to prepare. Instead of the standard full year of training, the Apollo 8 crew had only four months. As a result, every hour was spoken for. And yet as the number of days until launch dwindled and each day of the upcoming voyage came into sharper focus, they came to understand more fully the gravity of the Christmas Eve broadcast.

Borman was in the midst of training for one of the most critical aspects of the mission when he was informed of the scheduled Christmas Eve broadcast. In fact, he was learning the procedures for a blazingly fast reentry from the moon into earth's atmosphere at 25,000 miles per hour. Instead of distracting himself with what would be "appropriate" for the telecast, he mentioned it to a friend, Si Bourgin. Bourgin, who Borman described as "a sensitive, intellectual guy," didn't have an answer at hand. He did, however, mention the challenge to a journalist by the name of Joe Laitin. Laitin didn't have any suggestions either, but he happened to mention it to his wife, Christine, who is reported to have said, "Well, why don't they just read from the book of Genesis?"

Something about the suggestion seemed right to Laitin, and over the course of the next few days it made its way from Laitin to Bourgin to Borman, who in turn shared it with Lovell and Anders. This, they agreed, was "something appropriate." Without telling anyone what they intended to do, the first ten verses of the book of Genesis, the Creation Story, were typed onto fireproof paper and inserted into Bill Anders' mission checklist log. Knowing that space and weight limitations on this flight wouldn't allow for a full printed copy of the Bible with them, this would have to do. So the first ten verses of the Bible made their way into the flight plan of the first mission to the moon.

A few weeks and a quarter-million miles later, the checklist now hung in weightlessness as the Apollo 8 spacecraft flew its next-to-last revolution around the moon. The crew did a bit of housecleaning, since the cabin itself would be the setting for the largest television audience in history. The telecast opened with live pictures of the earth and moon as seen from the command module. A few hours earlier at Mission Control, the giant screen of the earth suddenly vanished and in its place was the moon. For the first time ever, the ground crew was tracking the trajectory of a spacecraft around a celestial body other than the earth. The effect was overwhelming. A sign above the viewing room's glass partition began flashing "Quiet Please."

"Welcome from the moon, Houston," Lovell said.

Anders then pointed the television camera at the earth.

"What you're seeing," he said as the camera rolled, "is a view of the earth above the lunar horizon. We're going to follow along for a while and then turn around and give you a view of the long, shadowed terrain."

Frank Borman continued with his narration, remarking that the moon "certainly would not be a very inviting place to live and work."

"The vast loneliness is awe-inspiring," Lovell said, "and it makes you realize just what you have back there on earth." A moment later, Anders turned to the typed words inserted in the flight plan.

"For all the people on earth, the crew of Apollo 8 has a message we would like to send you."

NASA had no idea what to expect.

Just before they slipped behind the moon, in what was the most watched television broadcast to date, Anders began the reading that would captivate the world and grip those in Mission Control in awe-struck silence:

> *"In the beginning God created the heaven and the earth.*
>
> *And the earth was without form, and void; and darkness was upon the face of the deep.*
>
> *And the Spirit of God moved upon the face of the waters. And God said, Let there be light: and there was light.*
>
> *And God saw the light, that it was good: and God divided the light from the darkness."*

In an airport bar in Houston, Si Bourgin noticed the room suddenly fell silent. All eyes were on the grainy black-and-white image on the television screen.

In Washington, Joe and Christine Laitin lay in bed awestruck. Joe looked at his wife in disbelief. "That's the script I wrote!"

Christine looked back at him with a laugh. "You wrote?"

Back in the Apollo 8 command module, Anders now paused and passed the flight plan to Lovell, who continued the reading:

> *"And God called the light Day, and the darkness he called Night. And the evening and the morning were the first day.*
>
> *And God said, Let there be a firmament in the midst of the waters, and let it divide the waters from the waters.*
>
> *And God made the firmament, and divided the waters which*

were under the firmament from the waters which were above the firmament: and it was so.

And God called the firmament Heaven. And the evening and the morning were the second day."

Lovell then passed the flight plan to Frank Borman, who closed the reading:

"And God said, Let the waters under the heavens be gathered together unto one place, and let the dry land appear: and it was so.

And God called the dry land earth; and the gathering together of the waters called He seas: and God saw that it was good."

Borman then bade the entire planet farewell:

"And from the crew of Apollo 8, we close with good night, good luck, a Merry Christmas, and God bless all of you—all of you on the good earth."

Astronaut Gene Cernan had spent most of that day at Mission Control and was enraptured. "It made us feel as though we were all present at the creation of a new age," he said. It was a masterful touch. Afterward, Cernan and his wife left for a midnight mass at a little chapel at Ellington Air Force Base.

Walter Cronkite, a CBS television journalist who thought he had seen it all, was similarly moved.

"You know, I'm afraid that my first reaction was, 'Oh, this is a little too much, this is a little too dramatic.' I might even have thought, 'This is a little corny.' But by the time Borman had finished reading that excerpt from the Bible, I admit that I had tears in my eyes. It was really impressive and just the right thing to do at the moment. Just the right thing."

Gene Kranz was in Mission Control but off duty. "When they read from the book of Genesis, I cried," Kranz said. "And that's all there is to it."

TV Guide estimated that a billion or more people from around the globe watched the broadcast—nearly one in three of the 3.5 billion people on earth.

The mission's tenth and final lunar orbit came on Christmas morning.

To jettison out of lunar orbit and begin their long journey home, the crew fired the main engine, which once again brought a degree of anxiety among the crew and ground personnel. By now, however, there was much greater confidence in both the spacecraft and the retrofire engine that would get them home. Yet the possibility of a disastrous malfunction still hung in the air.

Meanwhile, back on earth, Rod Rose grabbed the recordings and headed to St. Christopher Episcopal Church with tapes of the prayer Borman had recorded and the Genesis reading. As the service was in progress, Apollo 8 swung behind the moon for the last time. For Flight Operations Director Chris Kraft, it was the tensest time in his entire tenure with the NASA manned space program.

Flight simulation supervisor Dick Koos was the resident expert at dreaming up hair-raising simulated flight failures on the ground, but this time the mission was for real. "It was at that point," Koos said, "that you really begin to realize it's no longer an academic enterprise. There are real guys up there in that thing. That's when your knees begin shaking."

On the back side of the moon the engine was ignited and the Apollo 8 command module came roaring back to life. Back on earth, Mission Control sat in silence in the midst of the communications blackout, waiting for Apollo 8 to re-emerge from behind the moon. As the minutes and seconds ticked by, flight controllers stood up and nervously began walking around the room.

Finally, an anxious flight director could take it no longer.

"Look, you guys," Chris Kraft announced over the intercom, "do what you want to do, but I'm going to sit here and I want to pray a little and I'd like to have it quiet here because this is one hell of a tense moment for me and those guys in the spacecraft. So, for God's sake, be quiet for me."

The room immediately fell silent. Astronaut Ken Mattingly was the CapCom for reentry. In the background hung the echo of his repeated transmissions:

"Apollo 8, Houston …"

"Apollo 8, Houston …"

No response.

After a few tense moments, a voice crackled through the headset.

"Houston, Apollo 8. Over," Borman reported.

A relieved Mattingly broke into a wide smile: "Hello, Apollo 8. Loud and clear."

As they curved around the backside of the moon and headed to earth, Lovell radioed, "Please be informed there is a Santa Claus."

"That's affirmative," Mattingly came back. "You're the best ones to know."

The crew of Apollo 8 were on their way home.

Rose had arranged to receive a phone call at the church when Lovell confirmed to the world that Apollo 8 had left lunar orbit and was homeward bound.

The call went out immediately upon Lovell's famous "Santa Claus" statement. "The timing was beautiful," Rose said, "because we could give that to the minister just as he was giving his final dismissal to the congregation. That was a great way to start Christmas Day."

In Mission Control, dozens of engineers and flight controllers in Florida and Houston tightened their headsets and gripped their consoles. The crew would need to position the spacecraft into a reentry corridor with only a five-degree margin of error—the equivalent of the width of a piece of paper viewed at arm's length.

As Apollo 8 passed over into earth's gravitational field, their speed gradually increased with each passing second. When they finally reentered earth's atmosphere, they did so literally in a ball of fire—faster than any other manned spacecraft. They splashed down in the Pacific Ocean just before dawn, completing one of the most historic and memorable space flights ever. As promised by Admiral McCain, the carrier U.S.S. *Yorktown* was standing by.

Borman had argued for a night landing to reduce the number of orbits around the moon and decrease the risk of anything going wrong. Now, because of the timing, they had to bob around for an hour and a half until dawn before Navy helicopters could drop frogmen into the water to recover the crew. As the astronauts waited, Don Jones, the helicopter commander, asked the beleaguered astronauts what the moon was made of.

"It's not made of green cheese," Borman replied. "It's made of American cheese."

Soon a light knock sounded on the hatch.

"I can remember this young frogman," Borman said, "a Navy SEAL, pulling the hatch back and poking his head in, and then, with a shocked look kind of falling backwards. I didn't have time to contemplate that very much because we had to hop in the life raft which was now tied to the spacecraft and then hoisted by the helicopter onto the U.S.S. *Yorktown*. But later after we had been debriefed by the doctors, we had a chance to go out and look at the spacecraft and to meet the rescue crew. So here were these Marines all lined up in their uniforms, and I recognized the young corporal there that had first stuck his head in. And I asked him, 'Corporal, thanks a lot and you know, we hadn't shaved, you know, we were dirty. We really must have looked bad.' "

And he said, without batting an eye, 'Sir, it wasn't how you looked, it was how you smelled.'"

⁂

Susan Borman, her blonde hair holding its own against the brisk Christmas Day breeze, clasped the sound tape of her astronaut husband's voice and smiled.

"These are my first Christmas presents," she said, standing outside St. Christopher Episcopal Church in League City. The small Christmas morning congregation had just heard the tape of Borman reciting a prayer to fulfill his assigned role in church from lunar orbit. They then listened to the tape recorded voices of Borman, Lovell, and Anders reading the story of the creation from the Book of Genesis. It was a Christmas present to everyone, beamed down from space on Christmas Eve.

"It's just what this small world was waiting for," Susan said.

Meanwhile, at St. John's Episcopal Church in La Porte, Lovell's wife, Marilyn, appeared wearing the new mink coat that had arrived mysteriously the day before. Four-year-old Jeffrey, the youngest of the four Lovell children, stood by her side holding a yellow helicopter, a gift from under the Lovell tree. It was his favorite Christmas gift.

"It flies like a rocket," he said.

Twice his big sister had to take him out of church to stop his fidgeting. The yellow helicopter was broken even before the service was over.

"I'm just so thankful my prayers were answered," Marilyn said after communion. "This is the happiest Christmas I think I'll ever have."

At Ellington Air Force Base, Valerie Anders attended mass with their five children at the Roman Catholic Church where the Anders' ten-year-old son, Glen, served as an altar boy. The Anders had opened most of their presents early, but there was one special gift yet to come. Bill Anders had ordered a pin to be especially made for his wife — a gold "8" crested by a moonstone with an azure lapis lazuli at its base.

"He's carrying it with him," she said, "I'll get it when he comes back next Sunday."

⁂

With the success of Apollo 8, NASA geologists were now enthusiastic at the prospect of retrieving rocks from the moon. But they also knew that fighter pilots and test pilots typically weren't very good geologists, so a team of geologists arranged for a prototype moon excavation site in the desert of Arizona which had formerly been Navajo land. Since NASA didn't do anything small, it became a major event, with helicopters in the air and fifty or so members of the training team and media hovering around two NASA astronauts wearing their space suits and all their gear. They had been training for two days when one morning around 9:00 a Navajo scouting party appeared on horseback over the horizon.

NASA Public Affairs Officer Ed Buckbee was onsite observing the training activity when the training supervisor approached him and pointed to the Indians, saying,. "Buckbee, you're in charge of Indians today." So Buckbee approached the Indians and attempted to communicate as best he could. One Indian in the group spoke a little English.

"He wanted to know what we were doing out in the desert," Buckbee said. "And so I pointed to these two guys that were suited up in gear and told him, 'We are training these two gentlemen to go to the moon.'"

The Indian said to Buckbee, "I want to return tomorrow with our chief." Buckbee told him he was welcome to do that.

The next morning, the same Indian returned with his chief, who was wearing Navajo face paint, a headdress, and carrying a spear. They rode up to the training site and dismounted. Buckbee introduced himself and spoke briefly to the chief through the translator. As the Navajos were about to leave, the chief walked over and handed one of the astronauts an audio tape, which the translator asked the astronaut to please take to the moon with him. The astronaut took the tape and agreed to do so. After saying their goodbyes, the Navajos rode off into the sunset and the geological training resumed.

Several days later, Buckbee pulled out the audio tape to listen to what the Indians had said. But the recording was in Navajo. So the next day he found a gentleman in Houston who could translate Navajo. The translator played the tape—then burst out laughing. "What did the chief say?" Buckbee asked.

The translator could hardly contain himself. "It says, 'Dear moon people, don't trust these white men. They are S.O.B.'s and they will steal your land.'"

Apollo 8's flight to the moon offered man's first look at himself from over 200,000 miles away, and the perceived enormity of our home planet shrank dramatically in the perspective.

Pope Paul VI later told Frank Borman, "I have spent my entire life trying to say to the world what you did on Christmas Eve."

Reverend Stout described it in these words: "The earth was a gorgeous blue orb suspended in blackness, but oh so small and seemingly insignificant. The idea of heaven being upward and hell downward no longer had meaning. Rather, God is of incomprehensible dimension and His spirit is within each one of us."

By this time the Apollo Prayer League had grown to nearly 40,000 members across the United States and was spreading into foreign countries. Yet even as the spellbinding vision of earth from space was capturing the world's attention, there were needs yet to be met on the surface of that blue orb itself. During that Christmas, the APL and hundreds of NASA employees joined with local agencies to distribute 2,000 boxes to aid needy children in the Houston area.

Meanwhile, atheist Madalyn O'Hair had taken note of the Christmas Eve scriptural broadcast from Apollo 8 and she was not at all happy. Her infamy for banning prayer in public schools had fed her insatiable ego and she continued to file lawsuits at a near pathological level. Like clockwork, almost every year she would find some event or issue that would outrage average citizens of American society and in the process generate nice-sized newspaper headlines. O'Hair was an astute, educated attorney, fully capable of exploiting the law to further her organization of American Atheists, and she was determined to stop "the religious exploitation of outer space."

But as outrageous as the woman was, she was about to meet her match in the form of one Reverend John Maxwell Stout. A tall lanky man with a mop of dark hair, his quiet demeanor concealed the convictions of a man with divine determination. His office at the Manned Spacecraft Center overlooked a burgeoning space agency intent on upholding the individual man's right to pray—anywhere.

"I thank you God for most this amazing day; for the leaping greenly spirits of trees and a blue true dream of sky; and for everything which is natural, which is infinite—which is *yes*."

- Rusty Schweickart, Lunar Module Pilot, Apollo 9
from a poem by E. E. Cummings

14
APOLLO 9: RENDEZVOUS IN SPACE

During the previous several months the nation's cities and college campuses had been shattered by riots and intense bitterness, pitting one American against another over a plethora of racial and political issues. Then, seemingly from out of nowhere, came this Christmas miracle. By year's end Americans were again drawn together by, of all things, a spaceflight. Unlike the war in Vietnam, Apollo 8 was a national effort that nearly all Americans were proud of.

The success of the U.S. lunar orbit thrust NASA into the indisputable lead in the moon race, although intelligence by the CIA suggested that the Soviet Union was not far behind. A man, whether American or Russian, would soon be landing on the moon. In remarks to the National Space Club in March 1968, Wernher von Braun declared that America was within striking distance of its goal.

"If all goes well," he said, "our goal of landing men on the moon and returning them safely by the end of this decade as planned in 1969 will be fulfilled." And then he added a satirical caveat: "That is, if we are not held up by having to pass through Russian customs."

Seven months later, after a successful lunar flyby of a Soviet unmanned spacecraft, von Braun was more pessimistic when asked whether the U.S. was going to beat the Soviets to the moon:

"I am beginning to doubt that we will," he said. "It is very important that we are there first, but in view of the spectacular performance of the Soviet spacecraft Zond 5, in late September, I am beginning to wonder. It will undoubtedly be a photo finish."

Dramatic as recent progress had been, an enormous number of critical technical problems remained unsolved. One of the most glaring was the lunar module. Apollo 8 had flown without one. To actually fly around the moon with a fully equipped lunar module and land it on the surface anytime soon seemed like a pipedream. But as preparation for Apollo 9 neared completion, that was about to change.

For the first time ever, one of the strangest vehicles conceived by man would be flown. The lunar module was a spidery, flimsy contraption covered in a material resembling aluminum foil. Apollo 9 Commander Jim McDivitt described the material as feeling "like tissue paper."

"If you took your finger and really poked hard at it," said Flight Director Gene Kranz, "you could poke right through the outer skin of the spacecraft. It's the kind of thing you were reasonably cautious that you didn't jam any pointed objects through."

It was so fragile that Grumman workers had trouble machine stamping the parts without creating stress lines and were finally required to stamp much of it by hand. Not only was the lunar module fragile, it was also huge, standing twenty-three feet high—a full twelve feet taller than the command module itself.

The comical looking contraption would be coupled to the rocket just to the rear of the command module during launch, then once in space, it would be released, the command module flipped, and docked to the top of the lunar module. Upon reaching the moon, the astronauts would open the connecting hatch door and float into the passenger compartment of the spindly apparatus, which would become their mode of descent to and ascent from the moon's surface. Testing to see if this was possible would be the primary objective of the upcoming Apollo 9 mission; however, the entire mission would take place in earth orbit. To accomplish this, NASA had placed the overweight lunar module on a diet program wherein it agreed to pay Grumman a bonus of ten thousand dollars for every pound of weight removed from the vehicle. A much lighter vehicle emerged.

This would be Gene Kranz's first official launch as a flight director. All flight control work shifts were identified by a color, and as head of the White Team, he wanted something special that would identify his group. Flight Directors typically wore three-piece suits, but Kranz's wife Marta had made him fighter pilot scarves during his Korean War fighter pilot days and suggested that she make him a white vest to make him appear solid. Initially, Gene balked for fear of looking ridiculous in front of the cameras. It was with some trepidation, therefore, that he slipped into the vest during a bathroom break in the midst of the Apollo 9 countdown. He returned to his flight desk and stood there in silence. To his immediate left, his systems flight controller was the first to catch a glimpse and wasn't sure if his new flight director had flipped his lid. "You know if you don't take that thing off they're going to come and haul you away," he quipped.

But it was too late. Kranz could hear the whir of the swivel cameras in Mission Control spinning around and zooming in on his vest, transmitting the image to thousands of tracking stations around the world. Like it or not, a statement had been made. And as the countdown resumed, the results were better than he could have hoped.

Apollo 9 launched with very little fanfare on March 3, 1969. No Apollo crew was better prepared than Jim McDivitt, Rusty Schweickart, and Dave Scott. They had trained together three years in 1966 as the initial backup crew for Apollo 1, before being replaced with Wally Schirra's backup crew. The gangly lunar landing module was appropriately named *Spider* and the command module *Gumdrop* on account of its shape and the blue wrapping in which the craft arrived at Kennedy Space Center. Though the flight was one of the most crucial to date, the public and the media, still satiated with the euphoria of Apollo 8, paid little attention.

Apollo 9 was the first flight of the lunar module, the first time it would operate in a weightless environment to see if it was capable of performing its job. It was important that its engine work and that the lunar module be able to maintain control of both the lunar and command modules should the unlikely need arise to use the lunar module, rather than the command module, as the driving force.

Once they were in earth orbit, McDivitt and Schweickart boarded the lunar landing module, being careful not to put a boot though its flimsy walls. McDivitt's initial attempt to undock the lunar module failed when the release latches hung up on one another. After some trepidation, the lunar module broke free with a distinctive *bang,* and for the first time ever was flying on its own, its spider legs silently unfolding as it drifted through the velvety darkness. By the end of the day they were ready for the final test. They lit up the main engine on the bottom of the lunar module—the engine that would eventually take two astronauts down to the surface of the moon— and it worked, just the way it did in the simulator.

Unlike the quiet hum of the command module, the lunar module was noisy, full of strange whirring and gonglike sounds. But it behaved superbly, the ascent and descent engines performing as advertised. In the first test of the lunar module flying solo, McDivitt and Schweickart separated from the command module by nearly two hundred miles, a maneuver intended to simulate Apollo 11's descent to the surface of the moon and back. It was a risky endeavor. If the lunar module was unable to re-dock with the command module, McDivitt and Schweickart would be permanently marooned onboard. But the funny-looking spacecraft proved itself trustworthy.

One of the scheduled tasks was a spacewalk by Rusty Schweickart to test the backpack life-support system in space and on the moon. Tethered to the lunar module by a twenty-five-foot nylon rope, he secured his feet in the gold restraints affixed to the surface outside the hatch. Scott and Schweickart had been scheduled to take pictures of the space walk, but while Scott was filming Schweickart from the command module's hatch, his camera jammed. So for a few brief moments, Schweickart was afforded an unplanned opportunity to take in the rare view. There, ripping through the vacuum of space at 25,000 miles an hour, he experienced a silence the depth of which he had never known. While perched outside the Apollo capsule high above the earth in complete silence, this typical macho fighter pilot had an ineffable encounter with his home planet:

> You look down there and you can't imagine how many borders and boundaries you crossed. At the Mideast you know there are hundreds of people killing each other over some imaginary line that you can't see. From where you see it, the thing is a whole, and it's so beautiful. And you wish you could take one from each side in hand and say, "Look at it from this perspective. Look at that. What's important?"

Political borders no longer had meaning to Schweickart. The rivers didn't take notice of either of their banks; the clouds didn't stop at the border between Russia and Europe; the oceans served communist and noncommunist worlds alike. National boundaries were nonexistent.

During those moments, something changed in Schweickart. He gained a deep sense of appreciation and responsibility for this magnificent earth.

"It's a feeling that says you have a responsibility. It's not for yourself. *The eye that doesn't see doesn't do justice to the body.* That's why it's there…and somehow you recognize that you're a piece of this total life."

The emotion became deeply personal for Schweickart and one that would lead him to spearhead a national scientific collaboration to explore means of detecting and deflecting approaching asteroids.

On final approach to the command module, the two vehicles docked perfectly. This delicate ballet of "rendezvous" was a hat trick the Soviets were continuing to struggle with and finding all but impossible to master. The Americans, it seemed, had it down cold.

Apollo 9's rendezvous in space was going so well and Flight Director

Kranz was so pleased with the reaction to his first white vest, that during the ensuing round-the-clock shifts, he asked Marta to make another vest to celebrate the Apollo 9 splashdown. This posed a dilemma for her since she had run out of white fabric. "Well, see if you can find some," Gene said. Being a woman of obvious creativity, she went searching through their home closets and found a white satin bathrobe with a pattern of gold and silver brocade, which she used to make one of the fanciest, most memorable vests of Kranz's career.

Overhead, Apollo 9 continued circling the earth every hour and a half, performing test after test. Finally, after ten days, 151 times around the world, 151 sunrises and sunsets, the crew lit the main engine for the last time and headed home for a splashdown in a fiery ball.

With this latest success under its belt, NASA was ready for the final dress rehearsal—Apollo 10. Commander Tom Stafford, Lunar Module Pilot Eugene Cernan, and Command Module Pilot John Young would head to the moon tasked with surveying a suitable landing site for the upcoming lunar landing.

One place of special interest was a region of the moon called the Sea of Tranquility.

"No one in their right mind can see such a sight and deny the spirituality of the experience, nor the existence of a Supreme Being… There were indeed moments when I honestly felt that I could reach out and touch the face of God."

- *Gene Cernan, Lunar Module Pilot, Apollo 10*

15
APOLLO 10: CLOSE ENCOUNTER

Those in the astronaut office knew all too well that Tom Stafford and Gene Cernan had made their way up the flight roster to a seat on Apollo 10 through an unfortunate twist of fate—an event the two astronauts themselves had witnessed first hand. They had originally been assigned as backups to the prime crew pilots, Charlie Bassett and Elliot See, on Gemini 9—an arrangement that was tragically altered on the afternoon of February 28, 1966.

The weather was nasty in St. Louis at 7:30 a.m. when the four hopped into NASA T-38s in Houston and headed for Lambert-St. Louis Municipal Airport, Cernan and Stafford piloting one, Bassett and See the other. The McDonnell Aircraft test facility housing their Gemini capsule stood just 500 yards off the St. Louis runway in Building 101. The weather in the area was the typical late winter mix with three miles' visibility, rain, and snow flurries. As they approached the airport, a thick white wall of snow reduced visibility to within 400 feet of the ground. They descended through the muck, wing tip to wing tip, almost losing sight of each other's plane in the squall.

As they banked their T-38s for landing and ground lights came into view, it became clear they had miscalculated the approach. They were too high, too fast, and too far down the runway. "We don't have a prayer of landing on this pass," Cernan told Stafford. Then, inexplicably, without warning, Elliot See wheeled his T-38 over into a sharp left turn.

"Damn," Tom barked. "Where the hell's he going?"

They were immediately lost in the fog. Cernan and Stafford never saw the crash, explosion, and fire that took the lives of their two comrades. The plane slammed onto the roof of Building 101, skidded to the edge, and cartwheeled into the parking lot, where it exploded.

There was immediate confusion on the ground over who had died. Cernan and Stafford came in smoothly. As they rolled to a stop, the tower asked for "the pilots' names of NASA nine-oh-seven." That was their plane. "Stafford and Cernan," Gene replied. Unknowingly, they had confirmed the deaths of Charlie Bassett and Elliot See.

Marilyn See was now alone with three kids to raise, and there was no role in the space program for widows and orphans. Two years earlier, astronaut Ted Freeman was killed when he flew into a flock of geese. One shattered his canopy. Now, two more astronauts were dead. NASA's backup program worked as designed, and Tom Stafford and Gene Cernan slipped seamlessly into rotation for the Gemini 9 prime crew positions vacated by the two lost astronauts.

The ripple effect would have downstream implications. In 1969 the flight roster once again cycled Stafford and Cernan into prime couch positions—this time on Apollo 10, headed far beyond the embrace of Mother Earth.

The mission would be the last one preparatory to a moon landing. Apollo 10 would take a loop around the moon on the same path cleared by Apollo 8. This time, however, two astronauts would descend in the lunar module to within a few miles of its craggy surface, the last test of the lunar module before the final touchdown on the surface of the moon scheduled two months later.

Some in Mission Control thought it foolish to go all the way to the moon, take all the risks, and then stop less than ten miles short of the surface. But Rod Rose, among others, disagreed. The more we knew in advance of a full lunar landing, he said, the better. It was important they be prepared for every eventuality—anticipated or otherwise.

The flight simulation trainers in Mission Control were charged with training the astronauts in the operation of the command and lunar modules and were adept at creating every combination of catastrophic vehicle failure imaginable. They spent days and nights dreaming them up. When all systems were operating properly the spacecraft was easy to fly. When things went haywire, it was an entirely different matter. As Apollo 10 backup lunar module pilot Edgar Mitchell, put it, "You don't train for things to go right, you train for failure…over and over."

The backup crew members and flight controllers trained alongside the primary crew. Once the crews and flight controllers mastered the basics of "flying," all of the training then became "what if" training for unexpected crises. During Apollo 10 pre-flight testing, the simulation engineers threw every possible trick in the book at the crew—every conceivable failure of every critical system the astronauts might take for granted. Typically, if the crew had difficulty recovering from a simulation failure, flight controllers would write up a detailed procedure describing the proper recovery action.

Such was the case on April 25, 1968, when the "sim guys" threw a hypothetical extreme failure at the backup crew. The cryogenic fuel tank in the command module, they postulated, had sustained a hydrogen leak fifty-one hours into the mission, resulting in the loss of all three fuel cells. As a result, the capsule had no power. Mitchell was in the simulator during the backup training exercise and couldn't believe the far-fetched scenario. "Guys, you just killed us," he said. "We don't need to practice getting killed. This is your problem. Go solve it yourselves. Don't waste our time because we think we'll be dead if that happened."

Mitchell wasn't alone in his belief. Unprepared for this kind of failure, the flight control team was also unable to determine a way to save the crew. Glynn Lunney, Flight Director during the simulation, minimized the significance of the loss and dismissed the entire scenario as an "unrealistic multiple failure case."

Regardless, Jim Hannigan, branch chief of the lunar module flight controllers, was concerned and wouldn't let it go. He authorized a thermal-electrical communications engineer, Don Puddy, to form a small task force to develop "Lunar Module Lifeboat Cases" to be used in the event all command module power or some other critical function was lost and the crew had to use the lunar module as a backup vehicle. In the months following the sim failure, recovery procedures were developed by Puddy's team and put on the shelf to be used "as needed."

Ironically, at the same time the recovery sim procedures were being developed, a bizarre sequence of events was setting into motion a dramatic example of this very situation. During the run-up to the flight, Oxygen Tank 2, one of the two oxygen tanks scheduled for Apollo 10, was accidentally dropped while being removed from the shelf for modification. When mixed with hydrogen, the cryogenic oxygen tanks, referred to as "cryos," provided the primary source of electrical power and water supplies for the crew during the mission. Granted, the tank only fell two inches—but it was dropped. A minor mishap, but enough to damage the tube assembly used to fill and empty the tank. The dropped tank, number 10024X-TA0009, was examined and retested, then tagged and recycled—not to Apollo 10, but to Apollo 13. For the time being, Apollo 10 was a Go for launch.

Driving up State Road 3 toward the launch complex at the Cape on May 17, 1969, Pad Leader Guenter Wendt saw ahead the flashing lights of a deputy sheriff's car on the shoulder of the road. The speed limit was strictly enforced along that stretch of road. Must be some unlucky soul, pulled over for speeding, Wendt figured. As he drove by, he glanced at the cop talking to a sandy haired man. It was none other than Gene Cernan, lunar module pilot for the upcoming Apollo 10 mission. Guenter wheeled the car over and backed up to where the two cars were stopped.

"Gene," Wendt said. "What are you doing out here? You should be getting ready!"

"Man, am I glad to see you," Cernan answered. Not wanting to be recognized as an astronaut who slipped out of pre-flight quarantine, Gene was having some trouble convincing the deputy to let him go. The last thing he needed was a standoff on the side of the road twenty hours before launch.

Wendt showed the officer his NASA identification badge and explained that Cernan was an astronaut, soon to be on his way to the moon. If management was called in to retrieve their wayward spaceman, it would probably get quite ugly. The cop had doubt written all over his face and wasn't sure what to believe. "Get out of here," the officer said, "and go on to your moon!"

With great relief, Cernan hopped into his car and sped off. The next time Wendt saw him, Cernan was encased in a spacesuit as he and the crew slid into their couch positions on a moon-bound spacecraft.

Together again three years after Gemini 9, Commander Tom Stafford and Lunar Module Pilot Gene Cernan, joined by Command Module Pilot John Young, set out on the path for the moon. Stafford and Cernan were slated to board the lunar module and descend to within ten miles of the Sea of Tranquility.

On May 18, only seven months before the end of the decade, Apollo 10 launched from Pad 39A with a fully outfitted lunar module named *Snoopy*. The command module was *Charlie Brown*. The image-conscious NASA public relations people argued that *Gumdrop* and *Spider* were not serious enough names for the historic Apollo 9 mission. They were even more underwhelmed when the crew obtained permission from *Peanuts* cartoonist Charles Schulz to name the command and lunar modules after his popular cartoon characters.

"The public relations types lost this one big-time," Cernan said. "Everybody on the planet knew the klutzy kid and his adventuresome beagle, and the names were embraced in a public relations bonanza." The intrepid, bubble-helmeted Snoopy had become a symbol of excellence around NASA, and before the hoopla quieted, the little dog's image was on decals, posters, dolls, kits, sweatshirts and buttons everywhere. The program had never seen anything like it.

Since *Snoopy* was already used as the logo of safety throughout NASA and featured on pins awarded to employees for excellence, NASA officials had difficulty denying its iconic significance.

Tom Stafford, a six-foot astronaut from the plains of western Oklahoma, was inducted with crewmate John Young in the second group of "New 9" astronauts. Cernan was inducted in the third group of fourteen. In September 1962, Stafford had been assigned as one of the few Air Force officers to Harvard Business School and had attended classes for all of three days before receiving the call from Deke Slayton. As if sensing the inevitability, upon arriving at Harvard he told the movers not to unpack the family belongings because "we might be leaving."

Tom, the only child of Mary Ellen Patten and Thomas Sabert Stafford, weighed into the world at only four and a half pounds. His mother had traveled to Cheney, Oklahoma, with her family as a baby in a covered wagon in 1901, not long after the Oklahoma land rush, and spent her early years living in a sod dugout her father had built. Twenty-seven years later, her life as a wife and mother unfolded when a tall slim man walked into the bank where she worked. The two married in 1921 and settled in Weatherford, Oklahoma. She bore her only son at the late age of thirty-six, a time in life when most other mothers already had six or seven children. Church was a staple in the Stafford home and Mary frequently read to her young son, Tom, from the Bible. Attendance at the local Methodist church was mandatory on Sundays and Wednesdays. During his final visit prior to the launch, she handed Tom a small black leather Bible to carry with him to the moon.

As he suited up for launch, he slipped it into the pocket of his space suit.

The ride up went smoothly; that is, until they fired their rockets headed into earth orbit. The spacecraft began gyrating like a giant pogo-stick. When the second stage ignited they were snapped back in their seats, and in a split second blasted through the first-stage fireball. The pogo stayed with them, worse than ever, as another million pounds of liquid hydrogen and oxygen fuel burned hot and hard for seven minutes. They accelerated skyward with breathtaking speed.

"We weren't done yet," recounted Cernan. "The spacecraft was talking to us, and we didn't like what it had to say. Low moans and a creaking groan indicated that metal was aching and straining somewhere in the back of us and the pogo wobbling indicated that big trouble was brewing down below. What the hell was going on back there? Then the escape tower, which we no longer needed, blew away with a loud thunderclap. The detonation was so sudden and fierce, I wondered for a bleak second if it would tear our spacecraft off the rocket."

After five minutes of suspense and gritted teeth, the burn ended. Eleven minutes after launch they were in earth orbit. One-and-a-half earth orbits later, they executed a translunar injection, ripping the spacecraft from earth's gravity and crashing into daylight.

The pogo was gone, but the astronauts were now suddenly shuddering to new and harder vibration from the booster rocket pressure relief valves. As their vision began to blur, it appeared as if the mission may end even before they left earth orbit. It was as if the bucking creature was shaking itself apart. Stafford's fingers reluctantly curled around the abort handle as he called Houston through gritted teeth. "Okay, we're experiencing frequency vibrations in the cabin." He literally spit out the words a syllable at a time.

"Stay with us now, baby," Cernan injected. "C'mon. Burn!"

They rode the bucking creature through the entire final burn. The third stage shut down right on cue, the bouncing stopped abruptly, and they were relaxed back in their couch like an easy chair.

During the run-up to launch, Stafford had proposed to George Low that the crew carry a lightweight color television camera. Apollo 7 and 8 had telecast only fuzzy black-and-white images. Apollo 9 did two live broadcasts, but these were also in black and white, and Stafford was determined to let the tax-paying American public share in the beauty of the missions they were funding. Undeterred that such a thing had not yet been developed, he garnered the help of NASA and Westinghouse engineers to concoct an imaging system similar to one developed earlier by CBS using a spinning wheel of blue, red, and green to encode signals which would then be converted on the ground into live color television images. With this, the group threw in a low-light-level TV camera tube, two French lenses, and an actuating motor from a Minuteman missile. The "skunk works" performed brilliantly and the camera was installed on Apollo 10 a mere ten days before launch.

Onboard *Charlie Brown*, as the command module moved away from *Snoopy* for reversal and docking, Stafford turned the camera toward the re-docking maneuver. As they slowly rotated, the earth came into view. What those watching from earth saw next were the first live Emmy Award-winning color images transmitted from space. Westinghouse and NASA engineers were ecstatic. Mission Control went crazy. After the rocky launch phase, they were on a free coast to the moon.

Now, in the quiet pull of the capsule toward the moon, Stafford took the small Bible from his pocket, the first ever of its kind to cross the abyss from the earth to the moon. Opening it randomly, he began reading the same beautiful King James scripture his mother had read to him during the early dust bowl days in Oklahoma. As they grew nearer the moon, he turned the camera and transmitted another live telecast view of earth, giving television spectators around the world a vivid look at their home planet, which by now had shrunk to the size of a grapefruit in the capsule window.

In the small town of Weatherford, Oklahoma, his mother pulled her chair close to the television. And Mary Ellen Stafford, the woman who had traveled in a covered wagon through the hot Midwestern plains and stayed to raise a family, watched from that same small town as her son transmitted breathtaking color images from space back to the television in her living room.

As the spacecraft slipped into lunar orbit sixty miles above the moon, Stafford and Cernan opened the connecting hatch and drifted into the lunar module, whereupon they bade John Young farewell and closed the hatch. When they disappeared around the moon, *Charlie Brown* and *Snoopy* were still docked. When they appeared on the other side they were fifty feet apart, flying in formation.

Stafford and Cernan fired a small burst from the engine and descended through the silence in an elliptical orbit toward the Sea of Tranquility where they would survey and photograph the terrain slated as the landing site for the first lunar landing on Apollo 11.

The homespun Americanisms used by the astronauts to christen features of the lunar terrain included Boot Hill, Joe's Crater, Wagon Road, and Dry Gulch. The astronauts had deliberately picked outlandish names during their long training period to make them stick in their minds. One that stuck in Stafford's mind was one he himself had named after his home state's terrain: "Oklahoma Hills"—a string of rills that would be just to the left of the moon's equator as Apollo 11 approached their lunar landing site.

As *Charlie Brown* fell away, Young called out, "Adios, we'll see you back in about six hours."

"Have a good time while we're gone, babe," Cernan replied.

"Yeah, don't get lonesome out there, John," Stafford said.

Although the to-do list was long, "the goals were only two," Cernan said, "reach our objective and survive." Survival, however, was far from a certainty. On May 21, Apollo 10 passed 47,000 feet above the moon. As they made their first pass over the southwestern corner of the Sea of Tranquility, Cernan was exuberant.

"I'm telling you, we are low." he radioed to CapCom Charlie Duke. "*We're close babe...* We is down among 'em, Charlie!"

"I hear you weaving your way up the freeway," Duke replied.

∽∧∧∼

The words came down loud, clear, and explicit from astronaut Tom Stafford aboard Apollo 10 when asked if he had swallowed too much chlorine in his water supply:

"You bet your sweet bippy I did," Stafford said.

"Bippy" is not in the dictionary. But just about everyone who heard Stafford knew he came very close to using another word.

It was a forewarning that this crew might set a precedent in language. If the Apollo 10 crew were not as articulate as their predecessors, they were certainly the most earthy. Stafford, Cernan, and Young used some basic Anglo-Saxon adjectives to describe the things they saw in outer space, nothing the average man doesn't use in the company of other men, but much unlike the jargon of previous astronauts who had managed to acquire a goody-goody image.

Two of the astronauts used a four-letter word meaning sexual intercourse. The first time it came ripping across the airways, a quarter million miles from earth, it so shocked the young girl typing the instant voice transcription that she substituted the words "damn it."

"We let that one go by," said a spacecraft center official. "It's the only one we let the transcribers change. If they don't want to type certain words, we'll get someone else to do it." He added that he doubted if the girl even knew the meaning of some of the words.

While skimming above the craters on the surface of the moon, the crew prepared to photograph the Sea of Tranquility. But as Stafford trained the lens of a specially designed Hasselblad camera, he encountered a glitch.

"This goddamn filter has failed on me," Stafford shouted as millions of people on earth listened in to the transmission. The transcriber in Houston jumped back from her typewriter. The Oklahoma Hills were fast approaching as Stafford quickly took out a substitute filter. It too jammed just as he caught a fleeting glimpse of the Sea of Tranquility passing beneath them. Traveling at 3000 miles per hour, all Stafford could do was report a clear visual that he knew Houston would understand: Apollo 11 was likely to have a touchy landing. There are "enough boulders around here to fill Galveston Bay," he reported.

So impressed was Stafford with the rocky terrain that at one point he enthusiastically announced, "There's Censorinus" referring to a large crater with gigantic boulders on its rim, "just bigger than shit."

In Mission Control, astronaut Jack Schmitt was asked by a reporter, "What did Colonel Stafford say?"

He said, "Oh, there's Censorinus…bigger than Schmitt."

Then, as Stafford and Cernan completed their "close encounter" and prepared to ascend for rendezvous with *Charlie Brown*, the unthinkable happened. Without warning, *Snoopy* suddenly wheeled over and dove for the lunar surface.

"Gimbal lock!" Stafford screamed.

"Son of a bitch," yelled Cernan over an open mike, once again startling the transcriber. "What the hell is happening?" They were suddenly bounding, diving, and spinning all over the place less than 47,000 feet above the rocks and craters.

"We were ass over tea kettle," said Cernan. "I saw the surface corkscrew through my window, then the knife edge of a horizon, then blackness, then the moon again, only this time coming from a different direction. We were totally out of control."

Instead of targeting the command module, *Snoopy's* computer had found a much larger target—the moon. A mis-set switch on the guidance system sent *Snoopy* into a premature search for *Charlie Brown*. The guidance box was located between the two astronauts, and Cernan had set the switch to the Abort Guidance System (AGS) mode, to simulate the ascent from the moon to be taken by Apollo 11. Stafford, not realizing the switch had been set, flipped it in the opposite direction for Primary Navigation Guidance System (PNGS) mode.

The lunar module was now hopelessly confused. For fifteen terrifying seconds, Snoopy arced toward the surface. The attitude indicator was nearing the red zone and they were tumbling into gimbal lock. Cernan and Stafford had every reason to believe they would be the first human beings to die on the moon. Stafford reacted instantly, which was almost too late. By flipping the guidance system to manual, he overrode the computers and grabbed manual control of the spacecraft, jettisoning the 20,000 pound descent stage to relieve weight and steering *Snoopy* away from certain doom.

"We've got our marbles," Stafford nervously replied once they were back on track, at which point Mission Control broke out a large cartoon showing *Snoopy* kissing *Charlie Brown*, saying, "You're right on target, *Charlie Brown*."

In Houston, the NASA official was caught once again having to explain the astronauts' colorful language to reporters. "It's no problem," he said in an attempt to smooth the situation over. "Those are three human beings up there and they acted like human beings. That's all. No more and no less."

But it had been a closer call than the astronauts realized. "After analyzing the data," Cernan said, "experts surmised that had we continued spinning for only two more seconds, Tom and I would have crashed into the side of the lunar mountains."

After they successfully regained control of *Snoopy* and their senses, the lunar module made its final approach to *Charlie Brown* and successfully docked. The crew said goodbye to *Snoopy* and sent their iconic lunar module into a heliocentric orbit on a one-way trip to the sun.

It wasn't until they were on their way home that it was determined by Mission Control that, as if jinxed by Glynn Lunney's comment about the "unrealistic" simulated fuel cell failure, one of the three Apollo 10 fuel cells had, in fact, failed due to a bad circuit. A second was operating at only half power.

Three days after leaving lunar orbit, Apollo 10 reentered the earth's atmosphere at 29,471 miles per hour, breaking the Guinness World Record for speed. After splashdown on May 26, 1969, the United States was ready to take the final bold step. While the crew of Apollo 10 had been busy orbiting the moon, NASA's massive "crawler" was inching slowly toward launch pad 39A at Cape Kennedy carrying the enormous Apollo 11 Saturn V rocket.

The scene was finally set. Neil Armstrong, Buzz Aldrin and Michael Collins now stood in the wings, ready to take the stage in the pageant of human history.

"Sir, my goal is to go to the moon."

- Buzz Aldrin, West Point Academy, 1951

"I want to be the first man to *come back* from the moon…"

- Neil Armstrong, Purdue University, 1955

16
MAKINGS OF A MOON WALK

By the spring of 1969, the energy surrounding the space program was palpable and building daily. There was a gathering realization that the United States was about to leave an indelible mark on human history. At a time when the nation was badly in need of hope, Apollo was on the cusp of delivering an antidote of national pride.

In March, NASA officials Chris Kraft, George Low, Bob Gilruth, and Deke Slayton had a decision to make: Who would be the first to set foot on the moon? Since Apollo 11 was scheduled as the first lunar landing, the first moon-walker would be Lunar Module Pilot Buzz Aldrin or Commander Neil Armstrong.

The first man on the moon, Kraft said, will be "an American hero beyond Lucky Lindbergh, beyond any soldier or politician or inventor." Kraft and Low assumed it would be the lunar module pilot, which meant Buzz Aldrin would be immortalized. During Gemini missions, the copilot always made the spacewalk, so this seemed a logical assumption. But this particular situation was without precedent—it was not just another spacewalk. After some discussion the final decision was put on hold.

Later that spring, Aldrin approached Armstrong about it. "Neil," he said, "you probably know I don't care very much one way or another about this. But we've got some tough training ahead of us and I think we have to settle this matter before it gets blown out of proportion."

Stoic as ever, Armstrong had little to say. "Buzz," he replied, "I realize the historical significance of all this and I just don't want to rule anything out right now."

The thought of being one of the first men to step on the moon was uniquely exhilarating for Aldrin. As a boy growing up in Montclair, New Jersey, he was smart, athletic, and exceedingly competitive. Born Edwin Eugene Aldrin, Jr., the nickname Buzz was bequeathed by his 18-month old sister, Fay Ann, who called him "buzzer" while trying to say "brother." The nickname stuck. The Aldrin family already had two daughters, and Aldrin Sr. was concerned they might never have a son. His wife, Marion, reassured her husband. "Don't worry," she said. "We'll just keep having them until we have a boy."

"Fortunately, the next child was Buzz," his sister Maddy said. "Otherwise, who knows how big the family would have gotten?"

Aldrin Sr. had a degree in physics from Clark University and was an early student of Dr. Robert Goddard, the rocket scientist instrumental in developing the technology perfected by Wernher von Braun, and whose rockets would now propel Buzz to the moon. He was a disciplinarian who set high standards for his son.

A review of family history revealed that a military career was imbedded in Buzz Aldrin's DNA. His parents met in the Philippines when his mother's father was a military chaplain to General Douglas MacArthur and his father a pilot and aide to Billy Mitchell. Aldrin Sr. later became an Army Air Corps pilot, a full colonel who fought in two world wars, and he had specific plans for his only son: Buzz would attend Annapolis, establish a career in the Navy, and become a successful businessman. Buzz, on the other hand, had ideas of his own.

"I always wanted to get into aviation. I knew that from the time I was growing up. My father was flying airplanes and I wanted to fly airplanes. So it sort of narrowed down that I would go to one of the academies. My father preferred the Naval Academy, but since I had gotten seasick as a youngster in summer camp, I thought going to West Point would be a better thing to do."

At West Point graduation, a classmate overheard Buzz deliver a startling comment. As he strode to the podium to accept his diploma, Aldrin was prepared for the standard question asked of each graduate. "The commanding officer asks them what their life's goal is," the classmate related. "As Aldrin got his diploma, he said, 'Sir, my goal is to go to the moon.' Imagine that—in 1951!" Aldrin graduated third in his class at West Point and second at MIT, an enviable ranking by anyone's standards. But Buzz Aldrin set high standards for himself. He was reared to be the very best that he could be, and there was an abiding sense that he wasn't doing that.

Technology and the sweep of history, however, would conspire with the young Aldrin in extraordinary ways. At the moment, he wanted to be a fighter pilot, which is what he became, flying the F-86 Saber. And he was exceedingly good at it. During the Korean War, he shot down two MiGs. The first he described as "a piece of cake." The second, as "the hairiest experience I've had flying machines in this planet's atmosphere."

The latter occurred the afternoon of June 7, 1953, when shortly after sighting a group of enemy MiGs below him, Aldrin peeled off in a "speed dive" and tore through the thick layers of lower atmosphere in hot pursuit. His plane began to buffet so ominously that he thought it might break apart, but the maneuver delivered him to the tail of his target. Barely in control of his aircraft, he fired off a short burst of .50 caliber machinegun fire. The canopy of the MiG popped open, and the pilot ejected, trailing "a red silk scarf." Aldrin snapped a quick series of photos of the pilot, then jammed the throttle of the F-86 forward and exited the scene at 500 mph. The photos of the pilot ejecting from a MiG in a flaming tailspin were featured in *Life* magazine's "Picture of the Week" and were vivid reminders to the American public of the realities of war.

On October 5, 1957, Buzz was stationed in Germany and in his room preparing for a routine flight when he heard the Russians had launched Sputnik. At the time, he was flying the F-100 with friend and fellow West Pointer, Ed White. Sputnik, they both understood, would change their lives forever. America wasn't going to sit idly by and allow the Soviets to rule outer space; it was going to send men there, too. Aldrin and White were determined to be among them.

One day in 1958, White told Aldrin that after the military he was going to earn a master's degree in aeronautical engineering at the University of Michigan through the Air Force program. Then he intended to join the pilot school at Edwards Air Force Base where the best pilots would be selected to fly the experimental, high-altitude "rocket plane" which was rapidly evolving into a genuine spacecraft. This kind of talk was catnip to Aldrin.

"What do you think they're going to call the guys that finally fly up there?" he asked White. "Rocket pilots?"

Ed smiled and tossed his parachute over his shoulder.

"No," he said. "The Air Force is bound to think up a fancier name than that."

The thought was obviously on Aldrin's mind in 1961 when he enrolled at MIT for a doctoral program. "I understood how to intercept airplanes in the atmosphere," he said, "but everyone was going to have to be able to join up spacecrafts, so I chose as my thesis subject 'Rendezvous in Space.'" It would prove to be one of the wisest decisions of his professional career.

In 1963 Aldrin was selected to join the third group of astronauts. Techniques for an orbital rendezvous were under active development then, and Aldrin's interest in rendezvous was nothing short of fanatical. He developed such a reputation for cornering others and talking incessantly about rendezvous that his fellow astronauts dubbed him "Dr. Rendezvous." It didn't bother Buzz—and it didn't stop him either.

Aldrin had already flown on Gemini 12 with Jim Lovell in November 1966 and performed a record-setting spacewalk. Flight rotation changes eventually cycled his name to the forefront for the coveted Apollo 11 moon landing mission, and by the spring of 1969, the media was speculating on whose boot print would be first on the moon.

Then, in late March 1969, a leak by an undisclosed top NASA official resulted in a premature news release blasted across the face of the *Chicago Daily News*: "FIRST MAN TO STEP ON MOON WILL BE BUZZ ALDRIN." The role of performing one of the most momentous events in the history of mankind "would fall to the brainy 39-year-old Air Force colonel by virtue of his role as lunar module commander." Forty-five minutes later, the release said, the Apollo commander, Neil Armstrong, would descend and join him. The report continued: "The disclosure of Aldrin as the choice comes as a surprise to many who had speculated that the top commander would be entitled to pull rank and take his place in the history books as the first man to set foot on the satellite of the earth."

But within NASA the issue remained unsettled. "I don't think we've really decided, yet," Slayton told a reporter. "I think which one is to step out on the moon first will depend on some further simulations."

Over the next weeks, rumors of a decision surfaced, but NASA officials were not yet prepared to announce it. Finally, on April 14, Deke Slayton called the two astronauts to Armstrong's office. A decision had been made. Neil Armstrong would be first. Slayton then gave their reasons: Armstrong was more senior, and given the confined area in the capsule, the procedure for suiting up, maneuvering, and opening the hatch would require the commander be the first one out. It was pure logistics.

The rationale made sense to Buzz. "The training workload on the commander was enormous," he said. "The net result of symbolically putting the commander up first was absolutely correct."

Neil Armstrong was pragmatic in his outlook for the mission and keenly aware of the challenge. When a college friend once asked him whether he wanted to be the first man on the moon, Armstrong replied, "No, I want to be the first man to come *back* from the moon."

As a child growing up in the Midwest, he had no hint of the extraordinary place in history that awaited him. He was the product of a middle-class family that moved from one small Ohio town to another. Like Aldrin, he harbored unusual interests. At age two his father took him to the National Air Races in Cleveland, Ohio, which seemed to spark a fascination with airplanes. His interest intensified when he went for his first airplane ride in a Ford Tri-Motor "Tin Goose" at age six. After graduation from high school in 1947, he enrolled at Purdue University in aeronautical engineering, but his education was interrupted in 1949 when he was called to active duty as a Navy aviator flying Panther jets over Korea.

Armstrong eventually returned to Purdue to complete his master's degree, and in 1955 found his way to Edwards Air Force Base as a test pilot for experimental aircraft. In April 1960, he was selected to join a small group of experimental pilots to fly the X-20 "Dyna-Soar," a delta-wing aircraft capable of being launched into earth orbit. His later flights in X-15 jets became legendary. This was just the beginning of Armstrong's career flying exotic aircraft.

In 1962 he received a call from Deke Slayton, who told him NASA needed pilots like him who were willing to fly machines that had yet to be invented. Neil Armstrong, it seemed, had the stuff to do that. Armstrong's hallmark was an uncanny ability to remain cool in the midst of a crisis. As he would soon learn, there were many ways to die as an astronaut, and you didn't need to be strapped atop a rocket to get yourself killed. The astronaut training itself was often life-threatening. Take the case of the Lunar Landing Training Vehicle (LLTV), one of the most dangerous flying machines ever contrived. Its purpose was to simulate the flight of the lunar lander, as it had four "legs" and a massive fan that allowed it to hover like a helicopter. It was so peculiar and clumsy that the astronauts dubbed it the "flying bedstead."

"The LLTV was the ugliest contraption in the world," said Gene Cernan. "You either landed, ejected, or killed yourself." But it was the closest thing they had to simulate a lunar lander and Armstrong had no trepidation about flying it.

On May 6, 1968, Armstrong flew the bedstead to an altitude of one-hundred-plus feet in fairly high winds when it began bucking and backfiring. The machine then pitched up and banked steeply as it aimed for the ground. In a last-second effort to save his life, Armstrong ejected, the parachute barely deploying before he hit the surface. Within four seconds from the first sign of trouble, the aircraft was in flames. Armstrong stood up, brushed himself off, and headed back to his office to finish some paperwork.

A few minutes later, fellow astronaut Alan Bean came into the office he and Neil shared. Bean hadn't heard about the accident, and Armstrong didn't bother to mention it. After a brief exchange of pleasantries, Bean left and met up with a couple of NASA employees who were vividly discussing something in the hallway. Apparently there had been an accident of some kind.

"What happened?" Bean asked.

"Well, the wind was high and Neil ran out of fuel and bailed out [of the LLTV]."

"When did this happen?" Bean asked.

"It just happened an hour ago."

"That's bullshit!" Bean said. "I just came out of my office and Neil's there at his desk. He's in his flight suit, but he's in there shuffling some papers."

"No, it was Neil," they said.

Bean still couldn't believe it and went back to his office to tell Armstrong.

"I just heard the funniest story," he told Armstrong, who was still at his desk.

"What?" he asked.

"I heard that you bailed out of the LLTV an hour ago."

"Yeah," Armstrong said. "I did."

Bean was stunned. "What happened?" he asked.

"I lost control," Armstrong explained, "and had to bail out of the darn thing."

That was all he had to say about the matter. The LLTV was grounded shortly thereafter, even though Armstrong himself described it as an "accurate simulation" of what it would be like to fly over and land on the lunar surface. In any case, Armstrong's near-death encounter said a great deal about who he was as a pilot and a man. In the face of danger, he had remained steadfastly calm and escaped with his life. He now had the unspoken respect of the entire Astronaut Corps.

"We felt very strongly the great amount of effort, the great amount of technology, and the great amount of teamwork that had gone together to make this. And I think anyone would be foolish to think that God wasn't the leader of this."

- Tom Manison, Elder
Webster Presbyterian Church

17
THE SILVER CHALICE

During his days as an astronaut, Aldrin, his wife, Joan, and their children, Michael, Janice, and Andrew, regularly attended Webster Presbyterian Church, a small parish near the space center where Buzz taught Sunday school and served as a church elder. The pastor, Reverend Dean Woodruff, and Aldrin became close friends. One of the topics Woodruff often spoke of was the meaning of the communion service. "One of the principal symbols," Aldrin said of Woodruff's sermons, "is that God reveals himself in the common elements of everyday life. The traditions of the church had it that these elements are represented in communion with bread and wine—foods that were common in biblical times, as they were products of a man's labor."

Now, as Aldrin prepared to go to the moon, an idea began to germinate.

"One day while I was at Cape Kennedy, working with the sophisticated tools of the space effort," he said, "it occurred to me that these tools were the typical elements of life today. I wondered if it might be possible to take communion on the moon, symbolizing the thought that God was revealing Himself there, too, as man reached out into the universe. For there are many of us in the NASA program who do trust that what we were doing is part of God's eternal plan for man."

When he returned to Houston, Aldrin spoke with Reverend Woodruff about the idea. According to Aldrin, Woodruff was "enthusiastic."

Right away questions came up. Was it theologically correct for a layman to serve himself communion under these circumstances? Reverend Woodruff felt it would be, but thought it best to clear it with the clerk of the Presbyterian Church General Assembly. Meanwhile, as a NASA chaplain and ordained

Presbyterian minister, Reverend John Stout was also caught up in the dilemma and worked internally to facilitate the matter through NASA.

"It was a big question," Stout said. "In the hierarchy of the church, a layperson has to have approval at senior levels in the Presbyterian Church." There was also the question of performing a religious service on a government funded moon landing. The mix of religion and space was destined to bubble up through the top ranks of the agency. The timing could not have been better for Dr. Thomas Paine, the newly appointed NASA top official, and in some ways, it could not have been worse. On April 3, 1969, Paine had just taken over from James Webb as the NASA Administrator responsible for the first moon landing. In addition to assuming the job during an exciting time in the program, however, Paine also had to deal with a lawsuit filed by Madalyn Murray O'Hair. O'Hair had filed against NASA in general, and Thomas Paine in particular as its figurehead, for allowing the Apollo 8 reading from the book of Genesis. She claimed the space organization was allowing the use of "government property" as a platform for religious acts, a clear violation of the First and Fourteenth amendments requiring separation of church and state.

In May 1969, a letter from the National Presbyterian Center in Washington D.C. concerning the communion issue was funneled through Paine to the desk of Julian Scheer, who deferred the decision to the Apollo Prayer League. Given the lawsuit brewing with O'Hair, NASA wanted to steer clear of any direct involvement. Meanwhile, Reverend Woodruff was striving to resolve the situation directly with the church. After a time, permission was granted by both organizations and the way for the lunar communion was cleared.

This hurdle behind him, Aldrin knew he would need a way to get the communion elements on board the spacecraft. As it turned out, NASA already had a method in place. Each astronaut was permitted to carry a Personal Preference Kit, or PPK, in which to stow personal items, a small white pouch measuring roughly 4" x 6" that could weigh no more than 18 ounces. Final approval of the astronauts' PPK contents was made solely by Deke Slayton. PPK items were logged separately from NASA-owned items, which ostensibly excluded them from public controversy. "I could carry the bread in a plastic bag," Aldrin surmised, "the way regular in-flight food is wrapped. And the wine also—there will be just enough gravity on the moon for liquid to pour. I'll be able to drink normally from the cup."

The idea began to take shape.

"I have this Personal Preference Kit," he told Woodruff. "Go find me a silver chalice—but it can be no heavier than two ounces." The chalice, together with the bread and wine, would go in a special "communion kit" inside Aldrin's PPK. With these weight restrictions in mind, Reverend Woodruff set out to find a suitable chalice, a task he suspected would prove difficult. It had to be exceedingly small, light, elegant, and befitting a sacrament of historical proportions.

The first place he visited was Corrigan Jewelers at Alameda Mall in Houston. He wasn't expecting to find a chalice, but hoped for advice on where he might find one. He described the item he was searching for to the sales clerk behind the counter. She thought for a moment, then reached below and drew a tiny silver chalice from the display case.

"You mean like this?" she said, placing it on the counter.

"That's exactly what I mean," said the Reverend.

The next week he showed Aldrin the chalice. It was light enough. So light, in fact, that it allowed for another item. At that moment Reverend Woodruff saw a unique opportunity.

"I just happened to have gotten a bronze seal of the church," he said. "And we decided we just might be able to get this in so we would have another symbol of the church. I weighed it and it was just .5 ounce, and I just *happened* to have two, so I threw them together."

Together they weighed too much.

Reluctant to sacrifice the opportunity, Woodruff asked for assistance from one of his parishioners, Jack Kinzler, head of the Technical Services Division at NASA's Manned Spacecraft Center credited for developing the "zip gun." Kinzler managed a large group of electricians, machinists, metal fabricators, and welders and was adept at calibrating and fashioning metal. Kinzler took a file, pared the two coins, and then glued them together. The resulting fused seals were then light enough to include in Aldrin's PPK.

The plans Aldrin and Woodruff developed called for two very special communion services at the Webster Presbyterian Church. The first would be just prior to the Apollo 11 launch scheduled for July 16, 1969. Such a service was deemed acceptable in the eyes of NASA officials, as no one could stop an astronaut from practicing his faith on his own time. The controversial aspect of the plan, should it become known, had to do with the second communion scheduled for Sunday, July 20, during the time Aldrin and Armstrong were slated to be on the lunar surface.

On that Sunday, the congregation in Webster, Texas, would gather to receive communion at a prescribed hour Texas time, while Aldrin took communion himself as close as possible to the same hour from inside the lunar module. The intent was to represent "not only our local church," said Aldrin, "but the Church as a whole."

Finally, there was the matter of which passages of the Holy Scripture would be read during the lunar communion service. Aldrin discussed it with Woodruff.

"I thought long about this and came up at last with John 15:5," Aldrin said. "Dean would read the same passage at the full congregation service held back home that same day."

Aldrin also sought advice from his father, who suggested the passage from Psalm 8, the Song of Praise Rejoices. Aldrin scribbled both verses onto a small card and stowed it in his PPK along with the communion elements.

With this, everything was set for Aldrin's pre-flight communion service. In order to avoid media attention, Reverend Woodruff advised the congregation not to discuss Aldrin's communion plans outside the church. In light of O'Hair's lawsuit, there was no way NASA officials would approve of an astronaut publicizing a communion on the moon—especially with a church congregation. After the Genesis reading on Apollo 8, Deke Slayton has expressly warned against any religious observance during a mission. NASA Administrator Thomas Paine, on the other hand, had ordered that no one within the agency coach the astronauts on what to say while on the lunar surface. Thus, the line between what could and could not be said was somewhat blurred. The astronauts could say whatever they wanted, it seemed, but not as part of any religious observance. All his life Buzz had been deeply religious and the impending services meant a great deal to him. He hoped to carry the plan off without a hitch, but problems inevitably arose.

It was Saturday, the day before he was scheduled to depart Houston for the launch site at the Cape, when Aldrin asked the flight physician's permission to receive communion at a Webster Presbyterian Church service the very next morning.

"You want to what?" The physician said.

Aldrin should have anticipated the doctor's reaction. The flight physician was taking no chances. A cold germ, a flu virus, and the whole moon shot might have to be aborted. In fact, to avoid contamination, Aldrin and his crewmates were required to appear at a pre-launch press conference wearing sterile masks and talking from behind a special partition. And now Aldrin was asking to join an entire church congregation in breaking bread. The flight physician would not hear of it. Aldrin called Reverend Woodruff with the news late that night.

"It doesn't look real good, Dean."

"What about a private service?" Woodruff asked. "Without the whole congregation?"

Aldrin figured this might be a possibility. He called the flight physician, who reluctantly agreed, provided there were only a handful of people present.

So the next day, shortly after the Sunday morning service, Aldrin's wife Joan and their oldest boy Mike—the only one of the three children who was a communicant—went to the church. There they met Reverend Woodruff, his wife Floy, and a close family friend, Tom Manison, with his wife, Robin. The seven went into the sanctuary. On the communion table were two loaves of bread, one for now, the other for two weeks later. Beside the two loaves were two chalices, one the small silver cup Woodruff had purchased for Aldrin's communion on the moon.

At the end of the service, Reverend Woodruff took the silver chalice, tore off a corner of the loaf of bread, and handed them to Buzz. A few hours later, Aldrin was speeding across the skies in a T-38 east toward Cape Kennedy.

At the Manned Spacecraft Center in Houston, Jack Kinzler was hurrying to complete his list of assignments for the upcoming Apollo 11 mission when Bob Gilruth, Director of the Manned Spacecraft Center, popped into his office. With the historic launch only weeks away, Gilruth wanted to know if Kinzler could "come up with something" to symbolize the country—something that would mark the historic event and remain on the moon for all time.

"Okay," Kinzler replied.

Even as Gilruth vanished around the corner, Kinzler had the answer. What could be more appropriate, he thought, than a flag that would symbolize the U.S.A.?

Kinzler's next idea was for a sign to be left on the lunar surface affixed to the spacecraft. He quickly jotted down a general message for a plaque and passed it to Gilruth the next day. NASA public information officials would undoubtedly come up with a more appropriate message, but this would give them something to start with. The message was reviewed and modified by NASA and forwarded to the White House for approval. It read:

> Here Men from the Planet Earth
> First Landed on the Moon
> July 1969 A.D.
> We Come in Peace for all Mankind.

As it turned out, NASA's final message wasn't exactly final. William Safire, a young speechwriter for President Nixon, received a copy of the proposed wording for the plaque, and some of the word choices troubled him. For instance, the word landed bothered him, as the C.I.A. suspected the Soviets might have already landed an unmanned vehicle. Nixon's chief speechwriter,

Pat Buchanan, suggested "set foot," which solved the problem. The words "We come in peace," sounded to Safire "like the sort of thing you'd say to Hollywood Indians." Safire suggested that they at least change the tense, "so that the message would not seem to be directed to lunar inhabitants." The fix was made to "came in peace."

Since the Nixon administration had a strong religious constituency, they left "July 1969 A.D." as it stood, since A.D. or Anno Domini, "in the year of our Lord," was a shrewd way of sneaking God in and would tell space travelers eons hence that earthlings in 1969 had religious inclinations.

During the run-up to the mission, considerable attention at the White House was directed to what President Nixon should say during a scheduled "telephone call to the moon" and how his image should be presented to the American public. During these discussions, astronaut Frank Borman, the Administration's NASA liaison, broke in with a sobering thought.

"You want to be thinking of some alternative posture for the President in the event of mishap—like what to say to the widows."

"Suddenly," Safire said, "we were faced with the dark side of the moon planning. Death, if it came, would not come in a terrible blaze of glory; the greatest danger was that the two astronauts, once on the moon, would not be able to return to the command module."

In that event, they would have to bid the world farewell and "close down communication" preparatory to suicide or starvation. It would hardly advance the cause of space exploration to force a half-billion viewers and listeners to participate in the agony of their demise. So Safire prepared an appropriate statement for the President about men who came in peace and stayed to rest in peace and placed it in his desk drawer in case of a tragedy.

The proposed wording for the plaque was then sent to the President for approval. By all reports, Nixon was excited by the prospect of being the sitting president of the country whose space travelers first set foot on another celestial body. The President approved the message—but only after adding his name at the bottom of the plaque. No one mentioned to him that his name wasn't entirely appropriate since he himself would not actually be landing on the moon.

Kinzler prepared a model of the plaque with the final wording, then inserted a small symbolic American flag in the upper portion and took it to Gilruth.

"You know, Jack," Gilruth said, "What you really could do is put an image of the Eastern Hemisphere and the Western Hemisphere on top. If any creatures from outer space should land on the moon later and look back toward earth, they would see this kind of an outline and they would know where this craft came from."

"Great idea," Kinzler said, and headed back to the shop.

―・・―

Lying on Jack Kinzler's desk in the Manned Spacecraft Center on the morning of July 15, 1969, was the final minted plaque. Next to it, folded neatly in its tubular fireproof fitting, was the first American lunar flag alongside an Allen wrench.

According to NASA Director of Engineering, Maxime Faget, "The U.S. had no intention of placing an American flag on the moon because it would look to the rest of the world like the U.S. was trying to claim it." But a few weeks before the launch, he said, "we changed our minds."

Before Armstrong and Aldrin departed for the Cape, Kinzler instructed them on how to deploy the flag, telescoping it like a portable umbrella to display the full illusionary "wind blown" effect. Kinzler had found a method to sculpt a wave into the flag when horizontally extended. Photos of the flag "waving" on the lunar surface would perplex viewers for years to come and elicit questions as to whether American astronauts had ever been to the moon at all, since there was no wind on the airless sphere.

"We had to package it so it would withstand heat from the descent engine when it was touching downward, firing the thrusters down to the moon's surface," Kinzler said. "We knew that the moon craft would be coming down with 3000 degree flames and the flag needed protection."

Since the only device the astronauts could use to pound the flag's shaft into the lunar surface was their geologic hammer, Kinzler was afraid they might tear it up, so at the top of the flag he installed a one-inch hardened steel tip. "I didn't want it all bent up when they left it on the moon," he said. He then had a red line painted near the bottom of the shaft so the astronauts would know how deep to pound the flag pole into the lunar soil. "We had data on unmanned vehicles on the moon," Kinzler said, "and we knew pretty much from the soundings what the surface was like."

Later that day George Low appeared at Kinzler's door.

"Are you ready?" he asked.

A Lear jet was waiting to take them to the Cape, where they would be picked up and driven directly to the launch site.

It was nearly dark when Kinzler ascended to the top of the fully-fueled Saturn V rocket. He affixed the plaque to the center step of the lunar ladder, then attached the tube containing the flag adjacent to the *Eagle's* lunar landing legs. The procedure for deployment would be simple: down the ladder, five steps to the left, a ninety-degree left turn, and the first American lunar flag would be within easy reach of the first man on the moon.

Apollo 11 liftoff was scheduled for 9:32 a.m. the following morning.

"I wish that everyone could see the earth from a hundred thousand miles distance, for it would certainly change our perspective. Perhaps if mankind could stop for a moment and see himself and the world—not from a hundred thousand miles out—but as God looks upon it, things would be a lot different. Our own perspective on His 'Divine Mission' would be a reality."

- Reverend John Stout
Director, The Apollo Prayer League

18
APOLLO 11: DESTINATION MOON

Before dawn on July 16, 1969, the crew of Apollo 11 was eating the traditional pre-launch breakfast of steak and eggs when a member of the support crew approached the flight physician with a concerned look. "The secretary just kissed Armstrong," he reported.

Under any other circumstances, the secretary's actions would not have been noteworthy. But as everyone knew, any physical contact with outsiders, including that of wives and family, could contaminate the astronauts and threaten the entire mission. The crew had even declined dinner with President Nixon the night before in order to avoid such a risk, and now an exuberant secretary had accosted Armstrong in the crew dining area and kissed him. After a quick assessment of the breach by the physician, it was decided the incident would be kept quiet and the launch would proceed on schedule.

At 6:00 a.m., the astronauts walked to the transfer van for the eight-mile ride to the launch pad. In the viewing area sat several members of Webster Presbyterian Church who Aldrin had invited to the launch. The astronauts emerged in their $100,000 space suits, helmets, and portable oxygen ventilators, all standing around 5'10 or 5'11 tall, all weighing roughly 165 pounds. In the pre-dawn light, it was impossible for the church members to tell them apart. Then, as the men passed the viewing area, one of them turned abruptly to the group and gave a thumbs-up. This one, they knew, was Buzz, signaling that everything—in every sense—was a Go.

Pre-launch access to the Saturn V rocket was tightly guarded by the ground crew and by one man in particular—Pad Leader Guenter Wendt. Because of his attention to detail and heavy German accent, John Glenn nicknamed him "fuehrer of der launch pad." A thin, smiling bespectacled technician, Guenter was the final word for the launch tower close-out team. Dressed in a white cape and overalls, he stood to the right of the hatch door, clipboard in hand, as the astronauts approached the capsule. His was the last earthbound face they would see before launch.

But as serious as "der fuehrer" was, he enjoyed a good prank. Gag gifts had become a pre-launch ritual, and Guenter had spent a lot of time thinking of a gag gift for the crew of this special mission. Remembering the ceremonial "key to the city" often given to visiting dignitaries, Guenter came up with the idea of a large "key to the moon" made of Styrofoam and wrapped in tin foil. A crescent moon made up the shaft. On one end of the shaft was an oval ring and on the other were teeth like those of an old skeleton key. In return, Armstrong presented Guenter with a small card he had tucked under the wristband of his Omega watch. It was a ticket for a space ride, "good between any two planets."

The crew named the lunar module *Eagle*; the command module was *Columbia*. Buzz was the last to board. Guenter dropped him off halfway up the gantry to wait until Armstrong and Collins were secured in their seats. The sun was just peeking over the horizon when Guenter returned for him, and the gantry elevator lifted them to the capsule. Buzz too had a special gift for Guenter. Since both were members of the Presbyterian Church, Aldrin presented him with a small condensed version of the Bible, "Good News for Modern Man." It was one of 130 copies Buzz had purchased for Webster Presbyterian Church in memory of his mother, who had taken her own life the previous year.

On the morning of July 16, only 169 days before the end of the decade, Apollo 11 rumbled to life with three American men aboard. Affixed to the lunar landing ladder was the plaque that read:

Here Men from Earth
First Set Foot Upon the Moon
July, 1969 A.D.
We Came in Peace for All Mankind

Aldrin carried in his PPK the tiny communion kit given to him by the church containing the silver chalice, a corner from Reverend Woodruff's communion bread loaf, and wine in a vial about the size of his finger tip. On the small card was scribbled the scriptural reading he had chosen for the communion. Also stowed in his PPK was a miniature copy of Robert Goddard's 1966 autobiography in recognition of Goddard as the "Father of the Space Age" and one of his father's first instructors. Among other memorabilia in his PPK were charms from his late mother's bracelet. Marion Moon Aldrin was seldom seen without the bracelet displaying fifteen charms of hearts and circles bearing the names of her children and grandchildren. Her maiden name, Moon, seemed a portent of the impending event.

The beaches and coastal roads were awash with people from all corners of the globe. Traffic slowed to a crawl and streets became all but impassable. The beaches themselves became parking lots. The countdown proceeded on time as hundreds of thousands pressed the gates and fences of the Kennedy Space Center, hoping to get a glimpse of the spacecraft in the distance. Three hundred technicians and controllers were involved in the countdown at the Cape, while hundreds more worked their way through the count at Mission Control in Houston. Thousands were on duty at tracking stations around the world, aboard tracking ships, and aircraft. Meanwhile, Walter Cronkite and Wally Schirra, now retired from NASA, provided live television commentary on CBS.

Excitement mounted as helicopters flew overhead and cars ferried people across the Cape seeking the best vantage point. Bleachers for VIPs, congressmen, celebrities, and foreign dignitaries were filled. Then came the final countdown, followed by an incredible roar. The crowds fell silent as the delayed rumble of the engine thrust grew into an earth-shaking eruption, vibrating the ground for five miles.

"It rattles your chest cavity," a Webster parishioner observed. "The heat comes up your pant leg and the decibel reading approaches the level of pain."

"With all the smoke we couldn't tell if it lifted off or blew up," said Jan Armstrong. Finally, in twenty seconds that seemed like forever, they saw Apollo 11 slowly emerge from the flames and smoke.

After an initial "rough ride" atop the Saturn V, the crew jettisoned the rocket and prepared for the four-day journey to the moon. Many of the usual mid-course corrections were deemed unnecessary by Mission Control, as the spacecraft trajectory was absolutely perfect. The crew attempted to rest, but it was difficult. "We had enough adrenaline pumping through our systems for several missions," Aldrin said.

As was traditional, Mission Control kept them apprised of the day's events on earth:

CapCom: Among the large headlines concerning Apollo this morning, there's one asking that you watch for a lovely girl with a big rabbit. An ancient legend says a beautiful Chinese girl called Chang-o has been living there for 4000 years. It seems she was banished to the moon because she stole the pill of immortality from her husband. You might also look for her companion, a large Chinese rabbit, who is easy to spot since he is always standing on his hind feet in the shade of a cinnamon tree. The name of the rabbit is not reported.

Aldrin: Roger. We'll keep a close eye out for the bunny girl.

CapCom: You residents of the spacecraft Columbia may also be interested in knowing that today is Independence Day in the country of Colombia. And Gloria Diaz of the Philippines was crowned Miss Universe last night. She defeated 60 other girls for the global beauty title. Miss Diaz is 18, with black hair and eyes, and measures 34-1/2, 23, 34-1/2.

But in the midst of the impending moon landing, another story was vying for headlines. On July 18, Senator Ted Kennedy attended a party on Chappaquiddick Island in Massachusetts. At about 11:00 p.m., he borrowed his chauffeur's keys to his Oldsmobile limousine and offered to give a ride home to Mary Jo Kopechne, a campaign worker. Leaving the island by way of an unlit bridge with no guard rail, Kennedy's car veered off the bridge and flipped into Poucha Pond. He swam to shore and walked back to the party. It wasn't until the next morning that police were notified. Kopechne had drowned and the senator was thrust into an onerous state of affairs that would plague him for years to come. The episode temporarily distracted Americans as sixty miles above the moon, Apollo 11 was completing its final orbit in preparation for the first moon landing.

Onboard the spacecraft, Armstrong and Aldrin made their way through the hatch. Once on board the lunar lander, they would descend upside down and would not see the moon until moments before touchdown when they pitched forward at 7,000 feet above the lunar surface.

"You cats take it easy," Collins radioed, as the lunar module undocked from the command module and silently floated away for its historic descent to the lunar surface. Armstrong radioed Collins a poignant and simple farewell. Collins, now circling alone in the cosmic dark, responded with a bit of irreverence.

"I think you've got a fine-looking flying machine, there, *Eagle*," Collins said, "despite the fact you're upside down."

"Somebody's upside down," Armstrong replied.

"You guys take care."

"See you later."

Collins looked over his shoulder at his crewmates and glimpsed a vision that he would never forget. As Neil and Buzz drifted away in the lunar module toward the moon, the earth suddenly rose into view directly behind them.

"So I could say to myself, look, that's all there are," Collins said. "Three billion people there and two people here—and that's all there are."

The two people in the lunar module descended silently toward the emptiness of the moon below.

"Apollo first placed human feet on a planet other than earth. It is a fact. And at the same time, it is a symbol. It is a symbol of humans reaching outward—a symbol of the human spirit. Yet, it was not accomplished by human endeavor alone. It was strengthened with the spirit of God within those who accomplished it."

- Dick Koos, Mission Control
Apollo Flight Simulation Supervisor

19
ONE SMALL STEP

When he arrived at Mission Control that morning, Flight Director Gene Kranz had already been to mass at the Shrine of the True Cross near his home in Dickinson. He brought with him a brand-new vest of white brocade that Marta had made especially for this day's shift. The vests had become Kranz's hallmark and a symbol of team readiness in Mission Control. Kranz worked all the odd-numbered Apollo missions and wore a different vest for each flight and for each shift.

Gene and Marta were a team. In fact, they were their own best example of teamwork. Years ago they had sought their parents' approval to marry and were summarily denied. They were just too different, they were told, and the marriage was doomed to fail. Kranz, an Air Force military man with a trademark flattop, was the son of a German immigrant; Marta Cadena was the daughter of Mexican immigrants who fled their country during the Mexican Revolution. Undeterred, the young couple resolutely proceeded to obtain a marriage license and exchange wedding vows at a Catholic church in Marta's home town of Eagle Pass, Texas.

Eleven years later the marriage was stronger than ever. Flight Director Kranz knew how to make a marriage work just as he knew how to make a team of flight controllers pull together. Now, as Black Team Flight Director Glynn Lunney closed out his shift, Kranz carefully removed the vest from its plastic bag and meticulously buttoned it up the front. The White Team was now officially in charge of the first moon landing.

When the Apollo 11 lunar module circled behind the moon, Kranz directed the doors to Mission Control be locked, a sign to the flight controllers that there was no turning back and no way out.

"We don't even think of tying this game," Kranz announced, "we think only to win... And after we finish this sonofagun, we're gonna go out and have a beer and we'll say, 'Dammit, we really did something.'"

Kranz was aware the team was nervous. Besides hosting the first moon landing, there was another reason. During a final simulation test two weeks earlier, just as confidence was soaring, flight simulation supervisor Dick Koos decided to toss one last unlikely "sim" at them. "Hey guys," he said, "open your books to Case No. 26 and have them load it in the simulators."

The technicians responded and Case 26 was loaded. Dick smiled and turned to his team.

"Okay gang, let's sock it to them and see what they know about computer program alarms."

The lunar module computer provided a series of four-digit alarm codes signaling varying degrees of severity. In the front row, 26-year-old flight engineer Steve Bales was ready. Bales was charged with a Go-NoGo call on program alarms. He had done well so far in training and was glad the test was coming to an end. Within seconds a 1201 alarm code flashed on his console. This was the first time he had seen this code and it was meaningless to him. As he watched, another series of 1201 and 1202 alarms appeared. The codes warned that the computer was overloaded and unable to complete its job during the cycle. Bales was desperate. After another burst of alarms and a check with his back-room support team, he delivered the news to Kranz that a Flight Director never wants to hear.

"Flight...something is wrong... Abort the landing... ABORT!"

On that note of alarm, Armstrong and Aldrin, in the simulator, confirmed the abort call and ignited the ascent engine in a simulated emergency liftoff from the moon. Kranz was *not* happy. He felt they should have ended the test with a successful simulated landing.

But if anyone was unhappier than Kranz, it was Dick Koos. During the simulation debriefing he erupted. "THIS WAS NOT AN ABORT. YOU SHOULD HAVE CONTINUED THE LANDING." He grabbed Kranz by the throat.

"If normal flight operations were proceeding and the crew displays all indicating that mission-critical tasks were getting done, you do not abort!" he continued. "You must have two cues before aborting. You called for an abort with only one!"

Case closed. A hard lesson had been burned into the brains of Mission Control personnel and into the brain of Steve Bales in particular. When Bales prepared the final abort code list, the 1201 and 1202 overload alarms were *not* included. Now, with the lunar descent imminent, the failed simulation was still fresh on the controllers' minds.

At approximately 4:00 p.m. EDT on July 20, 1969, Armstrong and Aldrin hovered above the surface of the moon. The terrain beneath them was unlike anything they had seen in training. While the simulator showed relatively few craters, maybe several hundred, what they were seeing were thousands of craters, some as deep as the Grand Canyon.

Five minutes into the descent came the first alarm.

"Program Alarm," Armstrong reported, "It's a 1201."

In Mission Control, hearts shot up into throats. It was up to flight engineer Steve Bales to say Go or Abort. His stomach knotted. Dick Koos's final sim test had become a reality.

"1201 alarm," repeated Kranz on the loop, awaiting an answer.

With minimum hesitation, Bales responded. "We're Go, Flight." His voice quivered slightly.

Forty-two seconds later, another 1202 alarm. Bales' response this time was immediate. "We're Go on that alarm. We're Go."

As the lunar module dropped its legs toward the moon's surface, the astronauts caught the first glimpse of the terrain through the bottom of their window. *Eagle's* computer was directing them toward a boulder field with craters the size of football fields. With little over sixty seconds of fuel remaining in the descent engine, Armstrong overrode the automatic targeting system that was maneuvering them into the crater-strewn terrain and took manual control, searching quickly for a smoother landing site. Standing on the left side of the lunar module cockpit with *Eagle's* fuel almost down to nothing, he used his joystick to spin the spacecraft about, hurtling forward, feet first.

The spacecraft passed through 3000 feet altitude. Five seconds later, another 1201. Thirty-four seconds later, with the lander at only 770 feet, yet another 1202 alarm. Bales called a Go in each case. Armstrong's heart rate rose from 120 to 150 beats per minute. Beneath his vest, Kranz's chest was swollen with pride in his team.

In Mission Control, CapCom Charlie Duke passed information to the crew about the alarms. The alarms, together with Duke's repeated explanations, eventually became distracting. Finally Deke Slayton punched him in the side and said, "Charlie, shut up and let them land."

An extraordinary moment was developing. Armstrong had to either land or abort within the next few seconds. A landing site was just below them, but he still had to get the *Eagle* down.

A voice came over the flight director's loop, "Low level." Fuel was now below the point where it could be measured. No one was breathing in Mission Control. Aldrin began a read out of the final descent.

From his small apartment in Brielle, New Jersey, Aldrin's father listened intently with family members as his son relayed the final moments of touchdown to his commander.

"Okay, you're pegged on horizontal velocity," Aldrin reported to Armstrong.

"Take it down...100 feet.

"Five percent [fuel]."

"Quantity light...75 feet...40 feet..."

With only seconds of fuel left, the descent engine began to blow moon dust for miles in every direction.

"I knew that no matter what I would say or do from now on, this crew is going to go in for the landing," Kranz said. "So we just shut up here on the ground."

"Contact light!" Aldrin called out as the lander legs touched the soft powdery surface.

"Shutdown," reported Armstrong. He shut off the engine and keyed his mike: "Houston, Tranquility Base here. The *Eagle* has landed."

When they finally touched down, the fuel tank was almost empty.

Bedlam broke out in Mission Control. Duke's voice spoke above the cheers.

"Roger, Tranquility. We copy you on the ground. You've got a bunch of guys about to turn blue. We're breathing again. Thanks a lot."

The time was 4:17:42 p.m. EDT, Sunday, July 20, 1969. In this year, in this month, on this date, and at this hour, minute, and second, NASA had accomplished phase one of the challenge set forth by President Kennedy in 1961. Phase two was returning the astronauts safely to earth.

Onboard *Eagle*, it was deathly silent. Armstrong and Aldrin looked over at one another and grinned. After a back-slap and a handshake, they set to work preparing *Eagle* for takeoff, in the event they had to leave in a hurry. Once the tasks were complete, their schedule called for—of all things—a four-hour nap. This aspect of the mission schedule was utterly unfathomable to the crew.

"Whoever signed off on that plan," Aldrin said, "didn't know much psychology." The two astronauts were charged with adrenaline. "Telling us to try to sleep before the EVA was like telling kids on Christmas morning that they had to stay in bed until noon."

Armstrong asked Houston's permission to start preparations to leave the *Eagle* earlier than originally scheduled. With pressure from the astronauts and the press, permission was granted. Now, Aldrin knew, was the time for communion.

He turned to his PPK where the communion elements were stowed. Mindful of what he was to do and say publicly, he cued the mike.

"This is the LM pilot speaking," he said to Houston and broadcast to the entire world. "I'd like to request a moment of silence. I would like to invite each person listening in, whoever and wherever they may be, to pause for a moment and contemplate the events of the past few hours and to give thanks in his or her own individual way."

For Aldrin, this meant taking communion.

He placed the communion elements together with the silver chalice and scripture reading on a small table in front of the abort guidance system computer. Then, during the planned radio blackout, he turned and offered communion to Armstrong.

"No, I don't believe so," Armstrong replied, "but if that's your thing go ahead."

Aldrin then turned his attention to the holy ritual.

"I poured the wine into the chalice," he said. "In the one-sixth gravity of the moon the wine curled slowly and gracefully up the side of the cup. It was interesting to think that the very first liquid ever poured on the moon, and the first food eaten there, were communion elements.

"And so, just before I partook of the elements, I read the words I had chosen to indicate our trust that as man probes into space we are in fact acting in Christ… I sensed especially strongly my unity with our church back home and with the Church everywhere."

Then Aldrin read the verse from John 15:5 he had printed on the card:

> *"I am the vine, you are the branches. Whoever remains in me, and I in him, will bear much fruit; for you can do nothing without me."*

There in the lunar module, in the faint lunar gravity, Aldrin ate the blessed fragment of bread and drank the wine. As *Eagle's* body groaned with the impact of micrometeorites flying through the vacuum of space, he offered thanks "for the intelligence and the spirit that had brought two young pilots to the Sea of Tranquility."

Meanwhile on earth, millions of listeners from around the world gave thanks in their own way. The moment occurred around 3:30 p.m. Houston time on July 20, 1969.

That evening, as Aldrin's father walked along the beach outside his apartment in New Jersey, he paused to look up at the quarter moon. He could barely see the corner where his son had landed. At that moment he could not have been prouder. Somehow he knew Buzz's mother was watching too.

─∽ ∧∽─

NASA routinely made the *Voice of America* network available to the Apollo Prayer League as a means of broadcasting to league members around the world, and Reverend Stout kept a red phone in the press room during missions as a hotline for communications to and from the APL. Although he had helped pave the way for the communion service, he could never have anticipated the role he would play next.

After Apollo 11 landed it was apparent that news reporters had not been told of the time lapse before the astronauts would open the hatch. They expected the crew would simply land and hop out. While reporters had backup plans if something went terribly wrong, they had nothing for the near seven hours that lapsed between the astronauts' touchdown and exiting the lunar module.

Stout happened to be walking past the pressroom when Pedro Kattah, a network production engineer, called out to him, "John, come in here quick." Stout walked in and Kattah phoned *Voice of America* headquarters in Washington, D.C. "John Stout just came by," he said. "What do you want him to do?"

"Give him the mike and let him say anything he wants," they said. So for almost two hours Stout explained the background of the Apollo Prayer League and its outreach programs. His talk was translated into several languages and broadcast to countries across the globe.

"For awhile the Apollo Prayer League was the hot news," he said.

The historic nature of Apollo 11 ensured a crush of media attention, and Deke Slayton had gone to great lengths to shield the crew from the onslaught. As a result, news reporters tended to swarm around anyone else associated with the lunar mission, including Pastor Woodruff and members of the Webster Presbyterian Church. Woodruff soon discovered he had to walk a fine line. He obviously needed to acknowledge his famous parishioner, yet he didn't want anything he would say, especially regarding the lunar communion, to be misconstrued or reported out of context.

Woodruff had witnessed the launch of Apollo 11 at Cape Kennedy, then returned to Houston the following Saturday. Immediately, reporters quizzed him about what his sermon the following morning would be about.

"Well, I really don't want to reveal this right now," he said. Nor did he disclose that the congregation would be sharing communion with Buzz Aldrin while he was on the moon.

"I got to church very early that Sunday morning," he said, "and the CBS vans were already out front. There were thirty or forty press people covering every aspect of that day—this time, this place. You couldn't escape them. There were press people in the trees and the yard, climbing over back fences, and sitting on the steps."

Joan Aldrin and her children arrived. As she made her way to the sanctuary, microphones were thrust in her face for an impromptu interview. Tensely, she responded, "Let's do this later."

In spite of the overbearing media presence, the church service proceeded on schedule. The sanctuary was jammed with people, and ushers placed folding chairs at the back and in the foyer to accommodate the overflow.

During the preceding weeks, Reverend Woodruff had asked Jack Henry, the church organist, to come up with some rousing music for the service. An excerpt from "Thus Spoke Zarathustra" by Richard Strauss was the final choice—the theme to the movie *2001: A Space Odyssey*, which was playing in theaters around the country. As the organ boomed through the halls of the little church, there was a palpable sense of the historic odyssey about to occur hundreds of thousands of miles above.

The music fit perfectly with the unconventional, emotional service that followed. Jack Henry himself was amazed at the effect. "When the time came for the offertory," he said, "I plowed into this huge piece of music, and I do believe it had an electrifying effect on the entire congregation—it certainly did on me."

Meanwhile, on the altar sat the communion elements, including the loaf of bread with a missing corner. Joan Aldrin sat in the front row, surrounded by family and friends. From the pulpit, Pastor Woodruff kept details concerning the lunar communion vague.

"We had communion," he said, "and it was really a very moving thing, because Buzz did not want us to publicize his prayer if possible, because he knew the news media would play it up in a very corny way. And Buzz didn't want that. So the congregation knew—but the news media people didn't know.

"So the press was in the congregation and the loaf was sitting there with a corner missing. The reporters kept looking at it and I kept referring to the loaf in Ecclesiastical terms that the congregation knew but the press didn't understand. And the press people were sitting there scratching their heads."

"We had our service and it certainly was emotionally charged," Woodruff said. "At the time, the spacecraft was in lunar orbit. They would be descending to the lunar surface. So at the end of the worship service, when we were partaking of the bread and the wine, I told the congregation what Buzz would be doing and explained the missing corner of the loaf."

At the end of the service, Woodruff announced there would be no benediction. "I said that Buzz would be partaking of communion sometime today, and all of us naturally will be watching and listening to what's going on in the spacecraft. When he finishes with his communion, each one of us, wherever we are, will say our own benediction."

Woodruff's announcement was immediately picked up by the national news networks. As ABC news anchor Frank Reynolds broke the communion story on world-wide television, millions of viewers around the world were tuned in—including Madalyn Murray O'Hair.

After Aldrin's lunar communion, he and Armstrong began suiting up for their moonwalk, an arduous task that took several hours. Then, at 9:56 p.m. Houston time, the *Eagle* was depressurized. Armstrong reached over and opened the hatch, which gave way to the utter blackness of the lunar sky. He worked his way backward on his hands and knees and slowly pushed his way out of the module onto the small "porch." With Aldrin's help, he manipulated himself onto the ladder. As he stepped down, he pulled a line to activate a camera to document the event. The camera slowly unfolded from its compartment on the side of the lunar module.

At that moment, televisions around the world captured the ghostly image of a man descending the last rung on the ladder onto the lunar surface. An estimated 600 million people in 49 countries watched and listened as Armstrong spoke.

"That's one small step for [a] man, one giant leap for mankind."

Less than three minutes after the *Voice of America* started broadcasting news of the Apollo 11 moon landing to the Soviet Union, the frequencies were jammed by Russians transmitting a music-talk cultural program on the same frequency. Unknown to many, an unmanned Soviet spacecraft, *Luna 15*, was orbiting the moon at the time. After the first of several orbits, it dropped to its lowest point directly over the Apollo 11 landing site in the Sea of Tranquility before crashing into the lunar peaks. Eventually Moscow television broadcast seven minutes of film of the American moonwalkers, but maintained a low-key treatment of the Apollo mission.

A few minutes after Armstrong descended the ladder, Aldrin followed.

"Okay. Now I want to back up and partially close the hatch…making sure not to lock it on my way out," he said.

"A pretty good thought," Armstrong laughed, as the second man on the moon backed out of *Eagle's* hatch.

Aldrin hopped down, descending through the faint gravity. They looked at one another through their visors.

"Isn't that something?" Armstrong said. "Magnificent sight out here."

"Beautiful, beautiful," Aldrin replied. "Magnificent desolation."

In every direction the horizon fell away at a steep grade. Armstrong turned to the plaque and read it to the worldwide audience nearly a quarter of a million miles away.

"Here men from earth first set foot upon the moon, July, 1969 A.D. We came in peace for all mankind."

Then Aldrin pulled the flag from its compartment, suddenly experiencing a keen sense of stage fright.

"Everything and anything we did would be recorded, remembered, studied for ages," he said. "It felt a little like being the young kid in the third or fourth grade who is all of a sudden asked to go up on stage in front of the whole school and recite the Gettysburg Address… The eyes of the world were on us, and if we made a mistake, we would regret it for quite a while."

Planting the flag wouldn't be easy. It took both astronauts to force it into the hard lunar soil, but it finally stuck and stayed erect in the one-sixth lunar gravity.

For the next two hours and twenty-one minutes, Armstrong and Aldrin collected rocks and soil, set up scientific experiments, and received a personal congratulatory call direct from President Nixon at the White House.

"This has to be the proudest day of our lives," Nixon said. "For one priceless moment in the whole history of men, all the people on this earth are truly one."

The astronauts snapped to attention and saluted as smartly as they could in their bulky spacesuits.

"Thank you, Mr. President," Armstrong replied. "It's a great honor and privilege for us to be here representing not only the United States, but men of peace of all nations."

Then it was back to work.

Not long into the walk, Aldrin remarked about an unusual rock: "Neil, didn't I say we'd see some purple rocks?"

"Find a purple rock?" Armstrong asked.

"Yep. Very small, sparkly …"

Houston made note and the moon excavation pressed on.

Armstrong and Aldrin had been walking on the surface two hours when Houston advised they had ten minutes before preparing to board the lunar module for launch. But before boarding the *Eagle*, they had one remaining task to carry out. In a shoulder pocket, Aldrin had stowed mementos to be left on the moon, one a small packet with two Soviet medals commemorating Vladimir Komarov, who had died so tragically, and Yuri Gagarin, the first man in space. He also carried a small gold pin in the shape of an olive branch and a small commemorative silicon disc that read "From Planet Earth." On the coin were etched messages from seventy-four leaders from around the globe, including a prayer and Bible verse from Pope Paul VI.

The pouch also held an Apollo 1 patch. At that moment Aldrin thought of his friend Ed White, whose life had shaped his own.

"In a way," said Aldrin, "Ed had come with me to the moon."

> "When I consider the heavens, the work of thy fingers, the moon and the stars, which thou hast ordained, What is man that thou art mindful of him?"
>
> *- Buzz Aldrin, Lunar Module Pilot, Apollo 11*
> *Reading from Psalm 8:3-4*

20
RETURN TO EARTH

Just as he was the first to step onto the moon, Neil Armstrong was the last to step off. He gave one last look at the stark, airless landscape as Aldrin guided him through the hatch and sealed it. Twelve hours after *Eagle's* landing, the television camera left behind would cease transmitting and Houston would lose all visual transmissions from the moon.

If the ascent engine failed to ignite, there would be no return to earth. Unfortunately, this became an immediate possibility. As Armstrong boarded *Eagle*, his backpack accidentally snapped off the switch to the circuit breaker. This was no small matter, as it could prevent them from igniting the engine for liftoff.

"In looking around at some of the lunar dust on the floor, I discovered something that really didn't belong there—a broken end of a circuit breaker," Aldrin said. "So during countdown, I jammed a ballpoint pen into the circuit breaker where the switch had been and held it down."

Then came another problem. Earlier, as the *Eagle* descended to the lunar surface, the monitors in Houston revealed a blockage of frozen fuel in the system; and there was serious concern that the fuel might explode like a small bomb when suddenly warmed during takeoff.

"There's no telling what it will do," a Grumman engineer told George Low.

This was the nightmare scenario Frank Borman had warned the White House of. Liftoff from earth was always an anxious time, but liftoff from the moon was an entirely different matter. According to Glynn Lunney, when launching from earth, the flight crew always had the choice to abort or go ahead. With liftoff from the moon, however, "there were no decisions to make, you just say, 'Well, let's light this sumbitch and it better work.' "

At twenty seconds to launch, Charlie Duke gave the signal: "Tranquility Base, you're cleared for liftoff."

"Roger," Aldrin responded. "We're number one on the runway."

Everything worked. The ascent engine fired and *Eagle* was soon making a beeline for *Columbia*.

"Hello, Apollo 11," Duke called over the radio. "How did it go?"

"That was beautiful," Aldrin responded. "Twenty-six feet per second up—a very smooth, very quiet ride."

"You are Go. Everything is looking good."

"I'm going right down U.S. 1," said Armstrong.

Less than four hours later, Collins made out the approach of *Eagle*, and a few minutes later felt its jarring arrival as Dr. Rendezvous brought the two spacecrafts back together. The crew bid farewell to *Eagle* and it was sent into lunar orbit, where it would slowly descend over time and ultimately crash into the moon. Armstrong and Aldrin had collected over forty-seven pounds of moon rock to be studied by scientists at the Lunar Receiving Lab. Not only was the mission of tremendous historical significance, it was a great scientific achievement.

"You're looking great. It's been a mighty fine day," radioed Duke.

"Boy, you're not kidding," Collins replied.

Back home, life in the astronaut's households was chaotic but celebratory. Even before the launch, Armstrong, Aldrin, and Collins were household names. Jan Armstrong and her children were especially overwhelmed by the media crush. *Life* magazine, which had been granted an exclusive on the Apollo 11 mission, had asked permission in advance of the flight for journalist Dodie Hamblin to move in with the Armstrongs to cover the personal aspect of the mission. Neil Armstrong was among the most private of the astronauts, and the situation would remove the last shred of personal privacy the family enjoyed. But in the end, the family reluctantly conceded.

During this time, Jan Armstrong and her friend Jeanette Chase coached a girls' synchronized swimming team together at El Lago Keys Club, and it became increasingly difficult for Jan to sneak past the photographers and press people who had camped in her front yard. The only way she could escape the house unnoticed was to slip through two hinged planks in the back fence adjoining the Ed White family's back yard and exit the Whites' garage on a bicycle she kept there. The rope that hung from a tree in the Whites' back yard, afforded young Eddie Jr. a means of catapulting himself over the fence to visit the Armstrongs' son, Mark, undetected, however, did not appear to be a likely option for Jan.

When the girls' swim team needed travel money for the U.S. National swimming competition in Toledo, they set up a lemonade stand on a lawn two doors down so as not to draw attention to the Armstrong house. Yet the team generated enough business from *New York Daily News* coverage of their activity to cover trip expenses.

In addition to their mutual interest in synchronized swimming, Jan and Jeanette shared common ties to the Apollo 11 mission. Jan Armstrong's husband, Neil, was about to become the first man to walk on the moon, and Jeanette's husband, Bill, was a member of NASA's Landing and Recovery Division assigned to monitor *Columbia's* reentry should it miss the splashdown area.

A few days after Jeanette's husband left to join the recovery team, she received a call from Jan asking if she would like to meet her at the Cape to watch the launch. Jan, it appeared, had been given access to a private yacht moored on the Banana River just five miles from the launch site, which she described as "numero uno spot." They would be joined by another friend, Pat Spann; astronaut Dave Scott's wife, Lurton; Jan's two boys, Rick and Mark; and *Life* magazine journalist Dodie Hamblin. All of this was to be kept under wraps, since NASA had an unwritten rule that wives of the crew were not to be present at the Cape during their husbands' liftoff. Certainly no one at NASA wanted an astronaut's wife to experience personal tragedy in the VIP stands in full view of a television audience. So this would be a first.

Jeanette quickly found a baby sitter for her children and was soon on a Lear jet heading to Florida. The crowd at the Banana River was thick and a barrier blocked the boats. It was a madhouse.

"We were allowed to cross the [barrier] line, and people couldn't understand why we were allowed to cross and they weren't," Jeanette said.

After the slow, agonizing launch, Apollo 11 finally emerged from the billowing smoke and flames, then vanished into space like a ballistic missile in the blink of an eye.

"Coming back in the yacht, I guess people had figured out who we were, because they started honking and raising Cain," said Jeanette.

Once back on shore, the seven crammed themselves into a small sedan. *Life* reporter Dodie Hamblin was behind the wheel as they sped down a narrow back road, only to be pulled over by local police. Hamblin tried to explain that this *really* was Jan Armstrong in the back seat and that they *really* needed to get back to watch the television coverage. But the police were having none of it. No one had been made aware that Jan Armstrong was going to be at the launch. The officers didn't believe them and asked Jan for her I.D. She searched through her purse but couldn't find her driver's license. The policemen grew more skeptical. When she was finally able to produce a checkbook with her name on it, they decided to let the group go on their way, but only after Jan promised to send them her husband's autograph. The officers quickly jotted down their name and address on a piece of paper and handed it to Jan.

"As we drove off down the middle of this road," Jeanette said, "we could see the cops standing there in a swirl of dust, dumbfounded—not sure whether they'd really been taken."

Neither Aldrin nor Armstrong had indicated what they planned to say during their lunar broadcasts, and NASA had clearly ordered the subject of religion off limits. In her suit, Madalyn O'Hair alleged that NASA was dictating what the astronauts were to say, and NASA had no intention of aggravating the situation.

So when during a television broadcast on the return trip to earth, the astronauts were asked by Mission Control to reflect on the meaning of the historic event, Aldrin took the opportunity to slip in the Bible passage from Psalm 8 suggested by his father.

"This has been far more than three men on a mission to the moon," Aldrin began. "More, still, than the efforts of a government and industry team; more, even, than the efforts of one nation. We feel that this stands as a symbol of the insatiable curiosity of all mankind to explore the unknown. In reflecting on the events of the past several days, a verse from Psalms comes to mind."

He then read from the communion card. Beneath the verse he used for communion, written in lighter ink, was the passage from Psalm 8:3,4.

> "When I consider the heavens, the work of thy fingers, the moon and the stars, which thou hast ordained, What is man that thou art mindful of him?"

Aldrin read the passage, and concluded: "...and to all the other people that are listening and watching tonight, God bless you. Good night from Apollo 11."

Network news reporters listening in from Houston were mystified by Aldrin's recital and asked Reverend Stout what he was quoting from.

"I told a reporter to go get the Gideon's Bible from his motel room," said Stout. "I marked a bracket around Psalm 8 verses 3 and 4 and gave it to the press people so they would know what it was. It was something they needed to get right."

Once on board the recovery carrier U.S.S *Hornet*, the crew was able to watch videotapes of the Apollo 11 news coverage. They saw CBS's Walter Cronkite and Eric Sevareid expounding on details of the lunar landing. They saw crowds around the world transfixed on television sets, witnessing man's first steps on another celestial body. For the first time, Aldrin was struck with the historical import of the first lunar landing. While the three of them were a quarter of a million miles away, much of humanity was mesmerized by a cosmic wonder. At one point Aldrin turned to Armstrong and lamented, "Neil, we missed the whole thing."

At a later news conference, Collins was asked what he believed to be the most dangerous part of the mission.

"The part which we have overlooked in our preparations," he said candidly, "[was] the comedown for the heroes to return to earth and then be put in quarantine in a sealed trailer due to worries about lunar germs. The feeling was accentuated when we entered the trailer and were greeted with the sign: PLEASE DON'T FEED THE ANIMALS."

NASA cosmologist Carl Sagan had almost convinced officials the astronauts would bring back a bug or virus from the moon, which was the reason for their 21-day quarantine before and after the mission, since it was known that earth bugs required that period to manifest themselves. However, according to astronaut Al Worden, when questioned about how long it might take moon bugs to germinate, Sagan had no answer.

The Apollo 11 crew underwent an even ruder awakening when their quarantine ended. Among the first scheduled appearances was a trip to Milwaukee to receive the Pierre Marquette Discovery Medal from Marquette University. It proved to be an eye-opener for Aldrin.

"We were walking up the street," he said, "and there were a lot of demonstrating students around and they started throwing eggs at us, because we represented the establishment and the success of the establishment in achieving something that evidently wasn't the kind of thing they felt they wanted to support. But it struck me as something that I certainly want to remember, that even at the glory of our first public appearance, the students threw eggs at us to give a sense that not everyone was so exuberant at what we'd been doing."

Aldrin had another surprise awaiting him in Houston, courtesy of astronaut Fred Haise, an incurable prankster. During the broadcast of his historic moonwalk, Aldrin had announced to millions that he had seen a purple rock—but the reflection of the sun through his visor only made the rock appear purple. When Aldrin pulled up his visor, he saw that it was just another gray moon rock. But he had already proclaimed it purple. Haise couldn't resist the opportunity. He painted a rock purple, affixed it to a plaque, and presented it to Aldrin. The inscription was a cross between Alfred Joyce Kilmer's poem "Trees" and Tweety Bird's "I did! I did! I taw a putty tat!"

> To Buzz Aldrin for keen observation and exceptional judgment in furthering geology while milling around on the lunar surface on July 20, 1969.
>
> > I never saw a purple rock,
> > I never thought I'd see one;
> > But when I reached the moon, I did!
> > I did! I did! I saw one!
>
> Freddo

During the rush of Apollo 11 publicity, Madalyn O'Hair accused Neil Armstrong of being an atheist. But Armstrong listed himself as a "Deist" on an application submitted to a Methodist church for volunteer work as a Boy Scout troop leader. Though he was not outwardly religious, his spiritual framework supported a universe governed through natural law rather than the doctrine of traditional religion. A November 1969 APL newsletter reported him remarking to a group of young people in his hometown that during his historic flight to the moon he certainly became more aware of the existence of God.

After his return from the mission, Mike Collins experienced an emotion common among early seafarers. Having spent so much time aboard the command module *Columbia*, this modern-day Columbus had grown attached to his ship. One evening shortly after splashdown, he quietly went to visit the only surviving vestige of the Apollo spacecraft where it rested in quarantine at the Lunar Receiving Lab. With no one around, he went to the lower equipment bay, took out a pen, and wrote on the conical module:

"Spacecraft 107—alias Apollo 11—alias *Columbia*. The best ship to come down the line. God Bless Her. Michael Collins, CMP."

There would be other emotional comedowns, especially for Buzz Aldrin. One night after the return of Apollo 11, Aldrin found himself gazing up at the night sky.

"You've been to the moon," he said to himself. "You did it. First. It cannot be done again, not by you, not by anyone else. Now get the hell out of here and live the kind of life you want." But that directive, simple as it was, would take many twists and turns—and eventually lead the troubled astronaut to a deeper, wider sense of spirituality than he had ever known. Finding the kind of life he wanted would not be easy.

The crew's families would also be affected for some time to come. At the Aldrin house, where Reverend Woodruff and others had gathered to watch the splashdown, astronaut Jerry Carr translated the Apollo 11 communications for Joan. After the splashdown, she turned to him and said, "Good, now life can get back to normal."

"Forget it," Carr replied. "Forget it."

> "We have no measure of God's work. We only see it through our experience."
>
> *- Dick Koos, Mission Control*
> *Apollo Flight Simulation Supervisor*

21
THE REST OF THE STORY

For decades the popular perception of Buzz Aldrin has been that of a troubled, hypercompetitive astronaut. As competitive as he was, beneath the surface there was far more to the man.

By his own admission, Aldrin was not a social animal and had never mastered the art of small talk. He rose to the top of his profession as a result of sheer intellect and skill, not necessarily by way of social aptitude. Like many brilliant people, nearly all of his conversations were about his astronautic passions and almost nothing else. When those around him were chatting about Joe Namath or the New York Jets, he wanted to talk about the physics of rendezvous.

Ordinarily this would not have been much of an obstacle to one's personal happiness. But Aldrin's destiny would be far from ordinary. In January 1968 he was an obscure astronaut who had yet to fly an Apollo mission. A year later he was a celebrity, his name familiar to hundreds of millions around the world. But fame was not something for which he was well suited. Public speaking, which he never enjoyed and was now frequently asked to do, became an anxiety-ridden ordeal. While Aldrin came across to many at NASA as abrupt and distant, there were those who saw him differently.

Tom Manison and his wife, Robin, came to know Buzz and Joan as fellow parishioners at Houston's Webster Presbyterian Church. Like many of the parishioners, the Manisons were also NASA employees. Tom and Buzz were ordained as elders at the same time and worked on several church youth projects together. Their relationship became one of a close personal nature, reaching beyond the usual bonds of friendship to include the Manison's four-year-old son, Tommy.

Aldrin's attendance at the Webster Presbyterian Church gave him a sense of grounding, and so did the friendships he developed there. He found the Manisons to be a clan of "robust, open, friendly, and outdoor-loving Texans" and felt a sense of ease in their company. But during this particular time, the Manisons had problems of their own. In the fall of 1968, little Tommy developed leukemia. Even as he was preparing for the Apollo 11 mission, Buzz "became somewhat of a fixture in the Manisons routine," Tom recounted.

It was on Memorial Day 1968 that the Manisons received the diagnosis. Busy as he was, Buzz couldn't put the news aside and would pick up the phone and call little Tommy every chance he got. Having an Apollo astronaut hanging around was a dream come true for the little boy. He worshipped Buzz and became enchanted with rockets and airplanes. He especially liked helicopters.

"I can remember when Buzz was getting ready to go to the moon," Tom Manison said. "They used to do a lot of helicopter flying, something to do with the controls of the lunar lander. They would get into the helicopter and fly over to the TV antennas near Alvin, Texas. Every time they'd come over, Buzz would drop down, not too close mind you, but close enough to wag the tail of the helicopter, and everybody would come out and wave."

Tommy loved it. Helicopters, it seemed, were everywhere in the boy's life—even when he was admitted to John Sealy Hospital in Galveston. As chance would have it, the Coast Guard helicopters routinely passed his window on the top floor of the pediatric unit. As each passed by, little Tommy would look out and say: "There's Buzz, he's coming to see me." When Aldrin stopped by the hospital to visit him, Tommy said, "You know, Buzz, I saw you in your helicopter."

Buzz didn't know what Tommy was talking about, so the next time he saw Tom in church, he asked.

"Well he thought that was you in the helicopter," Tom said, and didn't think much more about it. Without telling anyone of his plans, Aldrin asked NASA administrators if he could take Tommy for a ride in one of the agency's helicopters. Permission was denied. Too many astronauts were asking for special privileges for their family and friends, and NASA had drawn the line. But little Tommy was never far from Aldrin's thoughts, and he again asked NASA officials for permission. Nevertheless, the request was again denied. The weeks passed as Tommy's condition worsened. With the announcement of the upcoming moon shot, the name of Buzz Aldrin, the soon-to-be moonwalker, was suddenly front page news. One of the first people Buzz called after the announcement was his friend Tom Manison. The conversation, however, had nothing to do with the upcoming Apollo 11 mission.

"Tom, you're going to think I'm crazy, but Deke Slayton just called me," Aldrin said. "Back in June I put in for permission to land my helicopter in your yard and take Tommy for a ride. Here it is January and Deke called and said all the details are worked out. With your permission we'll do it."

Twenty minutes later Aldrin and a co-pilot were landing a helicopter in the Manisons' backyard. "We had him all bundled up," Tom said, "and Buzz brought someone with him so that the other fellow could fly the helicopter, and he could hold Tommy." Tommy lay cradled in Aldrin's arms as the co-pilot took the stick. The boy's parents and Joan Aldrin stood in the field "wringing her hands" as the copter lifted off and Buzz took Tommy up to show him the horses and the cows—all the things one could see beneath them. He flew over NASA. They saw a train running down the track along Highway 3. They passed over Galveston Bay for Tommy to see the fishing boats. Shortly after turning for home, Tommy drifted off to sleep in Aldrin's arms. As the helicopter landed, Aldrin gestured to the boy's parents that Tommy had nodded off. While the helicopter idled, he handed him back to his father.

"As he handed him back to me," Tom said, "and if I live to be a hundred years old I'll never forget it, he said, 'You know, Tom, that's the first real genuine person-to-person experience I've had since I've been an astronaut.'"

Meanwhile, Aldrin's pre-flight training intensified. Shortly after the lunar landing, the Apollo 11 crew left on a world tour to showcase America's victory in the space race. The itinerary was grinding. It called for them to visit twenty-three countries in forty-five days. Joan, who accompanied her husband, noted, "Three kings in two days! Can you believe it?"

The crew and their spouses met movie stars, heads of state, royalty, and thousands of ordinary admirers. Along the way, Buzz made a stop to present the miniature of Robert Goddard's autobiography that he had flown to the moon to Goddard's widow—a symbol of the long-awaited success of the man who newspapers once called "the mad scientist."

Each time the Aldrins returned from a trip, a mountain of letters awaited them. "A few were answered," Aldrin said, "but most were stacked around my office in what looked like insurmountable piles." He wondered if life would ever return to normal. Between ticker-tape parades, press conferences, award ceremonies, and fundraisers on "the fried-chicken circuit," there were high-brow cocktail parties and hotel rooms in far-flung cities.

"With some bemusement," Aldrin said, "I wondered if fame would prove as fleeting as the liquor supply."

Sandwiched in between all the accolades and appearances came news of a special honor. Organizers for the Army-Navy game, scheduled for Thanksgiving weekend, called Aldrin to say this year's game was going to be dedicated to him. As a proud West Pointer, he accepted the invitation.

But the hard realities of life remained close at hand. At every opportunity, the Aldrins attended services at Webster Presbyterian Church, which served as a touchstone with the Manisons and little Tommy.

"They were always having to go somewhere," Tom said. "He'd come over about two or three times a week. And every time he'd come with some new little trinket the crown prince of some African country gave him…he'd bring it for Tommy to wear at home…walking canes, you name it, they had something different every night."

By autumn of 1969 the news turned grim. Just as the Apollo 11 astronauts were scheduled to leave for their around-the-world trip, Tom saw Buzz at church. Robin was at the hospital in Galveston with Tommy. Buzz asked how Tommy was doing and Tom told him. Buzz was concerned.

"Boy, I don't know how we're going to do it, but some way we're going to get down to see that boy before we leave on Air Force One in the morning at eight o'clock."

"Oh, Buzz, don't worry about it," Manison reassured him. "We know you're thinking about us and little Tommy."

With that the conversation ended. Tom went home and didn't think any more about it. An hour later the telephone rang.

It was Buzz.

"Joan and I are changing clothes," he said, "and we're ready to go to Galveston. You want to ride with us?"

Soon they were on their way to see little Tommy in the pediatric ward. The Manisons sneaked the Aldrins into the hospital through the back freight elevator to avoid the press. For the next two hours they sat at little Tommy's bedside drawing airplanes and playing games. Aldrin did his best to conceal his emotions.

"We choked back our tears," he said, "as we helped the frail little boy unwrap his treasures from a world he would never know." Then it was time to go. From the hospital, Aldrin was off for an engagement with the Nixon family at the White House.

As Thanksgiving of 1969 approached, the boy's health took a final turn for the worse.

"Things were really bad," Manison said. "Joan and Buzz called, and I told them things were tough. It was at that time Reverend Woodruff, his wife, Floy, Buzz, and Joan helped us make the decision to stop the transfusions for little Tommy. They were right there by our side. He died that morning."

It had been a long struggle. Tom and Robin had known for some time how it would end. Nevertheless the ordeal hit them hard. After their Tommy was gone, they came home and went to bed, exhausted with grief. It was Wednesday evening, November 26, the day before Thanksgiving.

"We got up about 9:30 or 10:00 to get some coffee," Tom said. "And guess who was sitting in the living room? Buzz was there Thanksgiving morning, reading the newspaper, waiting for us to get up." They talked about little Tommy for a while, and Tom asked if Buzz would be willing to be a pallbearer at Tommy's funeral.

"Sure," Aldrin said without hesitation. "That's why I came, so we can talk about it."

Without saying a word to anyone, Aldrin later contacted West Point. He called to tell them he wouldn't be able to attend the Army-Navy football game dedicated to him that Thanksgiving weekend or to receive the award during the half-time ceremonies.

Aldrin had something more important to do that day. Instead of receiving an award at his beloved Academy before an adoring crowd of tens of thousands, he would be bearing the small casket of a four-year-old boy.

"The fool hath said in his heart, There is no God."

- Psalm14:1

22

AN ATHEIST VOICE

The entire moon landing episode took its toll on Pastor Woodruff and his wife. As Woodruff put it, the ordeal left them "wrung out." With the Apollo 11 moon landing behind them, they left for Acapulco to recuperate.

They had not been vacationing long when the church secretary, Regina Fischer, called to say that a woman by the name of Madalyn Murray O'Hair had filed a lawsuit against NASA, naming its new administrative chief Thomas Paine, plus a long list of other officials, including the Apollo 11 crew, for performing a religious act on Apollo 11. Pastor Woodruff was also named in the lawsuit. "It named Buzz Aldrin, Neil Armstrong, and everyone in NASA, and their dog," she said.

The *O'Hair vs. Paine* lawsuit made front page news shortly after the first lunar landing. Up to this point O'Hair had been viewed as a minor nuisance; she was now a force to be reckoned with. A heavy-set woman with a strong voice, O'Hair was an Army veteran and a law school graduate. More important, she had money. It was clear Madalyn O'Hair wasn't going away any time soon.

According to those who visited there, you would not drop by O'Hair's handsome tan-shingle house on Greystone Drive in the quiet Austin neighborhood expecting to be served tea and crumpets. A more likely offering would be a swigger of Scotch and a healthy dose of vulgarity from O'Hair, accompanied by frequent screaming matches between her son, Jon, and her granddaughter, Robin. Inevitably, one of O'Hair's ill-tempered dogs would nip you.

O'Hair's lawsuits had become a staple of American life. She appeared on TV all over the country—foul-mouthed, witty, and passionate—inciting culture wars over same-sex marriage and faith-based initiatives. She was accused of schizophrenia, alcoholism, and embezzlement from her own organization of American Atheists, but certainly never of cowardice or sloth. Her law degree instilled in her an educated sense of outrage at the faintest hint of Christian privilege, which she hurled at the judicial system like a legal bayonet. Her language became so profane that it could not be reported, and many in the media stopped interviewing her. Along with her outrageous public persona, the woman's past was a preview of what lay ahead. Her road to atheism revealed a path strewn with shattered lives and ruptured relationships—not the least of which was her own.

Born Madalyn May in 1919 in the small coal town of Pottsville, Pennsylvania, she was baptized in the Presbyterian Church as an infant and grew up in a poor, working-class household. After reading the Bible from cover to cover as a teenager and deciding the entire book was nothing more than fallacy, she became an atheist. In 1941 she married John Henry Roths, which made for a brief and unhappy marriage.

Madalyn and her husband separated when they both enlisted in the military—he in the United States Marine Corps, she in the Women's Army Auxiliary Corps. In 1945, while serving as a cryptographer in Italy, she began an affair with William J. Murray Jr. and gave birth to a son, whom she named William. Murray, however, was already married and refused to divorce his wife, a detail that didn't stop Madalyn from divorcing her husband and calling herself Madalyn Murray. Eventually she moved to Texas, where she earned a law degree from the South Texas College of Law in Houston; however, she failed to pass the bar exams. In 1954 she gave birth to a second illegitimate son, Jon Garth Murray, by yet another man.

As the years advanced, the unconventional mother of two became infatuated with the tenets of Communism and the "Workers' Paradise" of the Soviet Union. When Soviet Cosmonaut Yuri Gagarin became the first human to venture into outer space, Madalyn saw the feat as proof of Communist supremacy. She eventually packed up her family and traveled to Russia with the intention of gaining Soviet citizenship. But the Russians made it clear they cared little about her infatuation with Yuri Gagarin, noting that she did not speak Russian nor did she truly understand Marxist principles. The officials explained that if she did manage to immigrate, she would likely end up unemployed. The punishment for not being employed in Russia was hard labor at half pay. Never one to forego creature comforts, Madalyn gathered up her boys and returned to Baltimore, where she enrolled her younger son, Jon, in elementary school and William in the ninth grade at Woodbourne Junior High.

Although no one knew at the time, an historic moment was at hand. From the halls of this small school would begin Madalyn O'Hair's litigious conquest of prayer in public schools. On William's first day at Woodbourne, his mother accompanied him down the long hall in silence, following the signs pointing to the school office. The doors to the classrooms stood open. As they passed by they saw the students standing with their hands over their hearts, reciting the Pledge of Allegiance.

Madalyn's face reddened. "Do they do this every day or is this something special?" she demanded of William.

"Every day," William said.

Madalyn stopped dead in front of another open classroom door. Her eyes widened. The students were standing beside their desks with their heads bowed, reciting the Lord's Prayer. At this she let loose a stream of profanity and led William down the hall to the guidance counselor's office, where she approached a young clerk.

"Hello—may I help you?" he asked.

"You sure can," Madalyn shot back, "but first I want to know why those kids are praying." She advanced to within inches of the man's desk. "Why are they doing that? It's un-American and unconstitutional!"

"Is your son a student here?" he asked.

"No," she replied, her face nearly purple with rage, "but he will be starting today. I'm Madalyn Murray O'Hair, I'm an atheist and I don't want him taught to pray."

"Look, Mrs. Murray," the young man replied, "nobody's ever complained about this before. Besides, let me tell you this. There were prayers in the schools of this city before there was a United States of America."

"Nonsense, that's just a bunch of garbage and you know it."

"Madam, I don't have to take this! If you don't like what we do here, put your son in a private school."

"It doesn't matter where I put him. You people have to be stopped."

It was at this point that the counselor said something prophetic: "Then why don't you sue us?"

Madalyn stared at the young man. He had just given her an idea that would forever alter the landscape of American life and, most particularly, public education. O'Hair launched a court case against the school and the state of Maryland. The lawsuit drew a great deal of support from likeminded atheists from across the country and great sums of money, presumably going toward legal fees.

"The mail increased tenfold," William recalled. "Every misfit in America was sending my mother letters of praise with a check enclosed. The phone rang constantly."

The lawsuit also brought a firestorm of publicity. While Madalyn gloried in all the attention, her son was paying a terrible price. At school his life became a brutal game of survival. He had to sneak through the back entry to avoid being beaten up and was often called a "Commie." On one occasion he was pushed in front of a bus by a fellow student. "By this time," William said, "I was beginning to feel like shark bait."

The case made Madalyn a national figure. It also promised to make her an enormous amount of money, providing her with a steady revenue stream of donations. Even negative publicity, she learned, had a dollar value. The more outrageous her behavior, the greater the media attention and the faster the money rolled in. It wasn't long before O'Hair became the public face of atheism in America.

On June 17, 1963, after most public schools had let students out for the summer vacation, the Supreme Court announced its decision. By an overwhelming margin of eight to one, the court ruled that opening Bible reading and prayer exercises in school were unconstitutional, citing that "it is not within the power of the government to invade the citadel." This ruling applied not only to Maryland, but to all states. With that, prayer and Bible readings were banned from public schools.

In 1964, *Life* magazine published an article referring to O'Hair as "the most hated woman in America." The article quoted her as saying: "Everything I learn makes me realize I don't know a thing. But compared to most cud-chewing, small-talking stupid American women…I'm a genius." In response to a derogatory editorial about her in *Life*, O'Hair wrote a letter to the magazine defending atheism:

> We find the Bible to be nauseating, historically inaccurate, and replete with the ravings of madmen. We find God to be sadistic, brutal, and a representation of hatred and vengeance. We find the Lord's Prayer to be that muttered by worms groveling for meager existence in a traumatic, paranoid world… The business of the public schools, where attendance is compulsory, is to prepare children to face the problems on earth, not to prepare them for heaven.

The letter, which ran alongside her picture, revealed again Madalyn's unique ability to infuriate the American public. The suit itself was mired in a litany of hearings and appeals.

William, in his twenties at the time , moved out of the household and fathered a baby girl, Robin, whom he sent to live with his mother "for only a year" when she was twenty months old. Eleven years later, Robin had become Madalyn's clone both physically and spiritually. Eventually Robin became so overweight and widespread she had to purchase two airline tickets because she was unable to fit into one seat. She was by now a devout atheist.

In 1965 Madalyn fled Baltimore for Austin to evade prosecution for having assaulted five Baltimore police officers. Her son, Jon, and granddaughter, Robin, moved in with her at the family home on Greystone Drive, for which she claimed to have paid more than $1 million in cash. The three became accustomed to fine dining, expensive clothes, and fancy automobiles. They indulged in world travel, circling the globe together three times. In spite of their constant bickering and name-calling, they ate lunch together, dined together after work, and returned to the big house on Greystone Drive together. For all practical purposes, they were three peas in a pod.

In Austin, Madalyn founded the American Atheists and began a personal campaign to draw public attention to the special privileges continually granted to religious bodies despite the provisions of the Fifth Amendment which states: "Congress shall make no law respecting an establishment of religion, or prohibiting the free exercise thereof."

Years later, much to Madalyn's chagrin, William announced his conversion to Christianity on Mother's Day, resulting in their permanent estrangement. As she put it, "One could call this a postnatal abortion on the part of a mother, I guess. I repudiate him entirely and completely for now and all times. He is beyond human forgiveness." Mrs. O'Hair then legally adopted Robin Murray, making Robin her daughter as well as her grandchild.

In time, even some of Madalyn's most ardent supporters began to question her motives and dealings. Like many charismatic movement leaders, O'Hair, they said, had utterly lost the ability to distinguish between herself and her cause. She and her family lived lavishly off of contributions of supporters intended to support the cause, many of whom were actually quite poor. According to William, she had become "the Hulk Hogan of atheism."

By 1969 the stories surrounding the matriarchal head of the American Atheists were legendary. The group coalesced around Madalyn and dedicated itself to "the complete and absolute separation of church and state." Now, in addition to complaining to local post office officials about "Pray for Peace" stamp cancellations, they could object to broadcast readings of biblical verses from a government-owned spacecraft.

In August 1969, *O'Hair vs. Paine* was filed in Federal District Court against NASA, citing religious acts on an Apollo mission. During their second TV transmission from space, the Apollo 8 astronauts succeeded in showing all of humanity a picture of itself, the home planet afloat in the blackness of space. But it was the show broadcast from lunar orbit on Christmas Eve that brought a spiritual dimension to the telecasts and incited O'Hair to bring suit against Paine and the entire NASA organization. In her suit, Madalyn criticized the moon-bound astronauts of Apollo 8 for engaging "in religious ceremonies and in an attempt to establish the Christian religion of the United States government before the world while on scientific-military expedition to, around, and about the moon." She didn't mind the astronauts praying, or so she claimed. What she objected to was that the astronauts had been *ordered* to pray by NASA. Moreover, they had done it out loud, as she put it, and on "company time."

Frank Borman responded personally.

"I think the reason that Apollo 8 may have become spiritual, is because we read from Genesis… We were told by NASA, this wonderful organization, wide-open organization, something like, 'you'll have a TV appearance on Christmas Eve. You're going to be seen by more people than anybody, witnessed by more people than anyone has ever seen before, and you've got to be prepared.' And I remember asking NASA Public Affairs Chief Julian Scheer, 'Well, what do you want us to do?' And the answer came back, and I'll remember it to my dying day because to me this is the essence of America. The answer came back: 'Say something appropriate.'"

"Now if my name had been Leonov," Borman said, "they would have been saying, 'Extol the virtues of Lenin and the great Communist Society,' and all that baloney." But NASA had given no orders of any kind.

O'Hair wasn't at all placated. She amended her Apollo 8 complaint to include the fact that Neil Armstrong and Buzz Aldrin left a small commemorative silicon disc on the moon carrying messages from world leaders. Etched onto the coin, in letters no larger than one-fourth the width of a human hair, were seventy-four messages. The U.S. State Department had authorized NASA to solicit messages of good will from the leaders of the world's nations to be left on the moon. Many of the messages were not merely texts. Some included intricate artwork and, more egregiously to O'Hair, Pope Paul VI's message included a biblical quotation from Psalm 8. Among the United Nations flags taken to be brought back as souvenirs, she pointed out, was the Vatican flag. Nothing, it seemed, had escaped Madalyn's notice.

O'Hair's suit sought to enjoin NASA from "directing or permitting religious activities, or ceremonies" in the space program. Not only were NASA's actions unconstitutional, she asserted, but the Pope's Christian prayer discriminated against other religions and nonbelievers such as herself. She filed an eight-page complaint in Judge Jack Roberts' U.S. District Court in Austin, but asked for a special tribunal because Roberts had "a Christian, sectarian bias."

From its inception, the O'Hair lawsuit would prove to be a tangle of thorns for the judicial system. It was filed in a court that required Federal judges to be sworn in with a statement of belief in God. This, of course, was a problem for Madalyn. Each day the court opened with "God Save the Nation and this Honorable Court." During the hearing, it also required that Madalyn utter the words "so help me God" in her oath—yet another matter over which she sued.

It did not go unnoticed that the United States was in a Cold War with a country that essentially shared O'Hair's beliefs. To many Americans, an attack on Christianity by a self-professed Communist felt a lot like an indirect attack on the United States by the Soviet Union; and if anyone had any doubts about the Soviets' official position on religion, they could reference a Soviet dictionary published in 1951. In it, the Bible is described as "a collection of fantastic legends without any scientific support…full of dark hints, historical mistakes and contradictions."

Americans from across the country responded to O'Hair and her lawsuit in a fit of rage. A letter to the editor in the *Amarillo Globe-Times* summarized the public's general reaction:

> Atheist communism in the form of a Russian satellite or Russian cosmonaut in space has not hesitated or failed to support atheism and its repudiation of God. Our fathers and their fathers supported their beliefs and faith in the living God in each document written and each coin minted.
>
> Never was there a more appropriate time for Frank Borman, Jim Lovell, and Bill Anders to express their Christian American heritage as they did. How else can we account for our great nation and exploration of our universe except through the will of God and the blood and sacrifice of its soldier-patriots and believers? And God forbid that the disbelievers outnumber the believers.
>
> My humble thanks to the astronauts for their decision to pray for and read to the entire world. My thanks to Houston Control for their display of the stars and bars and their profession of the American spirit.
>
> To the devil, or to some other appropriate place as He may desire, with Madalyn Murray O'Hair!
>
> *- John W.*

Justice Department attorney James Barnes was having none of O'Hair's nonsense. "NASA has no plans to instruct astronauts what to say," Barnes replied. "The statements that the astronauts made are their own and not NASA's. At the same time, NASA has no intention of circumscribing in any manner the astronauts' rights in the free exercise of religion."

In November of 1969, U. S. Attorney Seagel Whatley asked the court to dismiss the suit on grounds that Mrs. O'Hair and her group lacked the legal standing to sue the government. After brief arguments from both sides, the court took the matter under advisement. The judges gave no indication how or when they would decide the case. Contrary to what O'Hair would have liked, the lawsuit spawned by the Apollo missions only served to unite men and women throughout the nation in opposition to her views, not the least of whom were the members of the Apollo Prayer League.

As preparations were underway for Apollo 12 to cross the abyss from the earth to the moon, 40,000 of those members would make themselves heard in a single, unforgettable voice.

> "Prayer can be a very vital and powerful part of your life. Do not get involved with it unless you are ready to have your life changed."
>
> *- Reverend John Stout*
> *Director, The Apollo Prayer League*

23
40,000 VOICES

The letter arrived at the Manned Spacecraft Center addressed to Reverend John Stout, Director of the Apollo Prayer League. It was not the kind of letter Stout enjoyed receiving. The overall tone was distasteful and several of the words were not the kind found in the Easton's Bible Dictionary. The author's identity, however, was easily recognizable. Madalyn O'Hair's letter was the most profane, filthy letter ever received at the space center. It read in part:

> I read with absolute disgust your plans to take your [expletive] so-called Bibles to the moon... [expletive]. If you persist in contaminating the Space Program with your foul, disgusting [expletive] Christianity, we must take steps.

It wasn't the first letter Stout received from O'Hair, nor would it be the last. As her lawsuit against NASA progressed, it became increasingly clear she was pitting herself not only against the space agency, but anyone associated with the Apollo program. The situation would have been humorous had it not been such a distraction at a time when focus and resources were needed for two back-to-back missions—Apollo 12 and 13.

Stout was astounded at O'Hair and her atheist organization of followers. They were the antithesis of the organization he founded, which by the summer of 1969 had grown to over 40,000 members worldwide. But he knew those 40,000 voices, rising in unison, could be a formidable force in dispelling the *O'Hair vs. Paine* lawsuit, now reaching its final stages. Stout was intent on taking steps to counter the O'Hair propaganda and had the unique resources at hand to do so. He had available to him a large mailing list, the Prayer

League newsletters, the APL News Wire Service, and the implicit backing of NASA, who dearly wanted Madalyn O'Hair to go away. Coordinating such a massive group of people scattered around the world, however, posed a formidable challenge.

Then one day, a similarly outraged Detroit radio talk-show host by the name of Loretta Fry came up with an idea.

"Madalyn Murray O'Hair said she had 28,000 signatures protesting the Bible reading," Fry related to a UPI reporter. "We could get 28,000 from Detroit alone!"

Fry began broadcasting an appeal for petitions supporting the astronauts' right to pray and to read from the Bible during their missions. Almost immediately listeners began calling in, asking to sign. It didn't take long for other broadcasters to hop on the bandwagon.

The Apollo Prayer League resolved to challenge O'Hair on her own terms. As news of the lawsuit flashed across the airwaves, an army of Christian activists and Apollo Prayer League volunteers canvassed the country wielding steno pads and clipboards.

From the very beginning the response was overwhelming as hundreds of thousands of people began signing the circulated petitions emblazoned with the header: "EVIL TRIUMPHS WHEN GOOD MEN DO NOTHING." Madalyn's effort, by comparison, was anemic. At 28,000 signatures, it abruptly died. The APL petition drive, on the other hand, was only beginning.

As they neared 500,000 signatures and prepared to deliver the massive list to NASA Public Affairs Officer Paul Haney, word reached Madalyn, who threatened to appear in person to protest the presentation. Although her home was only a few hours' drive north of Houston, she never appeared. Energized by her failure to appear, APL volunteers pushed the petition drive to well over a million signatures.

As the drive gained momentum and the pile of petitions continued to mount, Stout announced: "A certain Austin atheist has written several letters of protest to the Apollo Prayer League, We are proud to announce to you that to date more than five million signatures and letters have been collected opposing her and supporting our astronauts in their right to freely express their own religious faith."

Although the judiciary was normally insulated from the whims of politics and public opinion, such grassroots support for NASA and the astronauts was turning out to be a public relations bonanza for the APL. A nuclear scientist who joined the effort described it as "more powerful than the chain reactions in an atomic bomb."

Funds and letters of support came from such notables as John Glenn and Neil Armstrong, and APL member ranks began to swell beyond

NASA personnel, contractors, and subcontractors. They were now being joined en masse by ordinary citizens—mechanics, physicians, plumbers and homemakers—from across the country and around the world. The APL's national petition drive ballooned, and before Madalyn O'Hair could utter another vulgarity, the drive obtained nearly ten million signatures and letters supporting the astronauts' freedom of expression and contributing visibility to the court case poised to guarantee that freedom.

By the time of O'Hair's court hearing in November of 1969, APL members could be found in sixty percent of all regional postal zones in the United States and sixteen foreign countries. Indeed, Prayer League members could be found at virtually all NASA locations, on Pacific tracking ships, and remote arctic radar outposts. In one month alone, 408,000 letters were received supporting the astronauts' right to pray. Arrangements had been made to store the prayer petitions in a vacant office at the Manned Spacecraft Center and the room was now packed floor to ceiling and overflowing. NASA hoped that it was now a dead issue and asked Stout to destroy the petitions in such a way that the press wouldn't find out. The only way to do this discreetly, Stout decided, was to smuggle them out a few boxes at a time and burn them along with NASA confidential material in an incinerator at Ellington Air Force Base.

Before they did, however, a photographer asked to photograph the mass of prayer petitions. To do this, a group of APL members taped several thousand of them together and wound the scroll around a broomstick, which they took to Eastminster Presbyterian Church in Pasadena. When the photographer arrived, he was astonished to find the petition scroll unrolled and stretched down the entire center aisle of the church.

Along with the prayer petition effort, Aerospace Emergency Relief was formed to undertake other outreach projects. As a branch of Aerospace Ministries, it was quick to offer relief supplies to the victims of war-torn Biafra and joined the Joint Church Airlift effort to drop supplies into the region. A call was put out to NASA employees and subcontractors, and groups of volunteers turned out to load the Biafra-bound cargo planes. When passage in and out of the Biafran region was blocked by dissidents, NASA employee Jack Joerns volunteered to parachute into the area to coordinate distribution on the ground.

In August 1969, the APL also provided relief supplies to Mississippi victims of Hurricane Camille, at that time the strongest land-falling tropical cyclone ever recorded. Again, when Hurricane Celia wiped out an area east of Corpus Christi in 1970, APL members transported truckloads of supplies donated by Houston corporations to the parking lot of a Corpus Christi Presbyterian church for distribution to needy residents. In Stout's mind, such activities lent a balance to the lives of NASA employees and provided a meaningful outlet from the relentless pressures of the Apollo program.

The typical Prayer League member, Stout said, was "anyone who believes in the power of prayer and would pray for the astronauts' safety." In 1969 he reinforced the idea of acceptance of members regardless of their religious affiliations. Prayer can take many forms, he noted:

> The Apollo Prayer League does not seek to convert others from their own beliefs. What the Apollo Prayer League is trying to do is to find those who already have this power—or to find those who do not mind having this latent power activated.

Daily prayer and Bible study meetings held at the space center were attended by members from virtually every Christian denomination. The location inside a federal establishment was startling to some, but had been agreed to by Paul Purser, Assistant to the Director at the Manned Spacecraft Center, so long as the meetings were held during off hours. Stout saw the League as a new kind of fellowship in the "first interplanetary church." With this in mind, he directed the league's focus on the broader concept of prayer:

> What does prayer have to do with this? If we are going to be honest with ourselves, we must ask what prayer is in the first place. Prayer may mean many things to many different people and prayer methods may be varied from group to group and from culture to culture. In the Apollo Prayer League, we are not trying to teach methods of prayer, although we hope that we can share these with each other. We are looking for those, whoever they are, who believe in the *power of prayer*.

Stout wasn't interested in soliciting empty prayers from well-meaning Christians, and he had little patience with those who claimed to be believers but silently harbored doubts. He wanted those who *believed* in the power and knew how to use it, he said:

> Now, do you really believe in the power of prayer? It may be better to ask: Do you really believe in God—in a God who

talks to us through prayer? Perhaps we can learn a lesson from America's self-acclaimed No. 1 atheist, Madalyn O'Hair. She told me that there are three things that a true atheist must believe: (1) There can be no TOP BANANA; (2) If there were a TOP BANANA, we would not have the ability to pray to him; (3) If we did have the ability to pray to a TOP BANANA, he wouldn't have the ability to answer our prayers. If she had a fourth point, it probably would be something like this: If the TOP BANANA did answer our prayers, would we have the ability to know it?

One of the central tenets of O'Hair's legal complaint was that verses from the Bible had been carried to the moon on what was essentially government business. This, she claimed, was unconstitutional because it crossed the line between the separation of church and state. There were those who took her argument seriously. Like Madalyn, they knew how the judicial system worked and felt NASA should be forced to follow the letter of the law. Reverend Stout understood the threat but remained unswayed. Instead of backing down, he quietly raised the ante. Shortly after the splashdown of Apollo 11, he began hatching an elaborate plan.

For several years, Stout had nurtured a vision of a Bible that could be taken to the moon—the first "lunar Bible"—and that vision was nearing fruition. The idea had been borne of Ed White's comment to him three years earlier when he told Stout he dreamed of carrying a Bible into space and to the surface of the moon.

"Astronaut White never lived to realize that dream," he said. "His inspiration to all of us who knew him at the Manned Spacecraft Center in Houston and his dream continues to live today in our hearts and thoughts."

An unlikely set of coincidences began to emerge, all of which seemed to conspire to make White's dream a reality. As a NASA scientist, Stout was well aware of the legal and physical limitations imposed by NASA on items carried on board the spacecraft. A traditionally-bound Bible would not fit in the astronauts' PPKs. But in spite of what Madalyn O'Hair thought, the astronauts could indeed carry personal religious items on their missions. The astronauts' PPK was considered his private property. Here, Stout saw an opening.

During the months leading up to Apollo 11, he became aware of a new technology originally introduced at the 1964 World's Fair. This technology involved the photo-reduction of large documents onto a tiny piece of film called a microfilm. In fact, National Cash Register Company (NCR) had photo-reduced World Publishing's entire text of the King James Bible onto a microfilm and it weighed only a fraction of a gram. When Stout inquired, NCR provided him with several samples of the text, each smaller than a two-inch slide. This, he saw, was a version that could be easily stowed in an astronaut's PPK.

Stout now held in his hand what he hoped to be the first lunar Bible. On a single microfilm tablet were all 1,245 pages and 773,746 words of the King James Bible. Printed in the lower right corner was "The National Cash Register Co.", the manufacturer, and in the upper right corner, "Holy Bible." While it could not be read with the naked eye, under a 150 power microscope a person could read, in perfect form, the book of John, verse one:

> *"In the beginning was the Word, and the Word was with God, and the Word was God."*

As Stout and the Apollo Prayer League began to glimpse its full religious and historic significance, the project took on ever-greater meaning. The endeavor was a modern-day extension of a Christian tradition harnessed with technology in the interest of spreading Christianity, a tradition stemming from the time of Emperor Constantine in the fourth century when a small army of scribes painstakingly produced a small number of Bibles in Greek. The first Christians took the Scripture with them as they ventured into the known and unknown realms of the world, spreading their faith as they went. Now, centuries later, Reverend Stout and the Apollo Prayer League wanted that tradition carried forward as mankind set out to explore the universe. Indeed, a Bible on the moon would represent man's most eloquent symbol of God's omnipresence throughout the heavens.

Certain facts were becoming evident to Stout: landing a Bible on the moon would require the cooperation of several key players within NASA as well as the astronauts themselves. The plan must follow NASA's prescribed policy for separation of church and state in order to avoid repercussions from Madalyn O'Hair and the American Atheists. As a result, informal channels of cooperation would need to be developed and contacts within the space agency discreetly called upon. APL board member Bradford Jackson, for instance, called upon his wife, Estelle, a NASA Branch Administrative Manager in the Astronaut's Office. Estelle went to the "powers that be" in NASA to allow the Bible to be taken to the moon. The decision would not be publicized. There would be no grand announcement, no press release. Nevertheless, permission was quietly granted.

Meanwhile, during the busy run-up to the Apollo 12 mission, Stout turned to his typewriter and addressed the burgeoning membership of the Apollo Prayer League:

> Less than a year ago, we witnessed man's first flight to the moon and return. We were elated when our astronauts recognized the existence of God on that flight. Since then Apollo 9 has flown, Apollo 10 has flown a second round trip to the moon; and during Apollo 11 we witnessed man's first steps on another planet. Now, within a few days, Apollo 12 will take another crew to the moon to again walk upon its surface. We should not let our success in life or our success in our space program keep us from our dependency upon God.

Even with the lunar Bible in hand, Stout still needed a way to get it onboard the spacecraft. So in the fall of 1969, he sent a request to NASA officials asking to store the lunar Bible onboard the next mission. NASA quietly referred him to the Apollo 12 crew: Pete Conrad, Alan Bean, and Dick Gordon. Stout's other brother, James, was a personal friend of Alan Bean and felt that Alan would be amenable to carrying the Bible in his PPK as a personal memento. Since it was very small and added virtually no weight, Bean agreed. It would be stored in his PPK, along with a banner from the Clear Lake Methodist Church, where he served as elder. The banner, bearing the Methodist symbol of the Cross and the Flame, was part of an embroidery of Christian symbols sewn by a church member. With this he added a grain of mustard seed, symbolized in Matthew, 17:20: *"If ye have faith as a grain of mustard seed, ye shall say unto this mountain, Remove hence to yonder place; and it shall remove; and nothing shall be impossible unto you."*

In addition to the microfilm Bible, Bradford and Estelle Jackson mounted a concerted effort behind the scenes to locate a small printed Bible for Bean to take with him on behalf of the APL as a gift for Reverend Stout. What they found was a very thin red Bible measuring 3 x 5 inches. Bean logged the two Bibles, the banner, and the mustard seed in his flight manifest and handed them to the support crew for stowing onboard the Apollo 12 spacecraft.

After nearly two thousand years of life on earth, the Bible was on its way to another world. In early November 1969, Stout announced the plan to the APL:

> The Apollo Prayer League has requested that their home town boy and former Sunday school teacher be remembered. We

certainly will remember him in our prayers as well as the other members of the program. But for Commander Alan Bean, we have given him a very special title during a recent presentation ceremony here at the Manned Spacecraft Center. The Apollo Prayer League has designated him as 'Honorum Space Pastor' of the Apollo Prayer League.

In a few weeks Bean would be carrying the Word of God to the surface of the moon aboard the lunar module *Intrepid*. The Bible's destiny, Stout reminded the League, was far from certain. Now, more than ever before, prayer was vital, as each Apollo mission carried with it many new risks:

> There will be approximately five million separate parts on the spacecraft rocket system when it lifts upward from Cape Kennedy. Each of these parts or its backup system must work. Approximately three hundred thousand men and women will be supporting the launch. They must not only seek perfection, they must *be* perfection.

John's wife Helen shared her husband's deep abiding faith and love for the space program. In anticipation of the Apollo 12 launch, she authored an article for a Christian magazine, observing: "Like our astronauts we must prepare for a flight through life in a very hostile environment. We must know our rules book—the Bible." And now a man was going to carry a Bible through outer space to one of the most hostile environments mankind had ever encountered.

To Reverend Stout, carrying the Bible to another celestial world signified an evangelical effort of cosmic proportions. He hoped the endeavor would also usher in a religious renewal here on earth, which he believed was mankind's best chance for saving itself:

> Perhaps this gesture will cause a back to the Bible movement. If not, then, some day, after our earth has polluted, bombed, starved, or hated itself out of existence, some lone space traveler from a distant world might land upon the moon to study this earth which was, and by chance—or divine guidance—he might find it and take it home. By studying it, which we should have done, perhaps his world can succeed where ours had failed.

Not everyone in the general public approved of Stout's venture, including the author of a letter he received from Passaic, New Jersey.

"Did you ever check the word of God to find out if the Lord wants men to go to the moon and other planets?" the author asked. "Well I did, and I learned that God does not want man to go. Man was made for the earth."

Stout smiled, opened his desk drawer and placed the letter alongside the one from Madalyn Murray O'Hair. Apollo 12 would launch in less than four months and he had more important things on his mind.

> "I'm not a religious person...I do not intellectually believe religious things. However, I ought to believe them, because someone has made them work out for me more than just the odds."
>
> *- Alan Bean, Lunar Module Pilot, Apollo 12*

24
APOLLO 12: SPACE COWBOYS

"I'm going to be the first man to eat spaghetti on the moon," Alan Bean announced to his children, Clay and Amy. They knew their dad loved spaghetti—he ate it nearly every night of the week. It took a moment for the real message to sink in. It wasn't *what* he would be eating—it was *where*. Launch day was scheduled for November 14, 1969.

Commander Pete Conrad, Lunar Module Pilot Alan Bean, and Command Module Pilot Dick Gordon were as different from one another as any three human beings could be. Conrad was a free-wheeling take-no-prisoners sort, a smart aleck rocket-jockey with a degree from Princeton who kept the public information officers eternally on their toes. Public Affairs Officer Paul Haney observed that no matter how many times the gap-toothed Conrad was reminded to keep his language printable, it never seemed to register. "They had to bleep out so much of what Pete Conrad said they hardly had enough left for a complete sentence," Haney said. "Every third word was a bleep."

Dick Gordon, on the other hand, could have been a poster child for NASA, with the bravado and handsome looks that personified the Korean jet fighter and archetypical astronaut that he was. Growing up, he contemplated becoming a priest, then a dentist, then switched career goals several times before the Korean conflict interrupted his plans.

Alan Bean, the rookie lunar module pilot, learned the ropes from his two veteran crew mates as he rose through the ranks. He was like their younger brother. Together, the crew of Apollo 12 exhibited a camaraderie and unique friendship unparalleled in the lunar program. They not only shared friendship, they shared history. Each had spent time at the U.S. Naval Test Pilot School in Patuxent (Pax) River, Maryland, in the late 1950s, where as an instructor at the school, Conrad taught both Bean and Gordon. The three worked together as backup for Apollo 9 before joining up again on Apollo 12. They were a close-knit bunch who felt it was as important to have fun as to be good. "If you can't be good," Conrad would say, "then at least be colorful."

And colorful they were, right down to their customized Corvettes. Cocoa Beach car dealer Jim Rathman received permission from General Motors to continue an executive lease arrangement whereby the astronauts could drive a new Chevrolet car of their choice for $1 per year. Conrad arranged a deal for three matching Riverside Gold '69 Corvettes with black winged striping customized with front fenders, one bearing the initials "CDR" for mission Commander Conrad; another, "CMP" for Command Module Pilot Gordon; and the third, "LMP" for Lunar Module Pilot Bean. On a pilot's pay, none of them could have afforded such a car. "I thought it was manna from heaven," Bean said.

The cars were stock stick-shift 427/390 four-speed coupes, sporting head restraints, 4-season air conditioning, special wheel covers, and an AM/FM pushbutton radio—all the options a man with "the need for speed" could possibly want. The trio then attempted to photograph themselves together on the hoods of their respective vettes using a camera timer, rushing back and forth in rapid succession and creating a sequence of photos resembling stand-up/sit-down comedy. The resulting picture personified the nature of the crew and became a favorite of autograph fans.

As those in the astronaut ranks knew, Alan Bean was not originally scheduled to be the lunar module pilot on Apollo 12. At the time of the crew assignment, "Beano," a short, skinny, unassuming guy, had been relegated to a desk job in *Skylab*, the program scheduled to transform Apollo into a space station. On the surface, Bean was reminiscent of the kid next door who cut the neighbor's grass and delivered the paper. But underneath was a fierce determination inherited from his mother that could knock down flat anything that got in his way.

He decided to be a Navy pilot when he paid ten cents to watch *Twelve O'Clock High* as a young boy in Fort Worth, Texas. He was not a particularly bright student in grammar school. "I had the idea that people had to be born 'A' students or sports stars," he said, "and that being good at things was built-in." So if he wasn't naturally good at something, he didn't try. Consequently, Bean didn't work hard at spelling or geography. Instead, he dreamed about airplanes.

At the age of seventeen, in an act of defiance to his parents who felt he should stick to his studies, he signed up for the U.S. Naval Air Reserve.

"I was skinny as a beanpole. In school I hated being short," he said. "But you can't be too big if you want to be a military pilot because the cockpits of military planes don't have much room." For the first time in his life, he was just right—five feet, nine inches tall. In the end, his parents yielded.

"I started to understand that there is no such thing as being born a superstar," he said. "No, the answer to my old problem of how to do better was not that I needed to be born smarter. The answer lay in working hard." The only way to overcome ignorance, he decided, was through diligence. So in high school he began applying himself in areas where he lacked natural ability, winning a scholarship to the University of Texas where in 1965 he obtained a bachelor degree in Aeronautical Engineering.

After serving as an ensign in the U.S. Navy, Bean earned a place in the U.S. Navy Test Pilot School in Pax River. Only the best made it to Pax River. Pilots there had to fly test planes in bad weather to make sure the plane could perform even when skies were stormy. And as Bean soon learned, this meant finding the biggest thunderstorm and flying straight through it.

"Every time the plane was hit, it shuddered and shook," he said. "Sometimes, when I landed, I saw that lightning had blown off my plane's radio antennas or shattered its nose cone to pieces. Always there was damage to repair. Of course, I acted very cool about the whole thing—as if being struck by lightning happened every day."

The faster and more dangerous the planes, the better Bean liked it. He was soon given his first assignment in the Navy's Service Test unit. The assignment, however, was not at all what he expected. Rather than being assigned to a critical area of jet maintenance, Bean was assigned to—of all things—ejection seats. It was his job to service and reassemble the ejection seats on each test plane. The seats had been the source of constant malfunction and, in some cases, the death of the pilot. It was a tragic puzzle; and each time Bean came up with a fix for the problem, some so-called expert would tell him why his idea wouldn't work. He finally devised a technique that would work with almost any ejection seat on any aircraft. It involved a checklist of key failure components whereby service technicians could catch the most prevalent defects for nearly any aircraft. After Bean's procedures were implemented, there were no more ejection seat tragedies.

It was at Pax River that Bean met fellow test pilots Dick Gordon and Pete Conrad. The three Navy men's paths became tightly woven in the months and years that followed—all the way to their selection in the third group of astronauts in 1963.

By 1966, Conrad and Gordon had already flown together on Gemini, and Bean was still waiting for his first space flight assignment. When he eventually received a call from Deke Slayton, it was again not the assignment he expected. Instead of assigning him to an Apollo mission, Slayton assigned him as Chief of the Apollo Applications Office for *Skylab*. While the title was prestigious enough, it meant he would be sitting behind a desk with no time for space flight training.

"I didn't want to be there. *Nobody* wanted to be there," he said. "But there I was, and once again I had a choice. Am I going to go pout because I'm not in Apollo where I really want to be and where my friends are? Or am I going to try to do *Skylab* as good as I can do. Well, I decided to do *Skylab* as good as I could do."

Bean would not have made it to the Apollo 12 crew had it not been for a tragic accident that took the life of C. C. Williams. The original Apollo 12 flight crew had been selected two years earlier. The commander was to be Pete Conrad, already a veteran of two Gemini flights. The command module pilot was to be Dick Gordon, who had flown in space together with Conrad on Gemini 11. And the lunar module pilot was to be Clifton Curtis (C.C.) Williams, who had no previous space flight experience.

Conrad had lobbied Deke Slayton to assign Alan Bean as his lunar module pilot, but to no avail. Then, two years before Apollo 12 was scheduled to fly, C. C. Williams died tragically in a bizarre sequence of events. While piloting a T-38 from the Cape to Mobile, Alabama, to see his father who was dying of cancer, a mechanical failure near Tallahassee, Florida, caused the controls of his T-38 trainer to freeze, and the airplane went into an uncontrollable aileron roll. As the jet lost its lift, it became a ballistic projectile, accelerating earthward at a horrendous rate. Williams attempted an emergency ejection, but the T-38 was dropping too fast and its trajectory too low for his parachute to save him.

After the accident, Commander Conrad knew exactly who he wanted to replace Williams—it was a man working a desk job in *Skylab*. This time Slayton consented.

By now, Bean had resigned himself to his job, so didn't think much of it when Conrad strolled through the door in *Skylab*. Bean had no idea what Pete wanted. And then Pete asked if he would like to be part of his crew on Apollo 12.

"It took two or three minutes for it to sink in that maybe I was going to be a part of Apollo after all," Bean said.

"There was never any question about which mission launched in the worst weather," Walt Cunningham said. "Apollo 12 was born in a thunderstorm."

It was thirty-eight degrees, overcast, and raining at the Cape on the morning of the launch. The crew wore mission patches bearing three stars representing each of the astronauts and a fourth star in remembrance of C. C. Williams. As the countdown proceeded, Reverend Stout, listening from the Manned Spacecraft Center, heard the muffled words of a brief prayer from someone in the cockpit: "God give us the strength to do what we have to do."

But if Conrad and his crew thought their first space adventure would go smoothly, they were wrong. The first problem arose just thirty-six seconds after liftoff, only one mile up. The rocket had just disappeared into the base of the clouds when observers on the ground saw two bright blue flashes of light. Lightning had struck the Saturn V rocket, followed the flames from the engines down through the launch tower, and grounded itself on the launch pad. Sixteen seconds later the rocket was struck by lightning again.

Warning lights on the command module control panel lit up like a Christmas tree. The entire platform had been knocked out, along with the guidance system needed to navigate the moon. The fuel cells had gone offline and were operating on battery power. Everything pointed to a dying spacecraft.

"We just lost the platform, gang," Conrad radioed. "I don't know what happened here. We had everything in the world drop out."

People in Mission Control weren't sure what happened either. Telemetry screens read gibberish. Every guidance, navigation, and computer system was gone. In Mission Control, flight controller John Aaron remembered an event during an earlier test when technicians accidentally powered up the spacecraft using only a single battery, resulting in the same type of garbled telemetry he was seeing now on Apollo 12. He knew exactly what to do. He told the capsule communicator to have the crew flip the "SCE" (Signal Condition Equipment) switch to the auxiliary position.

CapCom Jerry Carr had no idea what Aaron was talking about. Neither did rookie Flight Director Gerry Griffin.

"What?" Griffin responded.

"SCE to Aux, Flight," Aaron repeated.

Griffin passed along the instructions to CapCom Jerry Carr, "CapCom, SCE to Aux."

Carr responded, "What?"

"SCE to Aux," Griffin repeated.

Finally, Carr radioed to Conrad, "Apollo 12, Houston. Try SCE to Aux. Over."

"Say again, Houston?" Conrad fired back.

"SCE to Aux," barked Carr.

"What the hell is SCE? And where the hell is it?" Pete exclaimed.

"I know what it is!" Bean yelled. He reached across the instrument panel on the Apollo 12 platform and flipped the SCE switch to Auxiliary. Immediately the screens in Mission Control blinked and streams of data appeared.

Meanwhile, Dick Gordon struggled to realign the guidance platform, and he was doing it the only way he could—from the stars. But he couldn't see any stars. Apollo 12 was beginning its final earth orbit and this was his last chance to recover guidance coordinates before Houston ordered them to return.

"I don't see anything, Al!" he said.

Bean grabbed a star chart and told Gordon to watch for the constellation Orion, which should be coming into view. By now, Gordon's eyes had adjusted to the darkness and—like clockwork—there was Orion in the telescope. Gordon made the alignment and the platform was restored. With that, the three space travelers laughed themselves silly on a course for the moon.

When Dick Gordon slid the spacecraft into lunar orbit four days later, Bean was exuberant. As if he had never expected it, the moon was right there in full view out his window.

"Man! Look, at that place. Outstanding effort there, Dick Gordon. Flash Gordon pilots again!"

"Good Godfrey! That's a God-forsaken place," Conrad responded, catching his first glimpse of the barren planet. "But it's beautiful," he added quickly.

As the time for descent approached, Conrad and Bean took their positions in lunar module *Intrepid*. Their destination was a specific crater where the *Surveyor 3* robot spacecraft landed almost three years earlier. Examining *Surveyor 3* after thirty-one months of exposure to the lunar environment and returning with parts of the robot was a primary objective of the mission.

NASA still wasn't sure of the exact location where Armstrong and Aldrin landed on Apollo 11, so Apollo 12 planners had worked hard to achieve a precise landing. Their efforts proved remarkably effective. Conrad came as close as anyone had a right to expect. Bean couldn't believe his eyes. *Intrepid* made a pinpoint landing within 600 feet of *Surveyor 3*.

"There it is!" he said, spotting the crater. "There it is!! Son of a gun, right down the middle of the road!"

"Outstanding! Forty-two degrees! Pete! Forty-two degrees," he continued.

"Look out there! I can't believe it!! Amazing!! Fantastic!!"

But as soon as they touched down, Bean grew cautious. Later, he said, "When we landed on the moon and stopped and turned off the engine, I thought, 'Wow, we've got to get going 6000 miles an hour again to get to

Dick Gordon up in the command module. And then we've got to come out of lunar orbit and burn our engine to go back home.' Well...we all know that with cars or airplanes, problems come *when you get ready to start 'em.*"

On November 19, Pete Conrad opened the hatch and stepped out onto the porch of the lunar module. He knew people wouldn't remember the third man to walk on the moon, so there was no need to make a momentous statement. But he did have a quote. In fact, he had a bet to win. The bet had its origins in a conversation with Italian journalist Oriana Fallaci after the return of Apollo 11.

"What are they going to have you say?" Fallaci asked Conrad at a reception for the astronauts. She, like Madalyn O'Hair, was convinced that NASA bureaucrats had instructed Neil Armstrong what to say when he first set foot on the moon.

"It's up to me, darlin'," Pete answered.

She didn't believe him. "Impossible," Fallaci said. "They'll never let you get away with it."

"Look," Pete said, "I'll prove it to you. I'll make up my first words on the moon right here and now. It's on TV real time. I can't lie to you."

Conrad had a good idea. Since he was nearly the shortest guy in the Astronaut Office, why not say...

"You'll never do it," Fallaci said.

"How about five hundred bucks?"

They shook on it.

Conrad reached the last rung of the ladder. The final step was a big one, almost three feet. He held the ladder with both hands and jumped.

"Whoopie! Man!" he yelled. "That may have been a small one for Neil, but it's a long one for me!"

A short while later Bean joined him. Then began a string of camera mishaps that would plague Bean in the days ahead.

To improve the quality of television pictures from the moon, Apollo 12 carried a portable color camera which was very sensitive to bright light. One of Bean's first tasks on the surface was to remove the television camera from the lunar module and position it on a tripod to capture their lunar surface activities. Houston wasn't entirely happy with the camera angle, so Bean attempted to reposition it to the other side of the lunar module. The camera was so new that Pete and Al never had the chance to work with it. They had received no instructions about the sun angle, and when Alan moved the camera, he inadvertently pointed it toward a brilliant reflection of the sun from the silvery lunar lander, destroying the light-sensitive Videocon tube. Television coverage of the mission was abruptly terminated.

Houston thought the camera color wheel was the problem and instructed Bean to fix the camera by tapping it with his geologic hammer. "I just pounded it on the top with this hammer that I've got. I figured we didn't have a thing to lose," Bean radioed.

"Skillful fix, Al," Houston joked.

But nothing worked. The world stared at a black screen with a crystal white smear across the center. The televised Apollo 12 moon exploration had just turned into a radio show. Networks scrambled to fabricate an image of the moon landing, switching quickly to contrived moonscapes. ABC showed employees romping about in a fabricated studio setting dressed in spacesuits that looked eerily like costumes from *The Twilight Zone*. A few people in Mission Control wondered whether the camera had been destroyed intentionally.

Other things made them wonder, too, like Conrad's frequent chortles.

The crew had been assigned to perform specific geological excavations over an area of the undulating lunar surface. Halfway through the checklist book, Conrad flipped to page seven. And there gazing up at him was Playboy Playmate Miss October 1968. Beneath it was a caption from the backup crew, "Seen any hills and valleys?"

"Hey, Al, you want to flip over to procedure seven?" Conrad yelled at Bean. "I might need your help on this."

The pages that followed continued the levity, like caricatures of Snoopy astronauts, and a second Playmate complete with a geological reminder: "Don't forget: Describe the protuberances ..." No one ever figured out how the backup crew sneaked the inserts past Deke Slayton.

The camera mishaps then resumed. There was the matter of the camera shutter self-timer device Conrad had smuggled in the pocket of his space suit. The plan was to mount the Hasselblad camera on the tool carrier, set the timer, and pose in front of it. Since the timer was not part of their standard equipment, such an image would throw post-mission photo analysts into confusion over how the photo had been taken. Conrad was convinced it would be on the cover of *Life* magazine with the caption, "*Who Took This?*"

As he left the lunar module, Conrad dropped the timer into the tool carrier. But after completing a successful geological excursion, the tool carrier was now full of dust and rocks. Using hand signals so that the whole world would not listen in, the moonwalkers decided that Bean would dig through the rocks to find it. His only choice was to dump the rocks out of the tool carrier onto the ground, which he was reluctant to do for fear of losing them.

After hopelessly rummaging through the samples, Bean gave up and the men moved on to other tasks.

After nearly four hours of work, Conrad and Bean headed to *Intrepid* for a meal and some rest. They were both hungry and tired, and Bean looked forward to his favorite meal—spaghetti.

When they emerged again from the lunar module, their destination was the *Surveyor 3* robotic probe, which stood well within sight of the lunar lander. The probe had landed very hard on the moon two and a half years earlier in April 1967. So hard, in fact, that the onboard television camera worked for only a short period of time. The camera was used in conjunction with a shovel extension to transmit television pictures to NASA to determine the make-up of the lunar surface before sending men there. Conrad and Bean were to retrieve the camera and other parts of the robotic probe.

But the probe, they discovered, didn't look anything like its supposed twin in Houston; and the bolt-cutter in the tool kit required extreme force and perseverance to grind the parts free. After completing their objectives, Conrad and Bean boarded the lunar module and prepared for the return trip to rendezvous with Gordon in *Yankee Clipper*. In a final comedic moment, Conrad pulled the Hasselblad camera timer from the bottom of the tool carrier. Bean took it and threw it out on the moon as hard as he could. Maybe the next guys there could use it.

Firing the lunar module ascent engine to lift off from the surface had been a matter of concern on Apollo 11, and it was equally so for Apollo 12, considering that the spacecraft had been riddled by lightning.

Under normal procedures, Conrad, as mission commander, was to fly the lunar module while the lunar module pilot assisted by observing and reporting instrument readings. But Conrad knew his buddy loved to fly. Taking one last look at the moon, he turned to Bean.

"You want to do it?"

"Do what?"

"Take us out of here. You're the lunar module pilot."

"But you're the commander."

"You want me to order you then?"

Conrad didn't have to.

"I can remember counting down and thinking, 'If this thing will just burn six minutes and three seconds, we're home,'" Bean recalled. "I can remember looking at my watch and the timer on the panel and watching it count down."

When the countdown hit zero, Bean pushed the launch button, lifting

Intrepid toward the orbiting *Yankee Clipper*, becoming the first, and perhaps only, lunar module pilot to actually fly the lunar module.

When *Intrepid* docked with *Yankee Clipper*, Conrad knocked on the hatch.

"Who's there?" Gordon asked.

When he opened the hatch, he saw Conrad and Bean covered from head to toe in lunar dust.

"You're not coming into my ship like that!" Dick Gordon maintained a clean ship and knew the dust could clog air flow filters and endanger the mission. "Strip down," he said.

"What?" Conrad asked.

Gordon demanded that the two strip out of their filthy space suits and clean up before they came aboard. And so it was that, 240,000 miles from earth, Conrad and Bean entered the command module the way they came into the world—naked. By the third orbit they were dressed and ready for the trip back to earth. They jettisoned *Intrepid* and sent it crashing into the moon. The impact reverberated for a full thirty minutes. The crew of Apollo 12 was on their way home.

There had been no ceremonial dedications or religious readings during the Apollo 12 mission.

"Had it done any good to have prayed for Madalyn O'Hair when I was on the moon," Conrad said, "I would have done it."

The descent stage left on the moon bore a plaque signifying the second moon landing of Kennedy's proclamation. The plaque did not bear words of momentous import or the name of President Richard Nixon, as did the Apollo 11 lunar plaque. Instead, it reflected the plain and simple nature of the crew, void of flourish: "Apollo 12 November 1969" and the names of the three crew members. Conrad had taken along a duplicate of the plaque to return as a gift to his good friend and next door neighbor, Jack Kinzler, the man who had made it.

Reentry through the earth's atmosphere was a violent process that released an enormous amount of kinetic energy, and there was concern in Houston that the lightning strikes might have damaged the parachute deployment mechanism, which would mean an uncontrolled splashdown and certain death for the crew. But there was no way to determine this. After some discussion as to whether they should alert the crew, Houston decided against it, since there wasn't anything anyone could do about it anyway. The parachutes would either work or they wouldn't.

Mission Control watched in relief as a rescue helicopter transmitted images of three parachutes ballooning above the Apollo 12 command module. As the module descended, Bean experienced his final bout of camera malfunctions. He had placed the 16mm data acquisition camera in the window of the command module so that NASA could have movies of rocket staging and reentry. The camera was supposed to be removed from the bracket during the parachute descent, but in the rush to get through the landing checklist, Bean forgot to remove it. As the command module hit the Pacific Ocean, the camera flew from the bracket and whacked Bean in the head, knocking him out cold. Fortunately, he regained consciousness in time for recovery by the U.S.S. *Hornet*.

The persistent camera malfunctions did not go unnoticed by the head astronaut prankster, Fred Haise. After the crew's release from quarantine, he arranged an informal ceremony where he presented Bean with the "Broken Brownie Award," a smashed up Brownie camera stapled to a board in recognition of Bean's "outstanding camera work."

In spite of the mission's near disaster and ultimate success, members of the Apollo 12 crew later said they experienced neither euphoria nor letdown after their return. Conrad was still Pete Conrad, Gordon was still Dick Gordon. Alan Bean, however, was a different sort of Alan Bean.

Before the mission, Bean might have admitted that he was a religious man, but he did not remain so after his return. The event caused him to question everything he had ever known about God, mankind, and the cosmos.

"I don't know how to explain it," he said. "It's kind of a paradox, feeling that somebody up there loves me, and yet not believing there's somebody up there to love me. It's crazy."

But one thing Bean did bring back was inspiration. During the next four decades he would develop as a painter, depicting Apollo missions in breathtaking displays of artistry. It was a skill he had nurtured most of his life, and now—driven by scenes that only a handful of men had ever seen—he undertook with a passion.

"When we actually did go in space and walk on the moon, we accomplished our lifelong dream—and then that let the next dream come to the surface," he said. "Some people had a next dream, like myself, to become an artist. Other individuals had other dreams. Some unfortunate ones didn't have a dream to replace going to the moon. So when they came back, they had a sense of emptiness because they'd accomplished what they'd always wanted, and nothing replaced it."

Pete Conrad was not one of those who suffered from a feeling of emptiness. He headed straight over to *Skylab*. The economy was suffering record deficits and at a billion dollars a flight, he knew Apollo's days were numbered. Alan Bean followed close behind. Dick Gordon, however, told Conrad he was going to stick around and wait for a shot as commander of Apollo 18.

Conrad wasn't sure that was such a good idea. "I'm not sure there's going to be an Apollo 18," he said.

After splashdown, the crew entered quarantine, accompanied by their PPKs and a vast cargo of moon rocks. Following their release, however, Bean had unfortunate news for Reverend Stout. While the two Bibles did indeed make a historic voyage to the moon, they did not actually land on the lunar surface. Due to an error in the log manifest, the support crew stowed the PPK on the command module rather than the lunar module, therefore the Bibles remained in orbit and never descended to the moon.

Although disheartened, Reverend Stout knew the Bibles' journey was nevertheless significant. The Apollo 12 spacecraft carrying the first microfilm Bible had endured two lightning strikes, yet survived to realize one of NASA's most successful moon landings, an event that could not be overlooked.

Bean retrieved the contents of his PPK and presented the small red bound Bible to Reverend Stout with a hand-written note: "It was a pleasure to have taken your Bible to the moon." Then, when no one stepped forward to claim the microfilm Bible, he tucked it away with his other flight memorabilia, where it remained for the next thirty-five years.

Stout now set his sights on getting the lunar Bible on board the next flight, Apollo 13, this time with multiple copies produced with yet a newer generation of NCR microfilm technology. He wasn't sure how he would persuade the Apollo 13 crew to do this, but he knew a conservative new congressman who might be able to help. Stout had been waiting in a barber shop when he ran across an article in *Esquire* magazine containing an interview with various aspiring young politicians. One in particular caught his attention.

"He was the only one in all the interviews that mentioned God," Stout said. "He had the kind of Christian personal commitment we wanted for the historic meeting…so he was the one chosen to present the Apollo 13 crew with the Bibles."

The politician's name was George H.W. Bush.

"Words cannot begin to express what experience can convince us of—how our experience can grow through prayer. First of all, it comes through hard work—the hard work of prayer. Have you ever seen a piano tuner at work on a misused or unused priceless work of art? Do it sometime. God is asking you to turn on and tune in. As it takes time for the piano tuner to do his thing, it will probably take time for you."

- Reverend John Stout
Director, The Apollo Prayer League

25
THE "ICY COMMANDER" IS BACK

When the Apollo 13 crew was assigned in July 1969, Alan Shepard was in his sixth year at a desk job in the Manned Spacecraft Center as Chief of the Astronaut Office, supervising the day-to-day activities of a team that had grown to sixty-five astronauts. He had every reason to believe he would die with a stapler in his hand.

Everyone at the center stepped lightly around Shepard, whose persona was larger than his five-foot, eleven-inch frame. He was so intimidating, the story goes, that his secretary kept two photos of Shepard in her desk drawer—one smiling and one scowling. For visitors' sake, she would display the picture reflecting Shepard's mood of the day.

As gatekeeper for the astronauts, Shepard coordinated travel schedules, parceled out the astronauts' media time, and arbitrated public appearances. He was fiercely protective of the astronauts and merciless with reporters. Once, when asked if a well-known reporter could interview one of the Original 7 astronauts, Shepard shot back, "No, we already gave that [expletive] reporter an interview. He misspelled Wally [Schirra's] name and thought reentry was some kind of sex thing."

Although there was no known cure for Méniére's disease, Shepard was determined to stay on and stick it out, and when Slayton was promoted to Chief of Flight Crew Operations, NASA officials agreed to assign Shepard Slayton's old position as Chief of the Astronaut Office. Slayton and Shepard had been friends since their selection together in the Original 7, and Slayton sympathized with Shepard's dilemma. Slayton himself had been grounded four years earlier due to an irregular heart beat. When Shepard asked the flight surgeon if he could at least take a plane up with Slayton, the answer was quick: "No, two half pilots do not make a whole."

The only thing America's first man in space would be flying was a desk.

All this changed one afternoon in 1964 when Tom Stafford stopped by Shepard's office. Stafford had learned from a high school friend of an ear, nose, and throat specialist in Los Angeles named Dr. William House, who reportedly had cured Méniére's disease through a delicate surgical procedure whereby a tiny silicon tube was inserted in the inner ear canal. It was an iffy procedure with only a 20% chance of success. It could, in fact, make the condition worse. But the risk of a surgery gone bad was small potatoes for Shepard. Within weeks he was on a plane to Los Angeles.

"Well, what if it doesn't work?" he asked Dr. House.

"Well, you won't be any worse off than you are now, except you might lose your hearing. But other than that …"

Raised a Christian Scientist, Shepard had learned that God would make things right with man if man was right with God. After his visit with Dr. House, he flew home to consult his wife, Louise.

"The doctor can't promise he'll be successful, but I'm burning up inside, Louise. I want so badly to fly again in space. I'm willing to try anything."

Louise Shepard was a woman of strong faith, herself a devout Christian Scientist, and she knew the odds of his success were much greater than a mortal man could profess.

"Do it," she said. "Go for it."

So in late 1968, Shepard phoned Dr. House to schedule the operation. If it didn't work, he would leave NASA altogether.

He checked into St. Vincent's Hospital under the assumed name of Victor Poulis, a name suggested by Dr. House's Greek nurse. Dr. House cautioned that the condition was often uncorrectable and the results, if any, would take time. But Shepard wasn't buying it.

"In my case it is going to be correctable," he said flatly.

And he was right.

Six months later, Shepard was back in Deke Slayton's office pushing for the commander slot on the next unassigned Apollo mission.

"Get me a flight to the moon," he said.

Gordon Cooper was next in line to be commander of Apollo 13 and

Cooper felt the mission was rightfully his. But Slayton rationalized that since Shepard had been bumped off the earlier Gemini mission, he was now entitled to the first available prime crew position. In the face of controversy, Slayton made the final decision. Cooper's name was scratched from the crew list and replaced with that of Alan Shepard as commander of Apollo 13.

Shepard wasted no time selecting his crewmates: Stuart Roosa would be his command module pilot and Edgar Mitchell his lunar module pilot. Roosa and Mitchell were part of the group of astronauts selected in 1966 who dubbed themselves the "Original 19." Some among the astronaut corps questioned Shepard's choice. Roosa, a freckle-faced kid with a rural upbringing, and Mitchell, a brilliant but intense intellectual, were both rookies with no space experience. Further, neither bought into the astronauts' social scene—which suited Shepard just fine.

Objections to Shepard's selection of Mitchell and Roosa were especially keen among the Original 7. A number of astronauts from the Mercury and Apollo ranks were imminently qualified with proven space flight records. Adding to the intensity, the Apollo program was falling victim to a government budget shortfall. Apollo 18 and 19 missions were already on the chopping block. The U.S. was sliding into a recession and President Nixon was having trouble shielding the space program. Chances to fly one of the remaining Apollo missions were rapidly diminishing. At a small dinner gathering with a handful of Mercury astronauts, Shepard was asked why he selected Edgar Mitchell as his lunar module pilot.

"Because I want to get back," he said simply.

No one doubted Mitchell's competency in the lunar lander. He and Fred Haise had been the designated astronaut representatives in the design and test of the lunar module, and the two had intimate knowledge of every nut, bolt, and computer circuit in the strange new spacecraft. In addition, Mitchell had a reported IQ of 180 and a doctorate from MIT with a dissertation in "Low-thrust Interplanetary Navigation"—a theory designed to propel manned spacecraft to Mars and beyond. For Mitchell, flying to the moon was merely an intermediary puddle jump. Shepard, who had little experience with the lunar module, wanted the best at his back during the tricky descent to the moon. Mitchell, he felt, was the man.

As for Stuart Roosa, "He's the best stick-and-rudder man I've ever seen," Shepard said. A country boy from Claremore, Oklahoma, Roosa, like others, answered NASA's call for astronauts from Edwards Air Force Base in 1965. Thousands of qualified pilots had applied, so Roosa wasn't especially hopeful when he was invited to Houston for an interview with NASA officials, including Deke Slayton. His interview had scarcely begun when Slayton received a telephone call and abruptly left the room. He never came back.

Roosa called his wife, Joan, dejected. He knew it was over when Deke didn't stick around. But Slayton had left the room to take a call from the commander at Roosa's test pilot school. "You want Roosa at NASA," the commander had said, so Deke had already made up his mind and saw no reason to return to his office.

Shepard was not one to mince words and not shy about expressing displeasure. Roosa, for one, was scared to death of him and was more than a little nervous that summer afternoon in 1968 when Shepard summoned him and Edgar Mitchell to his office. Roosa was literally quaking in his boots.

"I want to ask y'all something," said Shepard. "Do you mind flying a crew with an old retread like me?"

Roosa was taken aback. He had yet to serve as even a backup crew member for a mission. Roosa looked at Shepard. "You mean on the backup crew?"

Shepard gave him his icy stare. "I never mentioned anything about a backup crew—I said *crew*."

At that moment, Roosa realized he was going to the moon.

Shepard's euphoria as commander of Apollo 13 proved to be short-lived. Slayton was getting feedback from NASA officials that Shepard's selection posed a political problem, since the Apollo Saturn vehicle was several times more complicated than the Mercury and Gemini capsules Shepard had previously flown. The consensus: Shepard needed more time to train. Shepard, Mitchell, and Roosa had only fifteen minutes of collective spaceflight time, and all of that logged by Shepard during his 1961 *Freedom 7* flight.

In contrast, Jim Lovell, tentatively scheduled as commander of Apollo 14, had spent more time in space than any other human being, chalking up 572 hours and nearly seven million miles. Lovell had already flown two Gemini missions and was among the first men to orbit the moon on Apollo 8. Not only was Lovell eminently qualified on the Saturn 5—Lovell's crew was ready.

Slayton broke the news to Shepard. "We can't do [it]. You're too much of a political problem."

"I've been training along with all these other guys and I'm ready to go," Shepard argued.

"But the public doesn't know that," Slayton said. "So we'll make a deal. We'll let you command Apollo 14 if you'll let us have another crew for Apollo 13."

Shepard conceded, and Lovell's Apollo 14 crew was swapped with Shepard's Apollo 13 crew. Thirteen would now be flown by Commander Jim

Lovell, Command Module Pilot Ken Mattingly, and Lunar Module Pilot Fred Haise.

Jim Lovell was going to see Mount Marilyn one more time.

―∧∧―

The irony of the Apollo 13 crew swap wasn't wasted on Edgar Mitchell. Two of his closest friends and rivals had just elbowed him out. Mitchell met Fred Haise and Ken Mattingly when he first arrived in Houston and the three agreed to share an apartment. Since those early days, they had maintained a friendly rivalry as to which would be first to the moon. Mitchell was momentarily thrilled at the prospect of arriving at the moon onboard Apollo 13; but now, in a flash, both Haise and Mattingly were scheduled to arrive there ahead of him. Disappointed, he relinquished the honor gracefully. Apollo 14 would follow in short order—and he needed time to train his "icy commander."

Because Haise and Mitchell were married men, the quarters the three shared were temporary for them until their families arrived. Ken Mattingly, however, was a bachelor. "Ken had virtually nothing outside of the program," Haise said. "That was his life. We had families, children, and all of the things that took us away from the program. Ken had nothing."

Indeed, Mattingly's exclusive focus was on flying. His father had worked in maintenance for Eastern Airlines when he was growing up and took him on trips when he traveled. Eventually, Mattingly gravitated toward anything with wings and would eat whatever cereal had a cut-out of a plane on the box. His introduction to a flying career came when his high school advisor forgot to tell his students about an upcoming ROTC exam and offered each a six-pack of beer to show up at the last minute and take the test. The exam landed Ken in the ROTC and on his way to a military career flying exotic aircraft. The opportunity to fly to the moon on a new spacecraft still being invented was the very essence of his aspirations.

Mattingly and Mitchell met when students at Edwards Air Force Base, where they were the only two Navy participants in a nearly all-Air Force pilot school. In the summer of 1965, Navy men at the base were given a choice between applying for the Manned Orbiting Laboratory (MOL) program or NASA's new space program. Air Force personnel, however, were allowed to apply for both. After conferring with each other, Mitchell and Mattingly decided to opt for the MOL program. "We'd seen all the press releases and heard about all these heroes at NASA," Ken said. "We decided we didn't have a chance, so we checked the MOL box." The applications were then forwarded to the Air Force review board. Mitchell and Mattingly ranked first and second in their class and were confident of their selection.

When the results were posted, neither one of their names was on the list, although all of their Air Force classmates were. Fortunately, an Air Force instructor at the base, John Prodan, arranged for the Navy to submit their applications directly to the NASA selection board. "I think there might have been sensitivities among the ranks about 'these clowns who wear Navy uniforms,'" he said. Both Mitchell and Mattingly made the first interview cut and were put on the short list for selection.

Mitchell was having dinner at home when the pivotal call came from Deke Slayton. "Ed came back grinning from ear to ear," said Mattingly. "But no call came for me. So I figured I was dead meat and the guys in the class were all hanging around trying to keep my spirits up."

The group was undergoing medical evaluations at Brooks Air Force Base in San Antonio and decided to take Ken out on the town for drinks. When Mattingly returned to base quarters to change clothes for the outing, the phone rang. It was the call that would change his life. "The announcement will be out by the end of the week," Slayton said, "but don't tell anybody."

They couldn't tell anybody—but neither could they hide it from their friends. The group headed to the River Walk in downtown San Antonio to celebrate. They were sitting on the patio of a Mexican restaurant when the waiter approached Mattingly with a napkin.

"That woman over there would like your autograph," he said.

The group stared at each other. "Somebody recognized me?" Mattingly said. "How could she know?

"Why does she want that?" he asked the waiter.

"*Well, you are the comedian Dickie Smothers, aren't you?*"

After the group regained their composure, Mattingly signed the napkin and handed it back to the waiter. "Tell her to hold onto this for a couple of days."

⁓∧⁓

Elsewhere, Fred Haise was also receiving a call from Deke Slayton. He, like Mattingly, had been selected as part of the Original 19—or the "Excess 19" as the Original 7 astronauts called them, since they didn't feel they needed any help. The roommates were like slices from three different pies. Haise was loose and quick, Mitchell was philosophic and intellectual, and Mattingly was quiet and intense. But in the weeks that followed, the unlikely trio became fast friends. "We were just the same three guys who came from the same dusty desert," said Mattingly. And in time, they hoped to fly a mission together.

None of the three were much for party life, but they shared off-duty time together. Mattingly spent an occasional evening with the Mitchell family, and Ed Mitchell would take Mattingly's 16-foot Sailfish for a cruise on Clear Lake.

The "Icy Commander" is Back

Once, while navigating the boat on a family outing, Mitchell inadvertently turned it over, dumping his entire family into the lake and jamming the mast into the mud. "It was a little embarrassing for me at that point," Mitchell said. "I made a turn and a gust of wind caught me. I had to pull Louise and the two girls out from under the boat." He didn't know quite how to break the news to Mattingly, but based on Ken's nonchalant reaction, "suspected he might have done the same thing a time or two himself."

As for Fred Haise and Ken Mattingly, it became nearly impossible to keep the two out of an airplane. They spent every available hour flying together. "Fred and I considered any day we didn't fly at least twice a complete loss—and that includes weekends," said Mattingly. "Fred, Ed, and I sort of formed a triumvirate. The three of us kept very close relations right up to the time we had to go fly."

As the Apollo program progressed, the three were separated by their respective NASA assignments. Haise and Mitchell went to the Grumman development lab at Bethpage, New York to work on development of the lunar module; and Mattingly headed to North American Aviation in Downey, California, to work on development of the command module.

Haise and Mitchell shared a small trailer office at Grumman where they worked alternating shifts. Their primary job was to interface with Grumman engineers in the operational checkout of the lunar module features, since switches and other controls that may have appeared well positioned to a Grumman engineer might not be easily operated with the thick gloves worn by astronauts. The two wanted traceability of every part, every wire. But testing amid an ambitious development program with tight deadlines was challenging. A continuous stream of congressmen, politicians, and VIPs were permitted inside the spacecraft, along with their aides and secretaries.

"Changes got so hot and heavy with instruments and wiring that it became difficult to track," said Haise. "One time when we were in the lunar module testing, it got hot because someone was welding right behind us."

Haise became known as a light hearted practical joker, while Mitchell maintained a more stoic demeanor that masked a very wry sense of humor. On one occasion, when waiting for Haise to relieve him, Mitchell was asked by Grumman engineers to test a balancing platform designed to familiarize the astronauts with mobility on the moon. Mitchell found the operation to be unusually perilous.

"It was kind of like a Buck Rogers flying belt where you stand on a platform, and if you've got perfect balance, it would sit still. There were only two degrees of freedom back and forth along the rail, and if you tilted, you accelerated the platform very quickly." Although securely strapped into a harness, Mitchell repeatedly lost his balance and flew end over end off the platform before he finally mastered it.

When Haise showed up to replace him, Mitchell took him to the platform and explained the simple mobility procedure Grumman wanted to test. He encouraged Haise to go first. "Go ahead and give it a try, Fred. Just go ahead and do it. You go first." Then he stood back and watched as Haise repeatedly fell off the bizarre contraption.

"Oh c'mon, now, it's a piece of cake," Mitchell said. "Let me have it." He hopped up on the platform, strung up the harness, and performed flawlessly. After what seemed an interminable amount of additional time and trials on Haise's part, he finally realized he'd been had. And, as everyone knew, getting one over on "Freddo" was rare.

"I think you're a *fink*," Haise said, labeling his teammate a turncoat.

It was during this time that Mitchell received a grim reminder of the risky business of space. The Original 19 were returning to Houston after inspecting the command module in Downey when a page for Mitchell sounded over the Los Angeles airport speaker system. It was a call from Deke Slayton. There had been an accident on launch pad 34 during an Apollo 1 ground test—the crew had perished in a flash fire. Deke needed four men to fly the missing-man formation over memorial services at a Houston church and asked Mitchell if he could put together a team.

Mitchell knew immediately who he would ask to fly with him, and on January 30, 1967, Mitchell, Mattingly, and Haise flew a wind-swept formation over memorial services in Seabrook, Texas. As the signal came from the church for the flyover, the fourth missing-man pilot, Vance Brand, veered off into the heavens as the "fallen angel" in memory of their lost comrades, while Mitchell, Haise, and Mattingly remained in flight formation, vanishing into the horizon together.

It was an unfathomably sad time in American history. Yet as the turbulence of the Sixties persisted, around the globe a vast group remained dedicated to exploring the moon. On occasion, the tragic dimension of the decade penetrated the sealed world of NASA. Such was the case with Ken Mattingly. Shortly after the Apollo 11 moon landing, he received a letter from the parents of a friend notifying him that their son had been shot down in Vietnam.

"I can't do this," he told himself. "I'm sitting in NASA in the lap of luxury while my buddies are dying." Seeing no immediate prospect for a space assignment, he visited Shepard's office. "It's time for me to go," he said. "I've had fun, but it's time for me to go back and earn my pay in Vietnam."

"I can understand that," Shepard said. "Think about it for a week or ten days and then come back."

Not long after that, Deke Slayton announced the next prime crew: Jim Lovell, Fred Haise, and Ken Mattingly. Presumably, Shepard had known that.

The assignment was difficult for Mattingly to grasp. He hadn't known Jim Lovell very well and wasn't sure why Lovell had selected him. But Lovell had picked one of the best lunar module pilots in the ranks, Fred Haise. And Haise knew Mattingly very well.

"Lovell didn't know me from Adam's house cat," Ken said. "I think he must have asked Freddo."

On the previous Apollo 12 flight, the microfilm lunar Bible given to Alan Bean had never landed on the moon, and the Apollo Prayer League now intensified its efforts in the quest. The plan for Apollo 12 had been to land the lunar Bible on the moon and then return with it to earth. Stout's hope now with Apollo 13 was to leave a Bible on the moon so that even centuries later, other celestial travelers might find it. But he knew the harsh lunar environment was sure to destroy the fragile microfilm within days, if not moments. Stout had studied and taught chemistry in Brazil and knew of a gold amortization process that involved applying a thin protective layer of gold over the letters of the Bible. He was aware that the Germans had done a lot of work in this area, but he had difficulty locating an expert within NASA. Eventually he found such a person in the very building where he worked at the Manned Spacecraft Center.

After further collaboration with the co-worker, Stout discovered that the protective diazole used by NCR to encase the image would protect the fragile Bible just as well if only the borders of the Bibles were gilded. With this, he abandoned the gold amortization process. "If we were going for 10,000 years, then we probably would use an improved version of the gold amortization process," Stout said. "But at this point, 1000 years would do."

Marilyn Lovell agreed to work with Stout on a design committee for the Bibles and to serve as liaison between the committee and her husband regarding stowage and handling. Stout again turned to Norman Vincent Peale for advice as to an appropriate gesture for the occasion of landing the first Bible on the moon. Peale suggested that the crew leave the first lunar Bible on top of a small mound of rocks, similar to the altar built by Noah when his ark came to rest on the mountains of Ararat. Lovell was said to be amenable to the idea.

For this purpose, Stout envisioned a single microfilm Bible covered with a gilded American emblem displaying the stars and stripes. It would be accompanied by a small commemorative certificate containing a description of the microfilm Bible, its origin from planet earth, and a translation of the first verse of Genesis in sixteen languages, which were believed to represent the languages spoken by sixty percent of the world. The United Bible Societies provided the translations. The first verse on the list was in English, "In the beginning, God created the heavens and the earth," followed by the same verse in Spanish, Chinese, German, and so on.

With this would be a separate packet containing multiple copies of the microfilm Bible which would later be distributed to APL members and supporters of the effort. In view of the advancing microfilm technology, Stout obtained a roll of 1500 identical microfilm copies of the King James Bible from Edmund Scientific, an NCR distributor. Of these, he packaged 512 together in a fireproof packet, with the gold-edged Bible wrapped separately so it could be easily removed and left on the moon along with a small folded certificate reading:

> *This is the first book taken from PLANET EARTH to the surface of another celestial body. It was dedicated to this spot on the moon by James A. Lovell, Jr. during the Apollo 13 flight, in company with Fred W. Haise, Jr. and Thomas K. Mattingly II.*

The Bibles would take less than a square inch or so of space and weigh a little over an ounce. Reverend Stout was now prepared to approach another crew with the Prayer League's request to carry the lunar Bibles. In a newsletter to APL members, he officially announced the dignitary chosen to present the Apollo 13 lunar Bibles:

> Since the astronauts are doing this on a personal basis we wanted someone of national prominence to be present when the Bibles were presented to the astronauts. Finally from a hundred suggested names, from the Pope to the President, a Christian congressman, George Bush, was chosen.

Congressman Bush had developed a special interest in the space program and agreed to participate in the presentation. So shortly before the April 11, 1970 launch, Bush accompanied Reverend Stout to Lovell's house with the lunar Bibles.

Lovell, Haise, and Mattingly were on hand at Lovell's home to greet

Congressman Bush and Reverend Stout and accept the lunar Bibles for the flight. Reverend Stout was thrilled at the prospect but braced for a backlash from Madalyn O'Hair. By this time, the ongoing political battle between O'Hair and NASA was well known, and Stout was beginning to enjoy his lively exchanges with the woman. So to avoid any last minute consternation, he called her directly. The word of God was on its way to the moon again, he told her. And if all went well, it would be there for a thousand years. If she had a serious problem with that, she was welcome to go there and retrieve it.

Jim Lovell was eager to crown his career with a moon landing. His wife Marilyn, however, was skeptical about his reassignment to the number 13 mission.

"13? Why 13?" she asked.

"Because it comes after 12, Honey."

It was a short answer to a long story. And as things evolved, the story would only get longer.

"Houston, we've had a problem."

- James Lovell, Commander, Apollo 13

"We've had more than a problem …"

*- Larry Sheaks, Mission Control
Environmental Control Systems Flight Controller*

26
APOLLO 13: A SPACE ODYSSEY

Less than two months after Apollo 12 splashed down in November 1969, the Apollo 13 spacecraft stood ready on the launch pad. It was scheduled for liftoff on the thirteenth minute of the thirteenth hour. The lunar landing was scheduled for the thirteenth of April. Unfortunately, no one at NASA was superstitious.

At the Apollo Prayer League, Estelle Jackson submitted a poem that spoke directly to the matter, quoting from Romans 13:12:

> "13" is a long-dreaded number,
> Superstition has made it seem so.
> But cast off the workings of darkness
> Accepting the challenge to "go."
>
> There is no reason for doubting,
> For you'll reap whatsoever you sow,
> You're the 7th manned flight of Apollo
> And "7" is perfect, you know.
>
> Cast off your fears of the darkness,
> Put on the armor of light,
> Trust and the Lord shall go with you,
> That is my prayer for your flight

NASA had never experienced a disaster during a fully-fueled manned mission and were confident this flight would be no exception. They had an expert crew and a proven spacecraft. The crew named the command module *Odyssey* to symbolize the long voyage to the moon, and the lunar module *Aquarius* after the Egyptian Goddess, the water carrier who brought life to the Nile Valley, as they hoped she would bring them safely home to earth.

Although NASA was not ready for a lunar landing in extremely rugged terrain, the site selection committee had long been interested in a place called the Fra Mauro Hills. Of particular interest was a feature called Cone Crate, a comparatively fresh crater believed to contain material the scientific community needed to help better understand the creation of the rugged lunar highlands.

Even before taking command of Apollo 13, veteran astronaut Jim Lovell had announced this would be his last trip. "There are a lot of people in our program who haven't flown yet and it's appropriate that I step aside."

In the months following the crew swap, Lovell, Haise, and Mattingly became unilaterally focused on a mission that, by virtue of previous flawless missions, was sure to succeed. Even Fred Haise's practical jokes, like flipping the re-pressurization valve during simulations to create a startling bang, did nothing to undermine the crew's confidence.

It was a well-known fact that as an Apollo crew trained together they developed an intuitive sense of teamwork and, over time, began to function as a single cohesive unit. Such was the chemistry of Lovell, Haise, and Mattingly as the April 11, 1970, launch date approached.

The backup crew of Commander John Young, Command Module Pilot Jack Swigert, and Lunar Module Pilot Charlie Duke trained alongside the prime crew as standby. John Young was an experienced commander with three space flights behind him; Jack Swigert, a swinging bachelor known for his little black book of phone numbers. Swigert was serving his first time on a backup crew, as was Charlie Duke.

Working in unison with the prime crew, the backup crew became a part of that well-oiled team. But as Fred Haise had recently learned as backup for Apollo 11, the backup crew's intensity begins to wane a month or so before launch as the chance for a seat on the mission fades. And as the Apollo 13 countdown approached, this was beginning to happen. John Young, Charlie Duke, and Jack Swigert, while still performing their backup assignments, were slowly disengaging from the mission. But this was about to change.

A week before launch, Charlie Duke returned from a family picnic that included, among others, little Paul House, a friend of his young son,

Charlie Jr. Paul, as it turned out, had the German measles, and Charlie had been exposed. Before he was aware of it, Charlie inadvertently exposed the entire Apollo 13 prime crew to Rubella. Flight surgeon Dr. Charles Berry decided to conduct blood tests to check the prime crew's immunity "just as a precaution."

The following week, the surgeon called Ken Mattingly aside. "We need to take your blood again. It doesn't look like you've had the disease and built a natural immunity. Do you remember having it?"

"I don't know."

Mattingly called home. "Hey, Mom, did I have measles?"

"I don't think so," she said.

Mattingly then began a series of morning and evening blood tests. Mattingly's backup, Jack Swigert, was found to be immune to the Rubella virus, so the surgeons decided to give Swigert training time in the simulator, just in case. The other astronauts speculated it was purely political. "It's just for the public," they assured Mattingly, "to show that they've really done everything."

Meanwhile, after a countdown demonstration test two weeks before launch, Mattingly received a call from Dave Brooks in flight crew support. There was a problem draining one of the two oxygen tanks. As a matter of procedure, the tanks were emptied to about half full and refilled with liquid oxygen prior to the flight. Oxygen Tank 1 behaved as it should, but Oxygen Tank 2 dropped to only 92 percent of capacity. The tank had previously been installed in the service module of Apollo 10, but had been swapped out and cycled to Apollo 13. After several unsuccessful attempts to de-tank more oxygen, engineers decided on another approach, and Dave Brooks was now calling to clear the procedure with Mattingly. Even though the regular drain system wasn't working properly, he told Ken, they could use the electrical heater in the tank to boil out the excess oxygen.

"How do you know that you're not doing damage?" Mattingly asked. Brooks explained that this procedure had been used before on another tank and the temperature gauge and protective circuits would prevent a problem—it was no big deal. Ken passed the information along to Haise and Lovell. The response was a unanimous "yes." The launch was less than two weeks away, the spacecraft was already on the launch pad, and Brooks' recommended procedure would keep them on schedule.

Meanwhile, another problem was going undetected inside Oxygen Tank 2. In 1965 NASA had approved a design modification to change the spacecraft's 28-volt tank heater thermostat protection to match the 65-volt system at Kennedy Space Center. The tank manufacturer, Beechcraft, neglected to convey the change to its sub-contractor, and the thermostat in the cryogenic (cryo) oxygen tanks were never upgraded to operate above 28 volts. To complicate matters, the temperature gauge on the ground test console could only record a tank temperature up to 80 degrees Fahrenheit, since under normal conditions the temperature would never exceed that.

The support team now began the 8-hour boiling process. A technician monitoring the boil-off was told not to allow the tank temperature to rise above 80 degrees. The heaters were then connected directly to the 65-volt ground-power supply. As the electrical surge ran through the tanks, the thermostat fused shut, setting off an intractable chain of events. Now, instead of tripping open when the temperature reached 80 degrees, the heaters kept running as temperature inside the tank soared to an estimated 1000 degrees. As the temperature rose, the Teflon insulation around the wire of the tank fans used to do the oxygen-hydrogen stir cracked. A later attempt to de-tank showed Oxygen Tank 2 still failing to empty, and the heaters were once again left on for hours. Unaware of the problem, engineers proceeded once again with de-tanking, and ultimately Oxygen Tank 2, number 10024X-TA0009—the tank that had been dropped and reassigned from Apollo 10—was cleared for launch. The cracked wiring lay dormant in the cold liquid oxygen in the Apollo 13 service module as the rocket was prepared for the 250,000-mile ride to the moon.

In Houston, Mattingly's measles dilemma dragged on. Sensing his uneasiness, Deke Slayton advised, "Go down and fly. You know, get rid of your frustrations."

So two days before launch, Mattingly drove to nearby Patrick Air Force Base outside the Cape and hopped into a T-38 and flew. Sure enough, he felt better. On the drive back to the Cape, he turned on the radio just in time to hear an announcement that NASA had substituted Jack Swigert for Ken Mattingly on Apollo 13.

He pulled over to the side of the road.

"If this is a practical joke," he thought, "it's really well done. But I don't think this is a joke."

Lovell and Haise were quickly called into a meeting with Deke Slayton and NASA Administrator Thomas Paine.

It was a sad, awkward moment. "Lovell argued to just take Ken and go," said Haise. "The incubation period was such that we'd have been done with the lunar activity and rendezvous and might have been in lunar orbit before Ken came down with the measles."

But the doctor told them it would be a month before Mattingly manifested the disease. They simply couldn't take the chance. Lovell and Haise didn't like the decision any better than Mattingly, but they were caught in a dilemma. A one-month delay meant mean draining the rocket booster tanks—hauling the whole stack back, cutting the lines, and re-welding them. The decision was finally made to substitute Swigert and press on.

But Swigert, Mattingly's backup, had spent most of the preceding thirty days not training, but playing social secretary for the prime crew. With less than 72 hours before launch, Swigert was assigned to Mattingly's center seat position. In order to work him into the team as quickly as possible, the three astronauts engaged in an intensive simulation of all flight maneuvers to ensure unquestioned teamwork. The mission plaque mounted on the lunar lander was hastily overlaid with a plaque bearing Swigert's name. The decision came so quickly that Swigert had no time to even invite family and friends to the launch.

When Mattingly returned to the launch facility, nobody knew what to say. He packed up and flew back to Houston. Deke phoned ahead to alert Flight Director Gene Kranz that—after five years of waiting and nearly a year of intensive training—Mattingly would be watching the Apollo 13 launch from a seat in Mission Control.

"I remember sitting on the steps next to the CapCom console," Mattingly said, "feeling like a fifth wheel, feeling very down, very sorry for myself."

Lovell and Haise were somber during the traditional walk-and-wave to the courier van on their way to the launch tower. "I was not very excited," said Haise. "I was flat because of the change-up in the mission. It really cut the legs off our attitudes emotionally. Wanting to go but half not wanting to go because we had to leave him behind."

While Lovell and Haise were confident of Swigert's ability as a command module pilot, he was not the one they had trained with for the past year, he was not the one whose mind they could almost read, and more importantly, he was not Ken Mattingly.

As they boarded the Saturn rocket, Lovell and Haise were just days away from joining one of history's most exclusive clubs.

On the morning of April 11, 1970, the mood in the Mission Control room was one of quiet confidence. The first leg of the Apollo 13 flight went fairly smoothly with the exception of a slight administrative detail. In the rush to substitute for Mattingly, Swigert had forgotten to file his Federal Income Tax return. The sudden realization on board Apollo 13 threw the usually cool Swigert into a mild panic.

"How do I apply for an extension?" he asked Houston. Amid laughter from Mission Control, he explained, "Things kinda happened real fast down there and I need an extension. I'm really serious. Would you …"

In Mission Control the CapCom was unsympathetic: "You're breaking up the room down here."

Swigert continued, "…turn it in?"

Flight Director Glynn Lunney passed the word up that American citizens out of the country get a 60-day extension on filing. "I assume that applies," he said.

Thirty-one hours into the flight, the crew executed a course change to target the rocky part of the moon called Frau Mauro. The crew would need to reset the computer to adjust the course before it rounded the moon, due in a few hours. With this change they gave up the free-return trajectory which would automatically swing them around the moon and back to earth if all systems failed. If something went wrong now, getting home would require another course adjustment and a second rocket burn to re-set the course for earth. Otherwise they could drift forever in the void of space.

The mission had been so routine to this point that there had been little to break communications silence, and conversation had been reduced to mindless chatter. Astronaut Jack Lousma was on CapCom.

Lovell: We're all going to bed now, Jack, after we play the last rendition of "With Our Eyes on the Stars…" [tape recorder music]

CapCom: Sounds like all the comforts of home. Have you guys got a flower on your breakfast table?

Lovell: Yes, Jack.

At 7:13 p.m. on April 13, White Team Flight Director Gene Kranz was concluding third shift operations and Glynn Lunney's Black Team was already on the floor preparing for the transition.

Kranz, referred to as "Flight," sat in the upper tier of consoles at the flight director's desk wearing yet another white brocade vest Marta had made. Kranz, a devout Catholic, was ever mindful of the risks inherent in spaceflight. "I find a way to get to church and pray for wise judgment, and pray also for my team and the crew," he said. "Our pastor, Father Eugene Cargill, knows the risks and the difficulties of our work and the need for extra guidance."

Directly below Kranz's desk in the windowless room were two tiers of flight controllers seated in front of consoles, each with dual real-time displays of systems flight data transmitted directly from the spacecraft. The console display units had rack-mounted drawers that could be pulled out by handles on either side of the display for servicing. At times of stress, a flight controller would find his hands firmly attached to these handles, which came to be known as "security" handles.

At the back of the room in a glassed-in gallery overlooking the control room, NASA managers, news media, and members of the astronaut families watched a live television transmission of the flight on a 10' x 10' rear projection screen mounted at the front of the room.

Up to twenty-five flight controllers reported directly to the Flight Director. Sy Liebergot, a bespectacled 34-year-old, was seated below Kranz in the second tier of controllers, his console table littered with a tobacco pouch, tamping tools, and a glass ashtray that held a pipe and ashes. Liebergot was EECOM for the mission, the flight controller in charge of electrical, environmental and consumables systems for the spacecraft. His two sage-green consoles flashed a series of 200 individual parameters, each updated once per second directly from instruments onboard Apollo 13. Parameters that varied outside of a given range would be highlighted. To Liebergot's left sat Jack Lousma at the CapCom console. Lousma was winding down his shift and Joe Kerwin was standing by to take over as the Black Team CapCom.

In the first tier of consoles directly at floor level, the darkest and lowest, sat a small group of flight controllers informally referred to as the "Trench"—the engineers in charge of spacecraft trajectory. Among them was Jerry Bostick, the flight dynamics officer. Bostick, an affable guy with a distinct southern drawl, was in charge of spacecraft trajectory, guidance, and software operations. This was his eighteenth mission and he knew the process well. The Trench was a tight, cult-like team with its own monogrammed matchbook covers and business cards.

The Mission Control room was not a pleasant environment in which to work. In the early Seventies, smoking was acceptable and the atmosphere was often pungent with smoke. Cigarette, pipes, and burning cigars lay in cheap amber ashtrays in front of consoles. During especially tedious shifts, cigarette butts piled up in huge mounds.

Sy Liebergot wasn't alone in monitoring and tending to *Odyssey's* electronic and life support systems. He was in voice contact with three other controllers in a staff support room across the hall who had access to even more detail on their consoles than he did. On one occasion during the Apollo 10 mission, Liebergot had stumbled across an unused voice loop that he dubbed "Bluebird" and began using as a private line to ask "dumb questions" of his backroom support team. It was no secret that the ground engineers in charge of monitoring the spacecraft's vital systems knew more about the status of the flight than the astronauts who flew it.

For his part, Liebergot was no stranger to Apollo launches. Nor was he a stranger to the harsher side of life. He'd survived his mother's knife-wielding rants that sent him diving under his bed at the age of six and the harsh hand of his father who had driven his mother past the brink of insanity. Brutal beatings with the buckle-end of his father's belt left an indelible mark on Liebergot's psyche. His father, a chronic gambler, was constantly on the run from the mob, moving the family from city to city so much that Sy was enrolled in the first grade in three different schools.

"At some point I realized that I no longer cried," he said.

His childhood experiences left him with a certain degree of emotional callousness that would serve him well in the hours ahead.

It wasn't unusual to find astronauts gathered behind the CapCom console and surrounding desks during a launch. But for this particular flight, Mission Control was unusually empty other than a handful of news people and relatives in the viewing area. The flight had been uneventful and flight controllers were anxious to log off and cue their consoles to the Houston-Astros baseball game. The Astros had beaten the Atlanta Braves the night before and controllers were eager to catch the next game.

A handful of astronauts wandered in and out of the room during the day, including Edgar Mitchell, who was there to observe the flight he himself had originally been assigned to fly. Charlie Duke was there, as well. Things were going rather predictably and a break with Liebergot at the Singin' Wheel, a nearby roadside bar, was coming up on Duke's list of things to do.

But 200,000 miles from earth, trouble lurked. Ten feet below the crew was the cylindrical service module housing the oxygen tanks and fuel cells that provided life support to the ship. Inside the service module, Oxygen Tank 2 lay crippled. Unaware of the problem, the crew reconfigured their spacecraft for the journey to the moon. They turned *Odyssey* around, docked it nose-to-nose with the lunar module, then positioned the spacecraft to head for the moon. The crew could now open the adjoining hatch and move between the command and lunar modules through a short tunnel.

From this point forward, the timing of events moved quickly.

Monday evening, April 13, 1970: Televised broadcast to earth
At 9:00 p.m. Houston time, the Apollo 13 crew was finishing up a television broadcast aboard the ship. Before terminating the telecast, Lovell and Haise floated through the tunnel full of confidence, demonstrating the connection between the command and lunar modules,

Lovell: "And now Fred's engaged in his favorite pastime."

CapCom: "He's not in the food locker is he?"

Lovell: "That's his second favorite pastime. He's rigging his hammock for a sleep on the lunar surface and he's going to try it out to see what it's going to be like."

CapCom: "Roger, sleeping and then eating."

Things were going so well that the crew was asked to cut their TV broadcast short. Lovell closed it, saying: "This is the crew of Apollo 13 wishing everybody there a nice evening."

Still in the lunar module, Haise prepared another of his prank re-pressurization checks.

[bang]

Lovell: "... Stand by, one."

Haise: "Yeah, I got them with the cabin repress valve again there, Jack."

Lovell: "Every time he does that our hearts jump in our mouths. And Jack, anytime you want to terminate TV we're all set to go."

CapCom: "Okay, Jim. It's been a real good TV show..."

Unknown to the astronauts, the TV broadcast had been pre-empted on most major networks with routine programming. The moon missions had become so predictable, the world was no longer watching.

"The spacecraft is in real good shape as far as we are concerned," Lousma radioed. "We're bored to tears down here."

It would be the last time anyone mentioned boredom.

Apollo 13 was four-fifths of the way to the moon, and after 55 hours and 50 minutes of flight, the cracked wires of Oxygen Tank 2 were now exposed.

Six minutes after the TV broadcast, Liebergot had one last engineering housekeeping task for the crew. He asked Lousma to relay a request for the crew to initiate a routine extra stir of the cryogenic oxygen tanks by activating the tiny fans in each tank. Liebergot had less than an hour left on the third shift and wanted to leave a clean slate for his relief EECOM, Clint Burton. Lousma voiced up the request.

"Thirteen, we've got one more item for you when you get a chance. We'd like for you to stir your cryo tanks..."

"OK," Swigert replied, "stand by."

He threw four switches.

As the electrical current charged through the fans inside the tank, the bare wires shorted and the surrounding Teflon fueled a raging fire, exploding Oxygen Tank 2 with the force of seven pounds of TNT—enough force to blow up a 3000 square foot house.

The spaceship shuddered.

Liebergot's console display flickered.

Seconds later came a disturbing message from Apollo 13.

"Okay, Houston, we've had a problem here," radioed Swigert.

"This is Houston," replied Lousma. "Say again, please?"

Lovell was still in the tunnel between *Aquarius* and *Odyssey*, clutching a camera and making his way among the wires. Sliding into the seat next to Swigert, he saw what Swigert saw. The instrument panel was ignited in a blaze of flashing yellow warning lights. Alarms were sounding and voltage levels of the main power supplies were dropping rapidly. Lovell shot a look at Haise and could tell it wasn't another one of Freddo's tricks.

"Houston, we've had a problem," Lovell confirmed.

Across the hall in Mission Control, an alarming array of data flashed across Larry Sheaks' console. Sheaks was a backroom life support specialist for Liebergot and what he saw was disturbing.

"We've got more than a problem," he fired back.

Liebergot leaned forward in his chair to get a better look at the numbers; nothing much happened for sixteen seconds. He couldn't make sense of it.

"When I looked at the screens, the data told me I had lost two fuel cells, I had lost Oxygen Tank 2, and I was losing Oxygen Tank 1. I'm looking at not a single failure, not a double failure, not a triple failure, but I'm looking at a quadruple failure. There's no possible way that could happen in the spacecraft the way it was designed!"

Liebergot was convinced that if that much of the spacecraft had been lost, the crew would already be dead. Thus, he concluded it was an instrumentation problem, similar to that incurred during the Apollo 12 lightning incident—no real damage to the ship. Although the brunt of the problem fell on Liebergot as EECOM, other flight controllers were having trouble too. Flight controllers in the Trench scrambled for more stats. Five long minutes passed.

Then, looking out the hatch in *Odyssey*, Lovell saw something that made his stomach turn—the sheen of a clear piped gaseous stream. "It looks to me...that we are venting something into space," he reported.

A momentary hush fell over flight controllers as the new information sank in. They had all the symptoms of a total power failure. If power was completely lost, they knew the ship could rely on a battery backup which would last for about ten hours.

Liebergot reached for the console grips.

"I have never felt so alone in all my life... I didn't have the answers. And I'm sitting here...and I felt alone."

During the mission, the only way for Ken Mattingly to monitor activities in the spacecraft was to watch the TV broadcast from the glassed-in viewing area at the back of Mission Control. Seating was available for seventy-four VIP visitors who, no matter their background, had only "observer" status. Fred Haise's wife, Mary, then seven months pregnant, was seated next to Marilyn Lovell. Among those in the small crowd was NASA Assistant Administrator George Abbey.

The TV broadcast had just ended when Abbey put his hand on Mattingly's shoulder. "You look like you need a drink."

"You got that right," said Mattingly.

"Let me get my briefcase and we'll go over to the Singin' Wheel," Abbey said.

He never came back. The news from Apollo 13 had just come in.

Mattingly headed down to the CapCom desk.

"What did he say?" Mattingly asked.

"I don't know," Lousma answered, "something about he's got a problem and we've got this telemetry data that's all messed up."

Next to Lousma, Mattingly could see Sy Liebergot's desk. The situation had fallen squarely in his lap and all the mess was on his console. As Liebergot tried to rationalize the data, panic rose as a gorge in his throat. One oxygen tank and two fuel cells were lost, the command module was dying, and his screen was hemorrhaging data.

The voice loop came alive with the gibberish of flight controllers analyzing the problem. The shaky voice of a trainee flight controller could be heard in the background, "I don't know where to start."

It was twenty minutes into the crisis when Liebergot accepted what his instruments had been telling him all along: the spacecraft was venting oxygen. The service module was running out of power and oxygen—the *Odyssey* was dying.

"It was not an instrumentation problem but some kind of a monster systems failure that I couldn't sort out... I shoved the panic down and grabbed the security handles with both hands and hung on."

At that point, the future of the mission pivoted around a blithering amount of data and the gut feel of Sy Liebergot. He summoned a hard-won strength, his voice growing firm on the communications loop as he broke the news to Kranz.

"Flight, I've got a feeling we've lost two fuel cells...and it's not an instrumentation problem."

The problem was indeed real. And time was running out.

Without electrical power, the command module was dead. Without oxygen, the crew was dead. They were on the verge of losing everybody and everything.

Monday evening, April 13, 1970: Crisis in Mission Control
The prognosis from the back room was dismal.

"You'd better think about getting into the lunar module," Liebergot told Kranz, his eyes never leaving the console display. It was now 45 minutes since the explosion and Liebergot's team estimated that at the oxygen supply's current rate of decay, the last fuel cell would be drained in less than two hours. Even worse, there was only 15 minutes of battery life left in the command module.

"That's the end…right there," Liebergot said.

The only chance they had of saving the crew was to shut down the command module entirely to save what was left of the battery power needed for reentry.

Flight Director Gene Kranz stared intently at the console, outwardly unmoved. "Okay, now let's everybody keep cool… Let's solve the problem, but let's not make it any worse by *guessing*."

Suggestions poured in from flight controllers, each focusing on a particular system but none focusing on the overall crisis. Kranz knew that one misguided decision at this critical point could cost the entire mission and the lives of three men. The situation warranted careful consideration of Chris Kraft's well-known admonition: "If you don't know what to do, don't do *anything*."

When the phone rang in Marilyn Lovell's house she was cordial. She had seen her husband's television broadcast in Mission Control earlier in the day and knew everything was fine. She had no reason to expect anything other than a neighborly call.

"Marilyn, this is Jerry Hammock. I just want you to know that every country has offered to help in bringing the men in wherever they splash down." Jerry was a member of the Apollo 13 recovery team and assumed Marilyn already knew.

Marilyn was mystified. "Jerry, what are you talking about? Have you been drinking?"

Caught off guard and a bit embarrassed, Jerry tried to make it simple until someone else could explain it to her. "They're having problems with the fuel cell," he said and got off the phone.

More calls from NASA yielded the dismal reality. When Marilyn asked about the odds of saving the men, NASA officials were honest: About 1 in

10. Pete Conrad and his wife Jane went directly to the Lovells' house to be at Marilyn's side. The Conrads and Lovells were close friends, and NASA made it a practice to have the wife of an astronaut on hand when they were about to have a brand-new widow.

"The entire world was brought to its knees in prayer. For one week there was only one problem in the world and that was in space. For one week the world knew where to go to find an answer to that problem—to God."

- Reverend John Stout
Director, The Apollo Prayer League

27
THE WHOLE WORLD PRAYED

For the first time in NASA's history there was confusion in the control room. Two days into the flight of Apollo 13, the mission changed from a routine flight into a race for survival.

Glynn Lunney's Black Team was standing by to relieve Gene Kranz's White Team. Other astronauts poured in and the number of people on site doubled. The back room flight engineers, who could see more on their consoles than the main floor controllers, were equally confused. The situation was beyond their experience.

The original Apollo 13 crew—Alan Shepard, Edgar Mitchell and Stuart Roosa—crowded around the Capcom console, along with Ken Mattingly, Deke Slayton, and a host of others.

"Perhaps Stu, Al, and I prayed a little more fervently," said Mitchell, "as we knew how very easily it could have been us up there in a mortally wounded machine."

In the hour after the explosion, amidst incessant chatter on the loop and overlapping conversations, one voice was heard calm and steady—that of EECOM Sy Liebergot. As flight operations were transferred to the Black Team, his message to Clint Burton was blunt. They needed to shut down *Odyssey* and do it *now*.

"Listen," he told Burton, "we need to go through the whole fuel cell shutdown. Heaters, reactant valves, and the pumps. We're in danger of losing all three fuel cells…when we lose those cells we've shut the bird down."

With Lunney's team taking over, Kranz gathered his shaken flight controllers together in the back room.

"Okay, listen up," he said. "When you leave this room, you must leave believing that *the crew is coming home*. I don't give a damn about the odds, and I don't give a damn that we've never done anything like this before. Flight control will *never* lose an American in space. You've got to believe, your people have got to believe, that this crew is coming home. Now let's get going."

What the Apollo 13 crew needed was a life boat. Fortunately, their space craft was equipped with the ultimate redundant system, a second space craft—the lunar module. They had known they could use it as a lifeboat, but no one had ever done it. Even before Lunney's team radioed up the order, Fred Haise on board the stricken *Odyssey* had come to the same conclusion—they needed to power up *Aquarius*.

Lovell and Haise scrambled back through the tunnel into *Aquarius*. A module equipped for two men for two days would now have to sustain three men for four. Jack Swigert began furiously shutting down *Odyssey*. In fifteen minutes the command module would no longer support life. The lunar module would now have to do that. The problem was, the normal procedure for powering up the lunar module was complex and took almost two hours.

Liebergot remembered when the Apollo 10 training simulation had failed during a nearly identical simulation and neither the crew nor the flight controllers had been able to save the ship. The crew had died, albeit virtually. But he knew Jim Hannigan's Tiger Team had developed a lifeboat procedure for that failed simulation. Liebergot called his lunar module counterpart, Bob Heselmeyer, on the Bluebird loop. "Do you remember the lifeboat procedures that we developed on that sim?"

As the lifeboat procedure was plucked from the shelf, flight controllers began furiously paring down the lunar module start-up procedure. Jack Lousma lopped off entire blocks of nonessential tasks. Haise was already in the lunar module initiating emergency power-up and flipping switches.

Preserving *Odyssey's* batteries for reentry became a religion. John Aaron, the flight controller who had solved the Apollo 12 instrumentation dilemma, was called in to analyze the Apollo 13 power situation. What he saw was alarming. Amperage drains were bleeding precious power from virtually every circuit. Aaron's assessment was ruthless. He didn't want *Odyssey* powered down to only minimal systems. He wanted it powered down—as in *off*. No guidance, no heaters, no telemetry needed by ground controllers to diagnose the problem. *Nothing*.

Even at that, Aaron wasn't sure there was enough battery power left in the command module for reentry even if they made it back. So Bill Peters, the lunar module power expert, undertook the task of preserving enough battery power in *Aquarius* to charge the jumper cables needed to power *Odyssey* for reentry.

While Haise and Lovell powered up the lunar module systems, Swigert was powering down every conceivable circuit in the command module. Finally, he threw the last power switch and the command module went dark. He left, closing the hatch behind him.

With the lunar module's life support systems coming on line, the immediate threat of death was averted, and flight controllers now focused on getting the astronauts back to earth in one of two ways.

The first option was to turn around immediately. Engineers in the Trench calculated that if the crew used *Odyssey's* main engine and burned every last drop of fuel, they could turn the spaceship around and come directly back to earth. But this would require using the damaged service propulsion module, and there were serious doubts it would work.

The second was to let the moon's gravity pull Apollo 13 around the moon, accelerating the spacecraft and slinging it back to earth. But no one had ever anticipated lugging the lunar module along as part of the stack. And they certainly hadn't anticipated a failure of the service module. If the astronauts couldn't get home faster, they would run out of power and die. A NASA team calculated that the remaining life support for the astronauts and the electrical power necessary to operate the lunar module would last only twenty-four to thirty-Six hours—but Apollo 13 was eighty-seven hours from home. Something had to give. Either the spacecraft had to return to earth faster, or a means had to be found to extend the life support systems before the crew was dead. If they did nothing, the spacecraft would slingshot around the moon, miss the earth, and end up in a permanent orbit around the sun.

It was a critical call and Kranz' team was split down the middle. John Aaron was in the midst of the fray, trying feverishly to conserve power in *Odyssey* for the final entry phase, assuming flight controllers could make the life support consumables last that long. Engineers huddled in a back room calculating and commiserating. "It was easy to jump to the conclusion that maybe this was impossible," said Aaron.

Finally, Kranz made the decision: Apollo 13 was going around the moon. With this strategy, getting the crew back alive would depend on successfully circling the moon in a free-return trajectory, something they had given up. This meant they would need to pull off two key engine burns: the first to slingshot the spacecraft around the backside of the moon into a free-return trajectory for earth; the second, even more critical burn, to speed its return. The free return, however, would put the spaceship in a trajectory for splashdown in the Indian Ocean, and recovery forces had already been dispatched to the original splashdown site in the Pacific. But if they could pull off the first burn to get the spacecraft headed back to earth, they would worry about the splashdown later.

Such burns were normally done using the spacecraft's guidance system coordinates, but Aaron had earlier demanded that the guidance be turned off, along with everything else. Lunney and Kranz, however, had a proposition. They told Aaron that if he would leave the automatic guidance platform on for now, after the second burn, only fourteen hours away, the crew would be ordered to turn off everything in the lunar module and coast home in darkness and cold. "There was never any question in my mind," Lunney recalled. "I would have fought to the death to prevent anyone from trying to turn that guidance platform down."

With pressure from Kranz and Lunney, Aaron acquiesced. But even with the guidance system left on, the burn posed a problem, because the lunar module and command module were docked nose to nose. For navigation purposes, the center of gravity was drastically off. And the lunar module descent engine was not designed to do a burn while docked to the command module. Instead of navigating only the lunar module, the crew would be navigating an entire stack of three linked modules 60 feet long. Mission Control's IBM real-time computer complex had no program to compute such a burn, and they had only one hour to come up with an answer. A normal Apollo mission to the moon took roughly 250 billion computations. But this was far from normal. To spearhead the earth-return maneuvers for Apollo 13, they would need to work overtime. In theory, basic motions in space are simple, but only when an imaginary planet and an imaginary spacecraft are involved. In practice they are far from simple.

While IBM worked nonstop to compute the first burn, Ron Berry, NASA chief of Lunar Mission Analysis, worked feverishly developing a program of his own using off-line computers in an auxiliary computing room. The answers from the two computers were similar, though not the same. But they could now at least make a reasonable assumption about the burn equation. Ultimately, the Trench chose Berry's off-line formulae.

By this time, *Odyssey* had been completely shut down and ground controllers in Houston were faced with another formidable task. New procedures for start-up of the command module had to be written and tested in the simulator before being radioed up to the crew. A job that would normally take three months would need to be developed in three days. NAA engineers in Downey ran emergency procedure calculations through computers. A team of thirty engineers at the MIT instrumentation lab where the Apollo guidance system was designed worked through the night. Ten phone lines were kept open between Mission Control and the seventy lunar module experts at the Grumman facility in Beth Page.

Ken Mattingly worked furiously with John Aaron devising procedures to power up the command module and handing them off to astronauts for testing in the flight simulators. While Mattingly was busy firing up the command module simulator in Building 5, Edgar Mitchell was climbing into the lunar module simulator. Mattingly's job was to test procedures in the command module; Mitchell's was to test every feasible maneuver for executing a manual burn from the lunar module in case the automated guidance system failed. Their buddy, Haise, some 250,000 miles away on Apollo 13, was prepared to do anything to save the mission.

Haise, Mattingly, and Mitchell had wanted to fly a mission together, and they were about to get their chance under the most formidable of circumstances.

Tuesday morning, April 14: Approaching the moon
Before firing the rocket for the first burn to set the course on a free-return trajectory for earth, the spacecraft had to be pointed in exactly the right direction. To do this, the crew first needed to align the guidance system using the proper coordinates. With instructions from Mission Control, Charlie Duke on Capcom told Lovell to align the cross-hairs of a sextant in the window so they were just grazing the horns of the crescent earth.

Lovell attempted to sight the unwieldy spacecraft.

"When I pitched out, it went into some wild gyration because the center of gravity had moved way out," he said later. "It was doing something you had never seen before, something you had to correlate in your brain… I had to essentially learn how to fly all over again."

Even as Lovell struggled to aim the ship, Mission Control radioed instructions for the burn, the first of its kind ever attempted in space.

"Okay, *Aquarius*, are you ready to copy?"

"That's affirm," Lovell replied.

"Okay, here we go…"

But Lovell had serious concerns. Ground controllers wanted to begin the burn in thirty-seven minutes; and at the rate they were going, their course would take them within sixty miles of the lunar surface. If they made a mistake or burned in the wrong direction, they could plow into the moon. Lovell asked for and received an extra hour to enter the data and prepare the ship.

"Don't worry," said Houston. "The ocean will be there."

Tuesday morning, April 14: First burn to restore free-return trajectory
At 2:43 Tuesday morning, nearly six hours after the accident, Apollo 13 took the first step on its long homeward journey.

Lovell was now ready.

"Five seconds before the burn there's a little indication on the computer that says, 'Do you really want to make this burn?' The computer is asking if we know what we're doing. I hit the proceed button to say, 'Okay, computer, you've got it.' " The engine turned on. The seconds ticked down to zero and the engine turned off.

The burn was perfect.

Though still moving away from earth, Apollo 13 was now on a course for home. But life support resources were dwindling, and at present speed the astronauts would run out of power and water before they made it back. On top of this was the cold. With little power in the lunar module and no heaters to maintain cabin temperatures, the temperature in *Aquarius* was plunging to near freezing. Of the three, Swigert was the most worried. He stood in the window watching the earth recede behind them, wondering if he would ever return.

Mission Control now prepared for the second, bigger burn after the swing around the moon. This burn would shave twenty-four hours off the trip and move the landing location back to the Pacific Ocean where recovery forces had been deployed. But flight controllers worried that even this might not be quick enough.

In Building 5, Mitchell and Mattingly worked tirelessly in the simulators, going through power-up options, trying new things, failing, and starting again. Other astronauts were called in to test the procedures. Meanwhile, Ron Berry and the Trench computed trajectory run after trajectory run, slightly changing the parameters each time.

"We must have made close to 1000 computer runs manually in a couple of hours," said Berry. "Most of the trajectory team folks went the first thirty-six hours without sleeping. We were starting to get a little blurry in our thinking. I did fall back on my faith at that time, searching for strength and the ability to think clearly. I saw other people in the control center doing the same thing."

Tuesday at 7:21 p.m., the spacecraft slipped behind the moon and lost contact with earth. Haise and Swigert, who had never been so close to the moon, snapped photos like a couple of tourists. For Jim Lovell, one glimpse of the moon was especially memorable: "As we whipped around, one of the last things that I saw sitting there was this little triangular mountain, Mount Marilyn."

Joe Kerwin succeeded Jack Lousma at 7:30 a.m. and a weary Jim Lovell checked in: "Joe, I'm afraid this is going to be the last moon mission for a long time."

It was a message Mission Control did not want to hear—certainly not on an open mike.

∽ʌ ʌ∽

Back at the White House, President Nixon's speechwriters hurriedly drafted a public statement titled: *"Apollo 13 Contingency Statement in Case of Failure."* The President would make the loss of three brave astronauts as palatable as possible to the public:

> In this tragic moment, the nation's first thoughts are with the Apollo 13 astronauts and their families. All of us must hope that the memory of the astronauts' own courage, so abundantly displayed, will now help sustain their loved ones. These men were three of America's best. They dared greatly. They died bravely. The world will long remember the searing human drama of Apollo 13 – and also the calm, the self-control, the quiet heroism, displayed by the men aboard it...

As news of the Apollo 13 explosion circled the globe, people everywhere stopped to consider the fate of three men drifting slowing away from earth, perhaps never to return. From his red phone in the press room, Reverend Stout relayed news of Apollo 13's crisis. Around the world, members of the APL responded with prayer.

"Never in the history of the world have so many people prayed so fervently at the same time," said Reverend Stout. "During the flight, the news was live by radio and television to more than 80,000,000 people overseas. Our words were translated into several foreign languages and rebroadcast through the *Voice of America* and several foreign commercial networks. It was a trying time for all of us."

Tuesday, April 14: Second burn to speed return to earth
Because the lunar module had no computing capability or means of automated transmission, Houston made the calculations for the second burn and radioed detailed instructions back to the crew, who copied them exactly before losing radio contact behind the moon. At 7:40 p.m., twenty-two hours after the accident, Apollo 13, wounded and battered, emerged from around the moon

headed for the long journey home. The second bigger burn to speed its return was just an hour away.

At 8:40 a.m. the crew executed a powerful four-and-a-half-minute burn and ticked off the countdown to engine start. The second burn, executed perfectly, was the last to be accomplished with the aid of the computer guidance system. In accordance with Kranz's and Lunney's agreement, the lunar module was then powered down. The only thing left on was the radio and a light bulb. The lunar module was now running on less power than a vacuum cleaner, the command module had dropped below freezing, and the only heaters on board *Aquarius* were three bodies.

In the Lovell house in Seabrook, Texas, NASA officials, friends, astronauts, and wives filled the living room. In search of a quiet spot, Marilyn went into the bedroom, then the bathroom, where she locked the door and fell on her knees in prayer. Her daughter remembers her mother's tribulation.

"She was very nervous, very upset, probably never did stop smoking, didn't sleep…she didn't want to tell me what was going on, she didn't want to address it, because she didn't want to believe it herself."

Reverend Stout and Thomas Paine arrived to console Marilyn. But to Stout's surprise, Marilyn turned to console him.

"Somehow," he recalled, "she sensed our concern about her husband and tried to put us at ease by saying, 'It's too bad that Jim was not able to leave the Bible on the moon.' At the time, the Bible seemed the most unimportant thing in the world to me. The most important thing in the world was prayer. And the whole world prayed as they approached earth."

An estimated one-third of the world was now following the fate of Apollo 13. It would take the skill of NASA engineers, the fortitude of three astronauts, and the grace of God to bring the astronauts home alive.

> "The miracle of the Apollo 13 story was not in one person rising to the occasion, but in hundreds and thousands working at peak performance to do the impossible."
>
> *- Ken Mattingly, Command Module Pilot, Apollo 16*

28
SPLASHDOWN: THE GRACE OF GOD

When Kranz returned, Mattingly, Aaron, and the group were gathered in a room near Mission Control. The controllers were subdued but shaken. They had contained the crisis, but the crew was still in extreme danger. Kranz feared they would lose their nerve. No one would ever forget what happened next.

"It was a question of convincing the people that we were smart enough, sharp enough, fast enough—that as a team we could take an impossible situation and recover from it," said Kranz. His message to the group was simple: "This crew is coming home. You have to believe it. Your people have to believe it. And we must make it happen."

Mission Control calculated it would take Apollo 13 sixty-five hours to fly the 240,000 miles between the moon and the earth if the astronauts followed strict guidelines. But tracking data indicated they were significantly off course. There was only a two-degree margin for error in the narrow entry corridor above the earth's horizon, and the Trench had only hours to calculate the course correction that would get them into the corridor.

With trepidation, the biggest midcourse correction ever attempted by a returning Apollo lunar mission was scheduled. With the guidance computer shut down, the astronauts' skills would be dramatically tested. If the correction was too shallow, they would skip across the atmosphere and back into space, like a pebble skimming across a lake. Too steep and they would burn up on reentry.

"We didn't want it orbiting the sun," said Jerry Bostick. "But we sure didn't want it landing in Texas."

Wednesday night, April 15: Mid-course correction
The mid-course correction was dicey. The crew would need to keep the earth centered in the lunar module window—visually. To compound the risk, they couldn't afford to power up the guidance platform again and use the remaining battery power. The spacecraft alignment would normally have been done with star sightings, but debris from the exploded tank surrounded the spacecraft, shining so brightly the crew couldn't see the stars. In ten minutes following the explosion it had expanded to thirty miles in diameter.

Flight controllers instructed the crew to sight using the cusp of the sun's disc, which would be accurate within two degrees. The crew was allotted fourteen seconds from the moment of burn ignition to sight and lock in. The spacecraft would need to be pointed in the right direction by using a portion of the surface of the earth as a reference. The manual burn would take all three men. Lovell laid out the plan. Over on the side were two buttons. One said "Start," the other "Stop." These buttons would directly connect the battery to the lunar module descent engine to power the burn.

"Jack, you're going to tell us exactly when to start the burn," Lovell said, "and you'll time it with your wristwatch. Fred, you have a backup attitude controller, I have a primary. I know that when that engine goes on, I will never be able to keep the earth in the window by myself, because there are three attitude controllers with pitch and roll and yaw. You take your attitude controller and keep the earth from going back and forth too much. I'll take my attitude controller and keep the earth from going up and down too much."

With no seats in the lunar module, the three stood like captains at the helm of a ship. At 10:31 p.m. they were ready.

Swigert said "Start." Lovell hit the start button, and the engine kicked in. The earth was in their field of view.

"Okay," said Lovell, "We got it. I think we got it."

"Yes" said Haise. "It's coming back in... Just a second."

The spacecraft had shifted out of position.

"Yes, yaw's coming back in. Just about in."

"Yaw is in..."

At that point Lovell struggled to hold the spacecraft steady. It kept slipping out of alignment, and he and Haise were juggling the earth up and down and sideways.

"What have you got?"

"Upper right corner of the sun ..."

"We've got it!"

Fourteen seconds after they started, Swigert barked, "Stop."

Lovell hit the stop button. And they waited. Back at Mission Control, flight controllers broke into cheers and pounded their fists. The crew had a good alignment. There was a chance they would bring them home alive.

But their troubles weren't over. Earlier on Tuesday, Mission Control recognized another problem. The crew was creating deadly carbon dioxide in the two-man *Aquarius*, just by breathing. Normally, oxygen-scrubbing canisters were used to cleanse the air, but they were running out of canisters and the square backups from the command module were useless since they didn't fit the round openings in the lunar module. The crew was slowly suffocating.

Ed Smylie, an engineer in the crew systems division, had been at home watching television when he learned of the explosion. Within minutes he was at the space center, where he and other engineers found a solution to retrofit the square canisters. A makeshift canister would be made of three feet of duct tape, a plastic bag, one of the astronaut's plastic checklist cards, and a sock. After constructing and testing the make-shift device themselves, they radioed instructions. It took the Apollo 13 crew a tense two hours to find and assemble the pieces. The final product resembled a rural mailbox. But it worked. "One thing a Southern boy will never say," Smylie said, "is 'I don't think duct tape will fix it.'"

For the next twenty-four hours the crew endured more cold and discomfort. Temperatures in *Aquarius* dropped to 38 degrees Fahrenheit. The cabin walls were perspiring and the windows were partly frosted over. Lovell and Haise pulled on three sets of underwear and their lunar boots; Swigert stood in his flight boots in wet socks. The shortage of water had led the men to drink less and less, to the point of dehydration. To add to the unsavory cabin conditions, they were no longer allowed to dispense urine overboard, since the small "rockets" it created might alter their trajectory. Fred Haise developed a bladder infection.

"Finally it dawned on me how sick he was," said Lovell. "And I ended up putting my body around him—I wrapped it around him to keep him warm, because he was visibly shaking with the chills."

Earlier, Haise had been rummaging through one of the storage areas and stumbled onto a little package from his wife—pictures meant to be found and opened when he was on the lunar surface. "So there was the feeling that I hope someday I'd get to see them again," he said.

Back on earth, plans of a political nature were underway. President Nixon's assistant, Dwight Chapin, shared with the President's chief of staff a discussion he had with Frank Borman "concerning a contingency plan for Apollo 13 in case of disaster." Should the crew of Apollo 13 perish in space, President Nixon would "proceed to the homes of Lovell and then Haise to pay his personal sympathy and that of the country to the families, and then to Swigert's parents." After this he would head directly to the airport and fly back to Washington. "In terms of presenting the Medal of Freedom posthumously to the Astronauts," the memorandum read, "it is felt that this should be done at a later time in the Oval Office."

Even the President was hedging his bets. And for good reason. Another problem was brewing in the Pacific. Tropical storm Helen was strengthening near the Apollo 13 splashdown area.

Thursday, April 16: The world is watching
By now, millions of people around the world were following the drama on radio and television in public squares, private homes, schools, offices, and factories. Pope Paul VI announced to an audience of ten thousand at St. Peter's Basilica, "We cannot forget at this moment the lot of the astronauts of Apollo 13. We hope that at last their lives can be saved."

From Jerusalem's Wailing Wall to the floor of the Chicago Board of Trade, people prayed. Newspapers published pictures of the world with hands reaching out for the astronauts or clasped in prayer.

Reverend Stout had watched everything leading up to this moment.

> Fragile man has broken the barriers of gravity and for the first time another celestial body other than this earth has been touched. On our TV screens, we were with our present day pilgrims as man first left his footprints upon the surface of the moon. We were with them again when the Apollo 12 crew romped around like happy children, leaping in and out of craters. But more than ever, we and the entire world were with them on our knees as the Apollo 13 crew guided their crippled craft back home.

On Tuesday, April 14, Congress adopted a resolution urging all businesses and communications media to pause news coverage at 9 p.m. local time, to "permit persons to join in prayer for the safety of the astronauts." An internal memo was routed to White House Chief of Staff H. R. Haldeman outlining the political sensitivities:

> We have received a number of telephone calls from Members of Congress urging the President to go on television this evening

to urge all Americans to pray for the safe return of the Apollo 13 astronauts... Congressman Jerry Ford has pointed to the Congressional Resolution just passed by both Houses calling on all Americans to pray for the astronauts at 9:00 p.m. this evening. Jerry feels quite keenly that it will be most desirable for the President to take to the air to plead for divine intercession.

Church and prayer groups around the world responded, ballgames were interrupted, and people around the world laid down their tools for minutes of prayer. On Thursday, work at the spacecraft center virtually came to a stop for three minutes as directed by the President. At Wrigley Field, Milton Berle, scheduled to make a seventh inning appearance, halted the Cubs game "for a moment of prayer for the crewmen of the Apollo 13."

Then, as the world watched and waited, the odds for the recovery began to rise.

Now, with millions of people watching, praying, and hoping, NASA personnel found inexplicable answers coming out of nowhere.

"We seemed to be running with some intuitive link that surprised our team members," Kranz said. "Through some miracle, some burst of intuition, something we had all seen, heard, or felt kicked in... To this day, I still can't explain it."

Flight controller Bob Legler found a way to transfer electrical power from the lunar module batteries back through existing wiring to top off the command module batteries needed for reentry. "I wasn't one who was praying in those days," said Charlie Duke. "But I found out later that many people prayed for the safety of the crew and for wisdom in these decisions we had to make."

Mattingly and the recovery team debated about powering up the command module for reentry. They knew the Internal Measurement Unit (IMU), a device at the heart of the navigation system, was very delicate. To get the precision needed, the units were allowed to operate only at a restricted temperature of 60 degrees, plus or minus one degree. No one knew what the temperature in the command module was, but they knew it was below freezing—a long way from 60. There were concerns as to whether it would work at all. Phone calls flew between developers at the MIT Instrumentation Lab and Houston. Finally, word came from an MIT engineer.

"We didn't do any testing at those temperatures, but we have reason to believe it's okay."

"Are you really sure?" Houston asked.

"Yes, we have some unusual data that confirms that. An employee left an IMU in the back of his station wagon in a parking lot overnight during a snowstorm. It got down to about 30 degrees. The next day, he took it inside and hooked it up and it worked just fine."

By now, Mattingly had worked incessantly for three days on command module power-up procedures and had recorded his results on a roll of 16-inch-wide computer printout paper that was now 20 feet long—*handwritten*.

"We weren't too swift by that time," he said, "so we called in other astronauts familiar with the vehicle to get into the simulator and redline the checklist to catch things that were either ambiguous or wrong. Things we might have overlooked." Tom Stafford and others joined the effort.

The final simulation for the power-up would take four more hours.

Friday morning, April 16, command module power-up
At 4:00 a.m., Apollo 13 was less than nine hours from reentry, hurtling toward earth at 20,000 miles per hour.

In Mission Control, Mattingly had just run the entire checklist through the command module integrator and knew it better than anyone else. So in violation of usual procedure, he took a seat at the CapCom console. Onboard Apollo 13, Swigert took his seat in the command module.

"Let me take it from the top," Mattingly began. But Swigert was fatigued and had nothing on which to write the procedures.

"They had no idea," Mattingly said. "They didn't even get a preview that said, 'Hey, we're going to give you the next twelve feet of paper, so be prepared to write it down.' All they had was, 'Here comes a lot and you can't afford to make a mistake. Now don't mess it up.'"

The crew ripped covers and backs from checklists—every scrap of paper they could find.

Mattingly began to read slowly, line by line. Swigert copied every word, often interrupting Mattingly to ask for a repeat. The spacecraft was rolling unpredictably and the antennas were losing contact. It took Swigert two hours—or 40,000 miles—to write down the instructions.

At 5:00 the next morning, the crew would leave the lunar module and return to the command module. Assuming the frozen electronics in the command module came to life, they would jettison both the lunar and service modules and the gradual force of earth's gravity would begin to pull them home. The carrier U.S.S. *Iwo Jima* was already en route for recovery in the South Pacific.

Sometime during the morning, Mattingly walked out of the Mission

Control building for some fresh air. The parking lot was full. He looked up at the buildings, and in each office he saw a light and the shadow of a head.

In the White House, Nixon's National Security Advisor Henry Kissinger and Secretary of State William Rogers attempted to keep the President informed through updates from astronauts Mike Collins and Bill Anders. On April 17, Kissinger and Rogers exchanged updates in a phone conversation at the White House.

Kissinger: I had a briefing at 3:00 from Anders and the big problem will come tomorrow. They will power up the systems tomorrow. They don't know how they will react after being cool for so long.

Rogers: I had lunch with Michael Collins today. He thinks if they get into this corridor—which is a little like hitting the edge of a piece of paper from 30 feet—once you are in the corridor you are good about it.

Kissinger: That's what they tell us.

Rogers: He thinks the oxygen and electricity are okay.

Kissinger: Yes, but since they have never flown in this configuration before and never made these adjustments before, there are questions. They don't think the heat shield is bothered by the explosion. I thought it was like the government after things are in bad shape, then you still have to go 70,000 miles before you can turn around. Let me leave you on that…

Friday afternoon, April 17, Powering up the command module
At six hours before reentry, Swigert crawled through the tunnel from the lunar module to the command module, took his seat, and started throwing switches. Even now, no one knew whether the command module would start.

At this point, Swigert was having trouble reading the instructions. "Would you give me that one again?" he asked CapCom Joe Kerwin. "I just can't read my own handwriting."

On the ground, Mattingly put his finger on the proper place on the checklist to help Kerwin, who read Swigert the correct instruction. "We knew early in the game that the powered-down levels would push survival of the crew and survival of the systems," Mattingly said. "There was a good chance that when we brought up the command and service module it would be nonfunctional…that was the end of the road."

As Swigert turned on the command module's telemetry and the electronic display, Liebergot saw his EECOM console flicker on. The spacecraft was alive.

But there was no time to celebrate. First the crew had to get rid of the damaged service module. Lovell and Haise were still in the lunar module. The switch for jettisoning the lunar module was right next to the switch for jettisoning the service module, and Lovell was worried that his rookie command module pilot might mistakenly jettison the lunar module and send him and Haise into the nether sphere.

"I'm in the lunar module with a camera," said Lovell. "And I tell Jack, not once, not twice, but three times, 'Jack, we're jettisoning the *service module*, not the lunar module. *Don't jettison the lunar module.* Put some tape over that switch.'" Swigert attached a piece of paper with the word "NO" in big red letters over the lunar module jettison switch. Haise, still afraid Swigert might press the wrong switch, floated into the command module to assist.

When the service module was cut loose, the crew would get their first and only look at the damage. Swigert hit the switch to jettison the service module and Lovell backed the remaining stack away. Then, camera in hand, he maneuvered to get a shot of the service module as it drifted away.

"One whole side of the panel is blown off," he reported to Houston.

Now Mission Control wondered if Apollo 13's heat shield, situated just above the explosion, was intact. Any penetration of the heat shield might allow a leak and eventually a burn-through to the command module during reentry.

With reentry just fifteen minutes away, Lovell and Haise took their seats in *Odyssey* and waited for the order from Mission Control to jettison *Aquarius*. Kerwin gave the instruction and Swigert threw the LM-SEP switch for lunar module jettison.

"Farewell, Aquarius, and we thank you," said Kerwin.

"She sure was a good ship…" Haise said softly.

"I was sad to see it gone," he said later. "If I could have brought it back I'd like to have put it up in my back yard."

Friday afternoon, April 17: Odyssey heading for splashdown
The command module was now on its own. But to everyone's dismay, it was drifting off course. Mission control ordered one more mid-course correction

Splashdown: The Grace of God

55,000 miles from earth. It went smoothly, but even so, the spacecraft was still venting something that Houston feared had caused it to drift. Controllers could only hope it would hold its course.

In a poignant moment, the astronauts' last words before reentry were not for themselves, but for Mission Control.

"I know all of us here want to thank all of you guys down there for the very fine job you did," Swigert radioed.

"I second that," said Lovell.

"I'll tell you, we all had a good time doing it …" replied Joe Kerwin.

"I wish I could go to the…party tonight," said Jack Swigert half sincerely.

"We'll cover for you guys," said Kerwin. "And if Jack's got any phone numbers he wants us to call, why, pass them down."

It was a moment of levity before the interminable silence that followed.

Houston: "Thirty seconds to go for blackout…"

Swigert: "Thank you."

Kerwin added one more message for the crew: "Okay, *Odyssey* …just for your information…battery C will deplete around main chute time."

A few seconds later, *Odyssey* vanished in a sea of radio static. Kranz and the controllers stood riveted to their consoles, watching the main display on the front wall in Mission Control. *Odyssey* was about to splash down, for good or ill, within sight of the live TV camera onboard the aircraft carrier U.S.S. *Iwo Jima*.

As if a final nod from heaven, Tropical Storm Helen had moved out of the area and was no longer a threat.

Apollo 13 entered earth's atmosphere traveling 25,000 mph. Nineteen seconds later, the craft was enveloped in a 5,000 degree fireball. The command module's heat shield, which had been protected by the jettisoned service module, was now exposed. The frozen module immediately thawed and condensation inside the capsule began to rain on the astronauts. Blackout would last three minutes. The spacecraft was wobbling but headed for splashdown.

The silence was deathly in Mission Control. Flight controllers stood frozen, staring at the clock on the wall.

A BBC newscaster stood by: "And now coming is the moment—the last moments of Apollo 13 as it comes in—as it begins its reentry. The best thing we can do now is just listen and hope."

An American broadcaster: "We'll only know whether that heat shield was damaged by the explosion three days ago when they come out of radio blackout."

Three minutes passed.

Then four minutes…five…six…

Hearts sank in Mission Control.

"I felt all the adrenaline drain out of my body," said Bostick. "It was the worst moment of my entire life. After we had solved all the insurmountable problems, we had brought them back to earth in water, in the Pacific Ocean where the recovery forces were, and we had squeezed consumables and figured out how to power up the command module—and it worked—and now they weren't coming out of blackout."

Suddenly, a rescue helicopter over the Pacific caught a glimpse of two small parachute drogues deploying on tethers. From the deck of aircraft carrier U.S.S. *Iwo Jima*, cameras began rolling.

"Iwo Jima Control…I have a visual bearing 182."

"Apollo 13, Apollo 13, this is Recovery. Over…"

"Apollo 13, Apollo 13, this is Recovery. Over…"

"Apollo 13 …"

The BBC newscaster exploded in triumphant joy, *"There they are! There they are! They made it! All three chutes out!"*

The blackened spacecraft broke through the clouds to splashdown, its three main chutes deployed in a brilliant flourish of orange and white. Less than fifteen minutes later, *Odyssey* hit the water, blunt end down, within eyesight of the U.S.S. *Iwo Jima*—one of the closest splashdowns of any spacecraft.

At the sight of the chutes, Mission Control remained strangely quiet. As was the tradition, everyone waited until the crew appeared safely on the deck of the aircraft carrier, at which time wild cheering broke out, congratulatory cigars appeared, and the world map lit up at the front of the room.

Pete and Jane Conrad were at the Lovell house witnessing the drama on television.

"I still feel emotions in my body that will probably never leave me," Marilyn Lovell said. "Because I didn't know for four days if I was a wife or a widow."

When the phone rang a short time later, Marilyn didn't hear it. Pete Conrad had just popped the cork on a bottle of bubbly and she was still wiping tears from her eyes.

"Marilyn," a friend interrupted her, "President Nixon is on the line."

"Who?"

"It's President Nixon."

Several seconds went by as she registered the interruption.

"Marilyn, this is the President," said the voice on the end of the phone. "I wanted to know if you'd care to accompany me to Hawaii to pick up your husband."

"Mr. President," she said, "I'd love to."

Shortly after the crew's arrival, the President of the United States summarized the reaction of many: "The three astronauts did not reach the moon, but they reached the hearts of millions of people in America and in the world." Then, with the astronauts at his side, he produced the first of the two speeches he had prepared:

> "The men of Apollo 13, by their poise and skill under the most intense kind of pressure, epitomize the character that accepts danger and surmounts it. Theirs is the spirit that built America…"

The mission had weighed heavily on the President's mind. He knew what it meant to his constituency—the American people. "I had a man at the White House at a dinner who came up to me—I know he hasn't been in church for years," said Nixon, "and he said, 'I never prayed so hard in my life as I prayed these last three or four days.'"

Two days after splashdown, Thomas Paine reiterated President Nixon's strong support for a "vigorous" manned flight space program before a large crowd of employees at the Manned Spacecraft Center.

"We will press on," Paine quoted the President as saying, "in the exploration of the moon and will eventually explore the planets."

Nixon mentioned the reality that there were those who would "seize upon this accident as an opportunity to call for a slowdown or a turning back" of the Apollo program. Paine reassured NASA personnel "we are not the kind of people whose purpose is diverted by adversity or setback …we will not falter in our resolve." However, the launch of the next mission would be moved back pending "a thorough review." He then concluded, "We in NASA heartily agree that [the crew of Apollo 13] have earned all the vacation time we can give them."

For most in Mission Control, the ordeal had come to a victorious end. For Sy Liebergot, however, the nightmares began almost immediately. He began reliving the oxygen tank explosion over and over, berating himself for not recognizing the Oxygen Tank 2 indications sooner, for not averting the disaster. Finally, after weeks of mental turmoil, he came to realize that no matter what he might have done, no matter how quickly or how shrewdly he could have reacted, the outcome would have been the same. His life lessons had come full circle.

"The zenith of the human experience," he said, "is when a man takes tragedy and is able to convert it into something positive and fulfilling."

In the aftermath of the crisis, Edgar Mitchell had come to a realization of his own. At the podium before a crowded auditorium in Orlando, Florida, he concluded a description of the ordeal with an emotional tribute to the skill and ingenuity of NASA. Then, in a surprising moment of reflection, he acknowledged that a palpable underlying momentum seemed to carry the rescue forward. In Mitchell's mind, there was more to the recovery of Apollo 13 than human intervention. The power of millions of praying minds singularly focused on Apollo 13 had somehow affected the outcome. And he was coming to believe that such conscious power applied to more than Apollo 13.

Reverend Stout was sure he understood the source of that palpable force. "Of all that I have ever done, in or out of the space program," he said, "that night was the most crucial, the most important, and the most meaningful for me, because that was when I learned about the assurance of answered prayer."

An article appearing in the *Christian Science Monitor* on April 20 underscored his belief:

> Never in recorded history has a journey of such peril been watched and waited-out by almost the entire human race... There will be those who ascribe the rescue wholly to human skills and courage. Others will see at work, affirmed in countless prayers, a power and providence which perceives "the path which no fowl knoweth."

Madalyn O'Hair, however, had another perspective. When Stout returned to his office in the Manned Spacecraft Center, he received a call. "You know why Apollo 13 exploded, don't you?" O'Hair jibed. "Because it was carrying all of those Bibles!"

Stout was now more determined than ever:

> God somehow seemed to have used this flight to bring the world down on its knees from out of the clouds of war. He humbled the

technological giants of our time. Did prayer really have anything to do with the flight? When one of the flight controllers was told that a certain Sunday school class had taught that it wasn't God, but those guys in Mission Control who pulled it through, this flight controller laughed, because he too was praying for guidance. Yes, the Apollo 13 Mission was an unqualified success—that is the Apollo 13 *Prayer Mission*.

After his arrival in Houston for a press conference, Jim Lovell turned to Stout. "I am sorry I couldn't fulfill our mission," Lovell said. The Apollo 13 lunar Bibles nevertheless became symbols of worldwide faith and answered prayer. Subsequently, a number of the Apollo 13 lunar Bibles were given away, one forwarded to George H.W. Bush, who later presented it to the Smithsonian Institution. Some were autographed by the crew as mementos and several were returned to Reverend Stout, who deposited them in the Apollo Prayer League archives.

As a result of the near tragedy of Apollo 13, the APL expanded its prayer request hotline, using the *Voice of America* radio network to promote membership in foreign countries. The League was now playing a pivotal role for NASA in interfacing with the press and religious dignitaries in matters "of a religious nature."

Apollo 13 was barely dry from splashdown when Stout turned his attention to Apollo 14. Alan Shepard, Edgar Mitchell, and Stuart Roosa would be assigned the same lunar landing site and mission objectives as Apollo 13; and Stout was intent on seeing Apollo 14 assume another one of Apollo 13's roles. On May 8, 1970, he sought Alan Shepard's cooperation in getting the microfilm Bibles on board Apollo 14. The request ultimately made its way to Edgar Mitchell, who agreed to log the lunar Bibles in his PPK for stowage on board Apollo 14.

But Mitchell had definite ideas of his own. In a preflight meeting with Stout, he questioned the use of the American emblem affixed to the cover of the Apollo 13 Bible to be left on the moon, as this might indicate the achievements of the Apollo program were intended solely for the benefit of the United States. Stout agreed with Mitchell's wisdom and removed the American emblem.

The Word of God was for people of all nations.

"Looking back on these times, I see how naïve I was. For several years I would continue to underestimate the power of belief in our lives because of the pervasiveness of my classical scientific training."

--Edgar Mitchell, Lunar Module Pilot, Apollo 14

29
APOLLO 14: RESTORING FAITH

After Apollo 13, the government, already shaky about the cost and risk of the Apollo program, had strong trepidations about Apollo 14.

"If Apollo 14 doesn't go well, we may not have a future at all," said Apollo launch director, Walter Kapryan. "Apollo 14 has got to be a perfect mission." A January 1971 headline in *Time* magazine stated it bluntly: "Future of Space Program: A Lot Depends on Apollo 14 Trio." Success or failure, it said, could set the course of future space exploration.

By now, President Nixon's political advisers were pressuring him to end the moon program. America had beaten the Soviets to the moon, collected 128 pounds of rock samples, and every astronaut who launched had come back alive. Quit while we're ahead, they urged him. "I went against every piece of advice I got and decided to carry on with Apollo," Nixon confided in Apollo 14's Stuart Roosa.

On the heels of the Apollo 13 near disaster, Nixon was grappling with a national dilemma of his own making. On April 30, 1970, he announced on national television that U.S. troops would be invading Cambodia, the country bordering Vietnam through which the North Vietnamese were supplying the Viet Cong troops in the South. The response to Nixon's announcement was one of shock and anger. Protests erupted across the country, including one at Kent State University where, on May 4, the Ohio National Guard opened fire on unarmed college students, killing four and wounding nine others. Hundreds of universities, colleges, and high schools across the United States closed due to a strike by four million students. The event further divided the country in an already socially contentious time.

In the spring of 1970, the Apollo program was yet one more slice of stew in the social boiling pot.

~*~

Given the near disaster of Apollo 13, Reverend Stout decided to split a separate group of 300 lunar Bibles into two packets: 100 to be carried on the Apollo 14 lunar module and 200 on the command module, in case one or the other didn't make it. There would again be a separate first lunar Bible, but this time that particular Bible would take on a unique symbolism.

During a preflight meeting with Stout, Edgar Mitchell had requested that the first lunar Bible be redesigned in such a way that it could not be easily copied. After some consideration, Stout approached the APL committee with the idea of a single, unique "multi-focal" microfilm Bible designed and dedicated to the memory of Ed White. This "First Lunar Bible" would consist of two microfilm versions of the Bible separated by a thin piece of glass—one, the same edition of the King James Bible used by NCR; the other, the Revised Standard Version (RSV) used by Ed White. The Bibles would be developed by APL members themselves. To create the two microfilms, they would first need to manually cut and paste together a master version of each Bible and then photograph and reduce each image onto a Biswanger glass slide for conversion to positive emulsion microfilm. Ed's widow, Pat, furnished Stout a copy of her husband's Bible.

With approval from NASA engineering director Maxime Faget, the way was cleared to use the skill of NASA employees working on their own time. The after-hours labor was long and tedious, inspired by *Hour of Power* sermons broadcast from the Crystal Cathedral by evangelist Robert H. Schuller. "He was an inspiration during the final weeks to get the Bibles ready for the flight," Stout said. "We had him playing in the background while everyone was in the room putting the Bible together."

As work progressed, the finished product began to take shape in the form of two master boards measuring nearly eleven feet square. Capturing them on film became yet another challenge, since the photographic lab at the Manned Spacecraft Center was located directly under the flight path of Ellington Air Force Base and the floor vibrated intensely each time a plane flew over. Eventually, several images were captured and reduced to microfilm. Amazingly, the resulting tiny multi-focal Bible tablets was crystal clear when viewed through a microscope.

This particular Bible, they knew, would be extremely difficult to copy.

As 1970 drew to a close, Stout developed yet another notion. The lunar

Bibles would be accompanied by another microfilm listing the names of persons and groups involved in the effort to land the Bible on the moon. To create this "First Lunar Bible Honor Roll," APL members were asked to list up to ten names of individuals they felt had made a contribution to their life or to a worthy cause. The names of astronaut Ed White and Stout's fallen brother, Joe F. Stout, headed the list of honorees. In addition to the names of family and friends, other names submitted included Jesus, Socrates, Billy Graham, Presidents Franklin R. Roosevelt, George Washington, Abraham Lincoln, Harry Truman, and John F. Kennedy, along with Albert Einstein, Martin Luther King, and George H.W. Bush. Someone submitted the name of Alabama Governor George Wallace, which was also added to the list.

Heading Stout's own list was the name of Texas A&M football coach Homer Norton along with one of the toughest teams to have ever played the game of football, the squad of 1939 to 1941. Stout played football for Norton and was a sophomore end in the squad of Norton's last Cotton Bowl team.

Since the honor roll would accompany the First Lunar Bible, Stout asked Reverend James McCord, president of Princeton Theological Seminary, to submit an early Christian symbol to appear at the top of the honor roll. McCord was active in the World Council of Churches (WCC) and obtained permission to use the organization's symbol depicting a ship with a mast in the form of a cross, representing the gospel call for followers to become "fishers of men." Since WCC was a non-denominational organization with members from countries across all seven seas, McCord felt the symbol would be appropriate for all religions.

The symbol was routed through channels from McCord to Senator Bob Dole, who then forwarded it to APL member Norman Durst in Washington for transfer to Reverend Stout, who inserted it in the upper-left corner of the First Lunar Bible Honor Roll. In the upper right corner, Stout inserted a modern Christian symbol designed by APL member Judy Bohac, depicting a planet orbiting an open Bible surrounded by the phrase: "God's Word for a New Age." The final list of 3,569 names was pasted together and sent to a printer in Pasadena, Texas, for reduction to microfilm.

Stout then issued a special request to pastors in the Aerospace Ministries:

> To help commemorate this special event, we are asking each church to personally involve itself and its members on Sunday, January 31, which is the day of the launch. We are suggesting, first of all, that special prayers be made during your regular worship service that day.

As the launch date neared, Stout and a small group of APL members packaged the lunar Bibles for the flight. To ensure all Bibles destined for the lunar surface were accounted for, Stout carefully inserted a tissue index marker between each group of twenty. Inside this packet he inserted the special First Lunar Bible wrapped separately so it could easily be retrieved by Mitchell if he chose to leave it on the moon. With this Stout tucked the First Lunar Bible Honor Roll and a copy of the certificate flown on Apollo 13 containing the translations from Genesis. When complete, the lunar module packet of 101 Bibles measured a quarter of an inch thick, and the command module packet of 200 Bibles, less than half an inch thick. The two packets were then wrapped in flight-safety beta cloth in accordance with NASA regulations and handed over to Mitchell for stowing.

Prior to launch, Mitchell dutifully logged the lunar Bible packets in PPK 1098 for the command module and PPK 1099 for the lunar module, marked simply "Large microfilm package" and "Small microfilm package" with no mention of their biblical content.

"I am without words to express our personal gratitude for your taking the First Lunar Bible," Stout wrote to Mitchell on January 5, less than a month before launch.

As the hopes of the Prayer League turned to Apollo 14, Stout urged members not to lose focus:

> For many of us here in Houston, we are very concerned about this specific flight. It comes at a time of general letdown in the space program which approaches questionable safety for the flight. We want you to be concerned also—and your church—and your friends. We want to find everyone who has that power of prayer to pray with us. Why should we wait until something goes wrong, as it did in the flight of Apollo 13, to pray?

Just as Apollo 7 had been called upon to restore faith in the Apollo program after the fire, Apollo 14 was charged with restoring faith for a nation shaken by the drama of Apollo 13. The upcoming mission attracted more widespread attention than previous missions, not only because of its predecessor's near disaster, but because after ten years its popular commander, Alan Shepard, was making a comeback. At age forty-seven, Shepard was the oldest astronaut in the program. Adding to this, Shepard's "crew of rookies" had never so much as orbited the earth.

"We got all kinds of flak from the guys," he said. "In the first place, I hadn't flown anything since 1961, and here it was ten years later, and the two guys with me had not flown before at all. So they called us the three rookies. We had to put up with that. And the fact that everybody said, 'That old man shouldn't be up there on the moon.' "

The lack of experience didn't go unnoticed by the media. Headlines in *The Albuquerque News* announced: "The Brain and The Hillbilly Ride Apollo." Mitchell, known as one of the "brainiest" of the spacemen, was regarded as an intellectual as well as an expert flier. Roosa was a red-haired down-home boy known to love country music.

"We were about as far apart as three guys could be," Mitchell said. "If you considered personalities in a 360 degree circle, the three of us were 120 degrees apart. But we were a compatible professional team when it came to flying a spacecraft. That's what made it work."

Apollo 14 Commander Alan B. Shepard, Jr. had the kind of storybook childhood that could have been illustrated by Norman Rockwell. His rural schoolhouse in East Derry, New Hampshire had one teacher, six grades, and one room. Alan's teacher was a knuckle-rapping disciplinarian who taught him to study and stick to it. He took to that lesson so well that he finished all six grades in five years.

Alan's boyhood hero was Charles Lindbergh—the first to fly across the Atlantic—and Shepard was one of those kids who hung around airports when aviation was just getting established in the thirties. After school and on weekends, he'd hightail it to the landing field to help push airplanes in and out of hangars. Sometimes the pilots would let him sit at the controls and work the stick.

Perhaps it was the excitement of those early days of airplanes, new frontiers, and brave pilots. Perhaps it was Lindbergh's triumphant landing that indelibly marked Shepard with the idea that the impossible is not impossible for someone who gets out there and does it.

To outsiders, there were two Alan Shepards: one "the utterly and, if necessary, icily correct career Navy officer," the other, "Smilin' Al of the Cape." "Smilin' Al" was a fun-loving guy with a fast Corvette. But at home under the spell of his wife, Louise, Shepard was transformed from "funny-caustic" to "funny-gracious." Louise Shepard was a serene, almost beatific woman who told her husband, "There is no evil in the world." Alan and Louise had met at a drinking fountain at the Morey Field house at Principia College in St. Louis over a Christmas break where she was a student and he was visiting his sister, Polly. They married in 1945. It was one of the few marriages in the astronaut corps that would last.

Those who worked with him knew that Shepard's lunar module pilot, Edgar Mitchell, was "schooled up" a bit more tightly than the rest. He didn't fall neatly into the good-guy character, nor did he emit the swagger and bravado of an astronaut. He was a unique mix of testosterone and brilliance, and while he learned the highly complex procedures of spaceflight quicker than most, by his own admission he had an impatient streak. An engineer recalled that a moon surface experiment once failed to deploy properly while Mitchell was practicing for the Apollo 14 moon walks. "He grabbed the experiment and shook it until finally it broke." In this regard he was uniquely compatible with his icy commander.

"Alan didn't really care whether he was a nice guy or not," Mitchell said. "He was what he was." There was obvious respect between the two.

Mitchell not only had first-hand experience developing the lunar module at the Grumman laboratory, he was an experienced Korean War veteran with expertise in nighttime rough-weather landings on aircraft carriers. And he was cool under pressure. Once when returning to Grumman in a T-38 with Fred Haise, the plane's engines flamed out in a thunderstorm and the aircraft started dropping like a manhole cover.

"It was hard to get an air-start on those engines above 20,000 feet," Mitchell said. "And this was up around 45,000. If you didn't get the nose down and keep your speed up, you're going into a flat spin." He nose-dived below 20,000 feet and after several failed attempts, finally managed to air-start the engines. "That's where you learn your cool," Mitchell said. "That's where you really learn to deal with fear. You've got to stay functional or you're dead—your crew's dead."

Mitchell was an intellectual as well as a scientist and, as such, enjoyed delving into fields far removed from aviation or space. He was particularly fascinated by psychic experience and extrasensory perception. He was a religious man, but not in the conventional sense. His father was a cattleman during the Depression years, a man of nature with a spirituality akin to that of Native Americans. Growing up in New Mexico, Edgar inherited his father's natural love of creation and eventually drifted away from his strict Baptist upbringing by his mother. By the time he reached NASA, his view of life had grown vastly beyond the dogma of traditional religion.

Mitchell's journey from the dusty town of Roswell to NASA had been intensely focused. Like other astronauts, his trek through multiple colleges and military bases dictated dragging a wife and two children across the country more than once, and his home life showed signs of strain. Long absences combined with his obsession with the space program were taking a toll on his marriage, a scene being played out in many astronauts' homes.

―∿∿―

Through 1970 and early 1971, the Apollo 14 launch date underwent repeated delays. Two modifications were made to the service module as a result of the Apollo 13 accident. A third oxygen tank was added at a location isolated from the other two, and a spare battery was added to support the electrical load of the spacecraft at any point if the fuel cells should fail. NASA wasn't taking any chances. A plethora of smaller changes were made as well, including blank paper added to the back of the flight book. The crew trained for nineteen months, longer than any crew before. "We were the only crew that was six months out from a launch four times," said Roosa.

The delays were accompanied by a continued deterioration of White House interest in the space program. Investigation of the Apollo 13 accident yielded a reincarnation of many of the problems uncovered during the investigation of the Apollo 1 fire. On June 15, an internal memo to Peter Flannigan, Assistant to President Nixon, recapped results of the Apollo 13 Review Board, noting: "The findings of the Review Board and its recommendations clearly indicated that NASA failed to take effective remedial action in the years following the Apollo 1 [204] fire."

On September 2, 1970, NASA announced it was canceling the planned Apollo 18 and 19 missions. Apollo 18's command module and lunar module would be used on Apollo 17.

As the remaining missions dwindled, the Apollo 14 launch date drew near.

> "Maybe if people had a chance to see this, they wouldn't be so interested in their own particular territories. Perhaps we could put the Security Council on the space station and let them try to see where their little bailiwick is."
>
> *- Alan B. Shepard Jr., Commander, Apollo 14*

30
LUNAR LANDFALL

The command module was named *Kittyhawk*, in honor of the Wright Brothers who flew the first powered flight at Kitty Hawk, North Carolina. The lunar module was named *Antares* after the star in the constellation *Scorpio* which would be used to guide Shepard and Mitchell in their descent to the lunar surface. According to Mitchell, *Kittyhawk* was intentionally spelled as one word, although many in NASA didn't seem to notice.

Through endless hours in the lunar module simulator and ground training exercises, Mitchell transferred his depth of experience to the newly invigorated Alan Shepard, until Shepard's seat in the lunar lander became as familiar to him as that of his own Corvette. "Okay, boss," Mitchell finally declared, "I think you're ready now."

Launch day brought the usual pre-launch gift exchange with Pad Leader Guenter Wendt, who felt Shepard warranted an age-appropriate gift. "I came up with a unique piece of 'lunar support equipment' for the old geezer—a walking cane." But Shepard upstaged him, producing a World War II vintage German helmet stenciled "Col. Guenter Klink" after the bumbling character in the TV sitcom *Hogan's Heroes*. Wendt loved it. He then escorted the crew to the capsule, wished them luck, and locked the hatch, waving farewell through the capsule window.

On January 31, 1971, at 4:03 p.m. EST, nearly ten years after Shepard's first space flight, Apollo 14 blasted off carrying a world-famous commander, two rookie pilots, and a tiny microfilm packet destined to impact biblical history. As Apollo 14 lifted off, Roosa yelled, "Go, baby go!"

The Apollo 14 backup crew of Gene Cernan, Ron Evans, and Joe Engle referred to the prime crew as "the old man, the fat man, and the cute little redhead." Shepard was destined to be the oldest moonwalker, and Roosa had red hair, which left lunar module pilot Mitchell with the moniker "fat man."

As a practical joke, the backup crew had a special patch designed using the Coyote and Roadrunner cartoon characters. The Coyote represented the Apollo 14 prime crew. It had a beard for the "old man", a fat belly for the "fat man", and it was colored red for the "cute little red head." The Roadrunner, embroidered with "Beep Beep," represented the backup crew. The patch depicted the Roadrunner, or backup crew, beating the primary crew, Coyote, to the moon. The backup crew secretly stowed the patches in nearly every storage place in the Apollo 14 command module. The patches, floating in micro gravity, would greet the Apollo 14 crew in orbit.

As Apollo 14 took the customary one-and-a-half orbits of the earth, the backup crew's joke became painfully obvious. An hour and a half into the flight, Houston heard repeated references to the patches.

"There's more patches," said Shepard.

"OK, I got it," replied Mitchell.

"I'll get this one out of the way," Shepard continued. "I'm over here ... beep, beep, beep, all over the place."

Roosa chimed in, "He who laughs last."

Shepard was having none of it. "Well, I'm a hell of a lot happier that we're flying and looking at their patches, rather than the other way around."

Later came more indications of patches. "Beep, beep, my ass," Shepard mumbled. Flight controllers understood.

Mitchell, alleged as the "fat man," demonstrated on a television broadcast that he moved easily from the command module through the narrow tunnel to the lunar module. Fred Haise on CapCom quipped: "You can get through there quite easily, Ed. I guess zero G's does help."

"If you're commenting on what I think you're commenting on, it is entirely uncalled for," replied Mitchell.

From its beginning, Shepard's hard-earned moon mission seemed as luckless as Apollo 13's. A series of malfunctions began to unfold, each with enough potential to cancel the lunar landing. The first problem appeared only three hours after launch, when the astronauts turned their command ship around for docking. On previous flights, the maneuver had been routine. The mother ship would edge forward from the stack, turn around, and hook into the lunar lander nose-to-nose. This time, however, the twelve small latches on the docking mechanism—shorter than a cigarette and not much wider—failed to lock into position.

Mitchell was mid-sentence to Houston when the problem arose. It took him more than four hours to finish the sentence. He was describing the lunar lander as the command module moved toward it to dock. "We can see all of the orange yellow thermal protective around the lunar lander... And I can look across Stuart..." But he didn't finish. The modules refused to dock.

Roosa had made it a personal goal during training to set the record for minimum use of fuel during the initial docking maneuver, but now that he was actually doing it, it wasn't working. The second attempt to dock also failed. If the command service module couldn't dock with the lunar module, the lunar landing would be scrubbed. Engineers on the ground began pouring over the docking system design. Three more docking attempts were made over the next sixty minutes. They also failed. The amount of fuel consumed during the attempts increasingly frustrated Roosa.

An alternative docking procedure was radioed to Roosa by Mission Control: They would skip the soft dock procedure and instead hit the hard dock switch in conjunction with ramming the docking probe into the drogue on the lunar module. At the moment of impact, Shepard would immediately flip a switch to retract the docking probe out of the way, allowing the final latches to engage. Following instructions, Roosa backed off again and brought the lunar module forward with maximum thrust. At the instant of contact, Shepard hit "RETRACT."

"It's not working," he reported hastily. Then, four seconds later, "Okay, I got barber pole!" and they heard the wonderful ripple-fire of the hard dock latches. While it was music to the astronauts' ears, the unsolved problem was worrisome to Mission Control. It was imperative that the apparatus work perfectly when Shepard and Mitchell returned from the lunar surface and attempted to dock with *Kittyhawk* for the ride home.

Four and a half hours after he halted mid-sentence, Mitchell nonchalantly resumed his transmission with Mission Control, "I was going to tell you when we got busy there a little while ago, if you look out our window past Stuart, we can see the earth receding..."

On February 4, after three days in trans-lunar orbit, Shepard and Mitchell climbed into *Antares*. Roosa deposited them in a ten-mile orbit for descent to the moon, then returned through the vacuum to a sixty-mile orbit for his work of mapping the barren world.

But as *Antares* passed into the lunar night, barely ninety minutes from the beginning of their descent to the moon, more trouble arose. The computer guidance software was receiving an errant interrupt, producing a signal that would automatically abort the landing once it began. Moreover, they would soon be passing behind the moon into a communications blackout, cutting off any instructions from Mission Control. The problem was thought to be a ball of solder lodged in the abort switch, forcing it on. By tapping the panel with a pen, Mitchell was able to loosen it, but this was only a temporary fix. Nobody underestimated the drama posed by the Apollo 14 abort switch problem. If they couldn't come up with a workaround, a premature abort would send *Antares* hurtling back into lunar orbit and put an abrupt end to the lunar landing. Once again, ground engineers had to carefully, accurately, and swiftly find a way to bypass the problem. But of the ninety minutes available to solve the problem, *Antares* would be behind the moon and out of radio contact for sixty of those precious minutes.

In Houston, it was two o'clock in the morning as controllers worked to isolate the problem. Meanwhile, John Stout, whose primary position in NASA was to support the missions' information systems, knew who might solve the problem: a young engineer named Don Eyles at the MIT Instrumentation Laboratory. Eyles had written the program code for the "abort operation" command, and Stout had met him during one of the engineer's orientation trips to the Manned Spacecraft Center.

"When the problem occurred, I knew he was the one who could solve it," said Stout. "But he had already gone back to MIT. My hunch was that he would be watching the flight from the lab. My hunch was right."

Stout made a phone call and Eyles was put in touch with Houston. The solution Eyles came up with was to modify the abort switch software code to ignore the signal. However, the program code for this was buried in a ten-inch-thick computer listing in a large binder. Eyles grabbed the binder and headed to the lab's "man-to-loop" guidance simulator on the second floor. The program workaround, he knew, would simply change a few registers, first to fool the abort monitor into thinking that an abort was already in progress, and then reset it for a normal descent.

But before he could fully test the workaround, an over-anxious coworker relayed a set of untested code to Houston. Astronauts scrambled into the lunar module simulator and began entering the code. To their consternation, the simulator "crashed." Fortunately, the erroneous code was quickly detected and Eyles transmitted the corrected code, which was entered in the simulator and executed perfectly. Haise at the CapCom console was then handed a printout of the code changes to radio up to the crew. Since the lunar module had its own stand-alone computer and no means of data communications, the only way for Haise to communicate the code changes was verbally—and he would have only minutes to do this when *Antares* emerged from around the moon.

Back at the Manned Spacecraft Center, Stout wrapped up his work late and headed to the Mitchell residence to watch the final moments of the landing. A small crowd had arrived and photographers were lined up outside the house. Inside, he took a place on the floor in front of the TV set so as not to block Louise Mitchell's view from the sofa. Gene Cernan arrived shortly after and took a seat on the floor next to him.

By the time *Antares* emerged from behind the moon, Eyles' abort code changes were ready. With only fifteen minutes left before the landing would have to be canceled, Haise began radioing up the changes. Mitchell was ready. The crew had expected a few code revisions to get around the problem, but what they got was much more. The procedure was complex. Shepard and Mitchell now had only a scant ten minutes to complete the pre-descent checklist, enter navigation updates, and manually program the remedy into the computers. Because of fuel constraints, time was not on their side; and they simply couldn't afford another trip around the moon.

Shepard had counted on Mitchell's expertise in operating the lunar module, but Mitchell was now busy entering abort code changes into the guidance system. "I was a little naïve at that point about the landing radar," Shepard admitted. "But we didn't have much of a choice. It was either try that or give it up." The procedure required that Mitchell enter 61 keystrokes at precise intervals during the descent. Haise began the read-up:

> After the NOUN 62 countdown starts, but before ignition, key in
> VERB 21 NOUN 1 ENTER 1010 ENTER 107 ENTER
> Exactly 26 seconds after ignition, manually advance
> the descent engine throttle to 100% ...

The procedures dragged on ad infinitum and included a notation: "Note that the numbers are in 'octal,' based on the root of 8, not 10, so that the number '107' actually means '71.'"

The program changes, flawlessly executed by Mitchell, would cause *Antares* to ignore the false signal and bypass an automatic abort when the engine ignited, now only moments away. But Mitchell knew there was a penalty in bypassing the abort code. In order to continue the mission, they were surrendering the help of the emergency automatic abort function that could provide a swift ascent from the moon at the push of a single button.

In the final minutes before touchdown, Mitchell's fears almost materialized when a third, more serious, problem developed. The lunar landing radar failed to initiate. Due to an anomaly in the sequence of events, the computer now couldn't recognize and lock onto the landing radar signals bouncing off the lunar surface. They were essentially flying blind. Not only could they not check *Antares'* descent altitude with the landing radar, they could not simply look out the window, as they were quite literally on their backs, the window displaying nothing but a striking pattern of stars. What made this particularly unsettling was that Fra Mauro was one of the most rugged areas of the moon, filled with hills, valleys, and craters. Even if the landing was perfect, they had lost the benefit of the computer abort system in case of trouble. More immediately, mission rules forbade them to descend below 10,000 feet without the radar—a height they would reach within the next minute.

Shepard broke the silence with the countdown: "10 seconds to go …"

Apollo 14 was now in landing sequence.

Mitchell implored, *"Come on radar; that's a lock-on…radar.* [Houston,] have anything to get the radar in?"

Mission Control broke in and instructed Shepard to recycle the circuit breaker powering the landing radar.

"Okay, it's cycled."

Mitchell pleaded, *"Come on!"*

The radar locked in just in time.

Reaching across to the camera and flicking it on, Mitchell took a brief look out the window to his right. As they slowed descent and pitched forward, the moon's rocky horizon rose up.

"Okay, there's pitchover," said Mitchell, as *Antares* began moving upright into landing position.

In Houston, Haise and flight controllers held their breath.

"There's Cone Crater," Shepard called. "Right on the money!"

Mitchell and Shepard knew exactly where they were. *Antares* was heading dead on target for Fra Mauro, the original landing site for the failed Apollo 13 mission—and now theirs.

"That's it, Al. Right on the money," said Mitchell.

Shepard liked what he saw. "Okay. Fat as a goose! Beautiful!"

Fred Haise finally gave the green light.

"Okay *Antares*…Go for landing."

The lunar module began its 3000 foot descent to the surface of the moon, dropping at 75 feet per second. With the landing spot still sitting pretty in the window, Mitchell watched the descent progress.

"2000 feet, 48 feet per second," he navigated his commander. "Coming down a little fast."

With Shepard's hand gripping the stick and the moon's surface nearing, Mitchell's vast knowledge of the spacecraft, procedures, and checklist of requirements for a safe landing were now coming in very handy.

"Still a little fast, not bad," said Mitchell. "Going by Cone Crater right outside to my right."

Haise cut in, "Fuel is running low."

Mitchell's slow, steady readout continued as the lunar module descended.

"... still at 170 feet. Six percent fuel, okay. One hundred fifty feet." Mitchell knew they were there. "You're looking great," he told his commander. "You're on your own."

Shepard: "Starting down, starting down."

Haise: "Sixty seconds of fuel left."

Shepard: "Okay, Freddo, we're in good shape."

Mitchell: "Forty feet...3 feet per second...30 feet...20...10."

The lunar module shuddered and came to rest. Mitchell turned to his commander and flashed a smile, his first of the entire journey.

"Contact, Al... We're on the surface."

On February 5, 1971, Alan Bartlett Shepard, Jr. became the fifth man to walk on the moon. On board *Antares,* at the behest of his wife, Louise, he carried a 35mm canister containing a microfilm of the first edition of The Christian Science Monitor dated November 15, 1908, along with 100 microfilm strips of the front page of the same issue.

His first words after setting foot on the moon's soil—"It's been a long way, but we're here"—seemed a tribute to his own persistence and sharp elbows. But a few minutes later, neither the Icy Commander nor Smilin' Al were anywhere to be found. Standing on the battered Fra Mauro highland, Shepard looked back at the earth—and seeing no evil in the world, he wept.

Ten minutes later Mitchell followed Shepard down the ladder. At forty years of age, he had left the dusty roads of Roswell, the aircraft carriers of the Pacific, and the academic institutions of America behind. He had left his wife, his children, and his parents behind. He had left his buddies, Mattingly and Haise behind. And he had landed on the moon.

In the modest house in Nassau Bay, Texas, there was a loud roar. Reverend Stout lifted his six-foot-three-inch frame from the Mitchells' living room floor. It had been a long way for him, too.

"There was a feeling of relief, perhaps more so for me," he said, "for I knew the chance that was being taken." While others were still celebrating, Reverend Stout shook hands with Louise Mitchell and excused himself.

"I have to get back to the office," he said. "There's work to be done."

While the rest of the world might have viewed the Apollo 14 touchdown as simply another Apollo moon landing, there were others who understood a different significance. On February 5, 1971, the Apollo Prayer League records made note that the history of the Bible took a detour to a place it had never been before—to the farthest reaches of the moon.

"All of my adult life had been as an engineer, a scientist, a test pilot, a naval officer, working daily with mathematics, exotic equipment, the hard facts of the technical professions, and the realities of war. But nothing prepared me for what was about to happen."

- Edgar Mitchell, Lunar Module Pilot, Apollo 14

31
FULL CIRCLE

Orbiting overhead in *Kittyhawk*, isolated from the drama, Stuart Roosa became the fourth human to solo around the moon. To some people, the loneliness of this phase might seem frightening. But Roosa enjoyed flying by himself. Before the flight, a reporter asked how he felt about the isolation. "As long as I've been flying, I always thought that more than one seat was too many in an airplane," he replied.

In the weeks before launch, he had gathered 500 seeds of numerous trees—sycamores, Douglas firs, loblolly pines, sweet gums, and redwoods—which were now stored in his PPK, a tribute to his early days as a smokejumper with the U.S. Forest Service. "Moon trees" from those seeds would grace lawns and parks from the White House to Japan in years to come.

Roosa had been the Capcom during the Apollo 1 fire that killed Grissom, White, and Chaffee, and the sound of his fellow astronauts' cries still echoed in his mind. To this day, he was unable to talk about it. Now, in the solace of lunar orbit, familiar songs drifted through the command module. A disk jockey in Houston, a friend of Roosa's, had arranged for the astronauts to take along a cassette tape of music by their favorite artists. As Roosa spotted a crescent moon filling one of *Kittyhawk's* windows, Sonny James' version of *How Great Thou Art*, one of his favorite hymns since childhood, began to play:

When I in awesome wonder
Consider all the works Thy hands have made
I see the stars, I hear the mighty thunder

The Apostles of Apollo

> *Thy power throughout the universe displayed*
> *Then sings my soul, my Savior God to Thee,*
> *How great Thou art, how great Thou art.*

"I heard people say that Stuart had to go to the moon to find religion," his wife, Joan, said. "It was ridiculous. Stuart was religious when I met him. And he took his faith seriously." Joan, a devout Catholic, was teaching school in Virginia when the two met at the Langley Air Force Base Officer's Club in Hampton, Virginia. Roosa was a test pilot at Langley and, like Mitchell, was raised a Baptist. He would fly in the mornings and pick up Joan when she got off work. One day she asked, "You picking me up today?"

"No," said Stu. His answer surprised her since he usually took advantage of every chance to see her. Then he added, "Joan, I've started taking lessons in Catholicism. I've decided I want to be a Catholic. I'm not doing this just to please you. I feel I was born to be a Catholic." Stu became a Catholic and in 1957 the two were married in a Catholic Church.

Seven years later, he was flying at Edwards Air Force Base in California when NASA put out a call for more astronauts. Joan mentioned it to him, but he balked. He already had the job he always wanted, he told her. But she persisted and finally Stuart conceded. As thousands applied, NASA kept cutting. Finally, a pilot friend of Roosa's, Joe Engle, got a letter stating that he was still in the running. But Stu didn't get one.

"I guess that's that," he told Joan.

"No, I don't think so," she said and handed him his letter. It had come to the house instead of his office. The letter turned out to be his ticket to a ride around the moon, complete with his favorite background music.

For the first time on the moon, the spacesuit for the commander had red stripes on the helmet, elbows, and knees. These stripes made it easier to determine which astronaut was which in photographs and on television.

"Yes, it's a beautiful day here at Fra Mauro Base. Not a cloud in the sky," said Shepard, as he and Mitchell stepped out across the lunar surface. "Beautiful day for a game of golf," Mitchell answered—perhaps an indication of things to come.

Apollo 14 would be the first and only lunar mission to use a Mobile Equipment Transporter (MET), referred to by Shepard as the "lunar rickshaw." The hand-pulled cart with two small wheels was intended to aid the crew with carrying their equipment on the lunar surface. One of the goals was to haul the MET to the rim of Cone Crater and sample the rocks there. It was believed that the oldest rocks thrown out from the bottom of the crater would be near the rim. But the journey to the rim of Cone Crater was three miles round-trip and more difficult than expected. The terrain was steep and littered with boulders, causing the wheels of the MET to jump almost a foot off the ground. Pulling the MET up the steep undulating grade was laborious and exhausting. Finally, Shepard and Mitchell took to walking single file so they could recover any equipment that might bounce off the cart. At one point Mitchell radioed Houston that he and Shepard had picked up the MET and were now carrying it.

The two were pushing themselves to the limits of exertion. They were breathing heavily and had to stop and rest from time to time, and they continued to be stymied in identifying their exact location. Finally Mitchell relayed to Haise, "It's going to take longer than we expected. Our positions are all in doubt now, Freddo."

The Fra Mauro highlands had also been Fred Haise's target area on his ill-fated Apollo 13 flight, and from his seat in Mission Control he was doing the best he could to guide their traverse. But the landmarks so obvious on the lunar surface map were difficult to distinguish when viewed from the moon itself. Large craters that would stand out on a reasonably flat plane were hidden behind other craters and ridges—or 100 meters below them.

"You'd say, well, this next big crater ought to be a couple of hundred meters away. And it just wasn't anywhere in sight," said Mitchell. "So you press on to another ridge…and all you would see would be another ridge."

He and Shepard continued, but Cone Crater eluded them. Shepard doubted the wisdom of continuing and proposed stopping the attempt to reach the rim to allow more time for geological sampling. Mitchell protested, "Oh, let's give it a whirl. We can't stop without looking into Cone Crater."

The discussion continued between the two until Haise interrupted: "Okay, Al and Ed. In view of your location and how long it is going to take to get to Cone, the word from the back room is that they'd like you to consider where you are to the edge of Cone Crater."

The implication was clear. If they didn't know where they were, they needed to turn around and go back. The message pushed an already exhausted Mitchell to utter a sarcasm he knew his friend Haise would decipher quickly. "I think you're all *finks*," he snapped.

Before the flight, Shepard and Mitchell had made a bet with Deke Slayton that they would reach Cone Crater with the MET in tow. But Mitchell now suggested to Shepard that they leave the MET and continue up without it. "We could make it a lot faster," he said. Shepard disagreed. He didn't think the MET was slowing them down.

Fred Haise cut in: "Al and Ed, Deke says he'll cover your bet if you drop the MET." But Shepard rejected it flatly. It was just a question of time, he felt.

Time, however, was their enemy. Mission planners had underestimated the drag the MET would place on the timeline. Routine tasks were taking longer than expected and cutting into the time allotted to collect and tag rock samples. Moreover, most of the interesting rocks were too large for the sample bags. There were so many big rocks, in fact, that they got very few of the sampled rocks into the allotted bags. Shepard knew the geologists would not be happy with the results. "Well, you can put a little rock in a big bag," observed Mitchell, "but it's sure hard to put a big rock in a little bag."

Mission Control had given them a 30-minute extension, but that extension was now eaten up, and they didn't have enough time to document all the rock samples. Mission Control urged them to just pick up random samples on the way back to the lunar lander and hold to the mission schedule.

After returning to the lunar module, the astronauts pointed the TV camera back towards *Antares* so Mission Control could watch the close-out activities. At this point, Shepard had a surprise up his sleeve—or quite literally—in his pocket. Standing in full view of the TV camera, he reached into his pocket and produced the head of a specially crafted Spalding six-iron golf club, which he attached to a lunar module scoop. The entire collapsible implement weighed a little over sixteen ounces. Moving stiffly in his puffy space suit, he produced the first of two golf balls.

"In my left hand, I have a little white pellet that's familiar to millions of Americans," he said.

When the ball plopped to rest at his feet, Shepard tried to assume a conventional golf stance, only to find that the girth of his space suit made it impossible for him to grip his makeshift six-iron with both gloves at the same time. He had no choice but to attempt man's first extraterrestrial golf shot one-handed.

"I'm going to try a little sand trap shot here," Shepard said, then jerked his club back and followed through.

The ball popped almost straight up in a cloudy divot of moon dust and seemed to hang in mid-flight as if suspended on a string. Then it tailed off to the right and fell back to the lunar surface less than a hundred yards away.

"That looked like a slice to me, Al," jibbed Haise, watching the broadcast on a monitor.

Shepard then whacked the second ball in a high arc as far as it would go. "Miles and miles and miles," he said.

Not to miss out on the fun, Mitchell attempted a javelin throw using a staff left sticking in the ground nearby from their solar wind experiment. His javelin throw exceeded Shepard's second golf ball by roughly six inches, thus making Mitchell the winner of the first lunar sports spectacular.

As *Antares* ascended from the surface, Shepard reported, "Okay, Houston. Crew of *Antares* is leaving Fra Mauro base."

"Roger, Al," said Haise, "You and Ed did a great job. Don't think I could have done any better myself."

As *Antares* approached rendezvous with *Kittyhawk*, the docking problems on the outbound journey raised renewed concerns in Mission Control. This time Mission Control instructed Roosa to forget the earlier hard-dock procedure and use the standard docking procedure. It worked, all docking latches tripping closed in cadence.

Up to this point, the astronauts had been busy, minute by minute. But now, in the quiet of the command module, drifting silently in lunar orbit and rolling slowly in "barbeque mode" to maintain thermal balance, Edgar Mitchell found himself gazing out the window at the celestial panorama unfolding before him—the earth, the moon, the Milky Way, the sun—appearing in sequence as if controlled and orchestrated by divine authority.

As he hurtled earthward, he became engulfed by a profound sensation, a sense of universal connectedness. He intuitively sensed that his presence and the small blue planet in the window were all part of a harmonious, universal process—and that the glittering cosmos itself was in some way conscious.

"This new feeling was illusory, its full meaning somehow obscured; but its silent authority shook me to the very core," said Mitchell. "Here was something potent, something that could alter the course of a life."

Mitchell's upbringing had ingrained him with theories about religion and the Bible that eventually became too dogmatic for a prying scientific mind. From his perspective, the Bible had been used more as a basis for war than peace, and one he could no longer accept at face value. But his journey to the moon had molded his lack of belief into something new. "This is a sight that bores deep into the soul and shakes the foundation of your being."

Mitchell had an awakening—a feeling that there was a greater being, a guiding hand, and that science and spirituality were not two different dimensions of reality. They were, in fact, one and the same. This transforming experience sent him on a search for what he now saw as a more likely truth: the presence of a cosmic conscience and order in the universe. This journey toward the intersection of science and spirituality would consume the rest of his life.

Three days after their return from Cone Crater, Apollo 14 entered earth orbit and descended in a fevered rush through layers of life-giving atmosphere. On February 7, 1971, one of the most critical lunar landing missions came full circle to a smooth splashdown in the Pacific, accomplishing everything Apollo 13 was unable to do and more. After splashdown, the crew was debriefed and placed in three-week quarantine, the last crew to endure the post-flight ordeal, as the moon was deemed to be void of any life, including extraterrestrial viruses.

It was during quarantine that an experiment conducted by Mitchell during the flight came to light in the press. In the months prior to Apollo 14, while on vacation in the Bahamas, Mitchell met two physicians, Edward Maxey and Edward Boyle, who shared his curiosity about paranormal studies. It occurred to them that an unusual opportunity for a telepathy study was presenting itself.

In the past, they knew that parapsychologists had discovered that the time and distance between two participants in a telepathy experiment had no effect on the results. The transmission and reception of telepathic communications appeared to be instantaneous. A renowned psychologist at Duke University, Dr. J. B. Rhine, had conducted experiments in the field for thirty years with positive results.

Now, the "three Eds," as they called themselves, saw an opportunity to test the results across the furthest possible distance ever traveled—more than 200,000 miles between participants. But because this would be a personal, ad hoc experiment, they would have to keep their plans confidential, because the press corps that blanketed the Cape could blow the experiment out of proportion.

During the Apollo 14 flight, before Mitchell retired to his hammock each "night," he pulled out a table of random numbers and five distinct symbols arranged on a clipboard. He would concentrate on a symbol four times for fifteen seconds. Meanwhile, through tens of thousands of miles of empty space, his collaborators in Florida would attempt to jot down the symbols in the same sequence that Mitchell had randomly arranged on the clipboard. Mitchell repeated the process twenty times on the journey to and from the moon.

Even before splashdown, the story of the experiment was leaked to a reporter and was all over the press. One morning during the first week of quarantine, Shepard came across an article in a stack of newspapers entitled, "Astronaut Does ESP Experiment on Moon Flight." He chuckled as he read, certain it was an absurdity dreamed up by some creative reporter. He leaned over the table to tell Edgar about it. After an awkward silence, Mitchell told him that he had in fact done it. Shepard looked up, nonplused. But a moment later, Mitchell thought he saw a glint of mirth in Shepard's eye as he turned his attention to his breakfast plate. The subject was never brought up again.

At his first opportunity after quarantine, Mitchell collected the results of his private experiment and submitted them to Dr. Rhine. At the Duke laboratory, Rhine's analysis pointed toward a positive result. Seeking verification, the participants sought a second analysis by Dr. Karlis Osis in New York, a well-known researcher in the field. Dr. Osis reported the statistics were such that there was only a one-in-three-thousand probability that the results were purely chance. The experiment indicated an underlying force at play. The results reinforced Mitchell's scientific interest in the power of the conscious mind.

In spite of the entire crew's superb performance during the mission, Alan Shepard had become so notorious for his previous flight on *Freedom 7* that few could recall the name of his fellow moonwalker. "It was always 'Alan Shepard and that other guy,' " Mitchell said. To remedy this, after quarantine Mitchell started growing a beard to distinguish himself. The result was only marginally successful, he said. "It then became 'Alan Shepard and the guy with a beard.' "

Given the overriding pressure of mission-critical tasks, Mitchell exercised his option not to leave the First Lunar Bible on the moon, and after his release from quarantine, he retrieved his PPK and headed for the Manned Spacecraft Center to meet an anxiously awaiting Reverend Stout. As Mitchell handed Stout the packets containing the lunar Bibles, a photographer captured a beaming Reverend Stout—and "the guy with a beard."

Upon hearing of the event at the Northwestern University Traffic Institute in Evanston, Illinois, Reverend Stout's brother, James, a Sergeant in the Ft. Worth Police Department, arranged for a press release and announced to law enforcement attendees from across the country that several of their names had just landed on the moon. West Virginia State Trooper W. J. Shaw, Lt. Bernard J. Roberts of Madison, Wisconsin, and Sgt. Benny R. Dickison of Rockford, Illinois, were among the many listed on the First Lunar Bible Honor Roll.

The news release hit the airways instantly. An American astronaut had landed a Bible on the moon. Newspapers and media across the country picked up the story. From Hollywood, a journalist reported:

> I was intrigued by the report in the *Los Angeles Times*, March 28,

1971, that Edgar Mitchell had carried the all-time best seller, a Bible, to the moon. Congratulations to all concerned on pulling this off. When zippy-do authors can publicly off-load bilge to the effect that astronauts should be given the razzle-dazzle treatment of playboy athletes, news of this nature is like a breath of fresh air.

More news organizations were contacted and a media bonfire ensued amid a whirlwind of inaccurate and sometimes fictitious reports that Mitchell had carried the Bibles in the pocket of his spacesuit or had left one on the moon, which he had not. The stories were as varied as they were incredible. Stout was overwhelmed and Mitchell was miffed. The stories would generate misconceptions and queries from the public for years to come.

Shortly after their release from quarantine, Alan Shepard and Edgar Mitchell were welcomed in Washington, D.C. by a delighted President Nixon to receive the Distinguished Service Medal. Nixon joked with Shepard, placed the prestigious medal around his neck, and then turned to Mitchell—at which point the President fumbled with the second medal and dropped it flat on the floor. The President's smile quickly turned to a wince as he covered his eyes in embarrassment. The fumble was captured on camera and the following day's newspapers displayed a photo sequence showing an obviously amused Mitchell taking the presidential gaffe in stride.

Apollo 14 would be the last spaceflight for Shepard and the only one for Mitchell and Roosa. The extraordinary skill of the crew saved the program from another failure and, in essence, redeemed the Apollo program for three more missions. Shepard and Mitchell retrieved 94 pounds of lunar soil and rocks, and Roosa's photography helped the selection of future landing sites that geologists were eager to explore.

Later analysis proved Mitchell had been right about the location of Cone Crater. After comparing data on their lunar trek against Fra Mauro photo maps, NASA scientists estimated the astronauts had stopped within 65 feet of its rim. As one of the participating geologists later put it to Shepard, "You weren't lost, and you didn't know it."

"And a great and strong wind rent the mountains, and brake in pieces the rocks before the Lord; but the Lord was not in the wind: and after the wind an earthquake; but the Lord was not in the earthquake: and after the earthquake a fire; but the Lord was not in the fire: and after the fire a still small voice."

- 1 Kings 19:11, 12

32
A STILL SMALL VOICE

"Apollo is a pagan god!" blurted Madalyn O'Hair in a letter to Reverend Stout at news of Apollo 14's successful lunar Bible landing. O'Hair—who had knocked prayer and Bible-reading out of public schools—had tried to stop the astronauts from putting a Bible on the moon, but her influence fell short. The Bible made it to the moon, anyway, although the woman opined that the Bible didn't amount to much anyway.

O'Hair's letter wasn't the only abhorrent one Stout received. With the news release of the lunar Bible landing came other letters, one in particular from a man in Tampa, Florida.

"He used the word 'God' thirteen times in the letter — in the *wrong* way," said Stout. "He called me every name in the book."

Edgar Mitchell received a similar letter from the same man but with more sinister undertones, stating that he would "get rid of him" if he ever did anything like that again. Since Mitchell was a government employee, the letter was an indirect threat against NASA, and the FBI was called in to look at the two letters.

Stout wasn't easily threatened, however, and chose a more ministerial approach. "I see you mentioned God thirteen times," he responded in a letter to the recalcitrant Tampa writer. "And if you did that, you must be a very religious person and you have come to us because you need to be prayed for. Just let us know and we'll pray for you." The Tampa man was never heard from again.

Madalyn O'Hair, however, remained undeterred. Around this time she began calling herself Dr. O'Hair. "I want money and power and I'm going to get it," she said. After Apollo 14, O'Hair, billed as the world's most notorious non-believer, infidel, and atheist, undertook a tour of tent debates wherein she solicited local pastors to debate Christianity vs. Atheism, splitting any profits with her debater.

O'Hair and a Louisiana pastor, Reverend Bob Harrington, met and debated several times. Most of their appearances were arranged by local radio personalities in Texas and Louisiana. They ran ads in local newspapers, beginning in Tennessee as the "demon-directed damsel" and "the Chaplain of Bourbon Street." A gospel band was hired to stir the crowds, and when the spectators were warmed up, Harrington and O'Hair were introduced like two prize-fighters. O'Hair would first incite the audience with statements that derided God and the Bible, prompting the crowd to cry out for a lynching. When it was Harrington's turn to speak, he'd yell, "You people here who don't know God had better get saved—look who you're going to hell with!" O'Hair would dash back on stage, enraged, and attempt to wrestle the microphone from Harrington. Then she would knock down the American flag and jump on it. At this point, Harrington would take up the offering and ask people to "vote" for him or O'Hair with dollars by checking the appropriate picture on the envelope. As soon as the show was over, the pair split contributions.

On one occasion at Rice University, O'Hair challenged anyone to debate her about religious faith. Stout took her up on it, inviting her to debate him at Herman Park in Houston during a conference sponsored by the International Bible Society.

"You're not of my stature," she informed him. "Now if you get Billy Graham to come, I might do it."

"Well we're going to have it anyway and there'll be a chair for you," Stout told her. "If you don't show up, the chair will still be there."

O'Hair wrote a quick sarcastic response: "Well, the chair will be empty unless you get Billy Graham to show up and charge a fee and I get to take charge of the money for the debate and get at least 50% of all the money and then have reproduction rights for the debate. And then I'll come." Across the bottom of the note, imprinted in bold letters, was O'Hair's standard correspondence footer: "TAX THE CHURCH."

But O'Hair had stated at Rice University that she would debate anyone and Stout intended to hold her to it. So he proposed that they allot debate time based on the percentage of letters received at NASA pledging support for one or the other. Only four letters arrived in support of O'Hair. "That would be about four seconds of time for her in the debate," Stout said. "She never showed up."

Instead, O'Hair appeared at a college in Ohio. In her speech she lambasted the clergy and church officers. She harassed the religious students and professors, harangued those who trusted in prayer and mocked religious Americans as stupid. In her address, O'Hair referred to God as "Big Daddy," Jesus Christ as "J.C.," and the Holy Ghost as "the spook." Afterward, she took questions, which she used to further rail about religious ideas and spiritual beliefs. The audience was stunned.

As the event was about to break up, from the back of the auditorium came the small voice of a college girl. She spoke quietly and lovingly.

"Mrs. O'Hair, I'm so happy you came to speak to all of us at our college tonight. We have listened with attention to your tirade on our beliefs. We thank you for showing all of us what an atheist is. We appreciate your concern for us, but now in turn we must be ever grateful for your visit, because now and forever we have been strengthened in our Christian beliefs by listening to you tonight… Thank you, because you have strengthened my faith. Now I'll have even more love for 'Big Daddy,' 'J.C.,' and 'the spook.' So again, I say thank you, and bless your soul."

It was the first time O'Hair was at a loss for words.

By this time Madalyn was in full-blown war with NASA, and the Apollo Prayer League was actively involved in *O'Hair vs. Paine*. The case was scheduled for a hearing by the U.S. Fifth Circuit Court of Appeals before Justice Roberts, who Madalyn told a reporter "simply hates my guts." Judge Roberts tersely dismissed O'Hair's complaint against NASA, stating "Mrs. O'Hair's contention that the judicial oath 'So help me God' systematically excludes agnostics and atheists from the judiciary approaches absurdity."

Having suffered defeat at all levels of the legal system, O'Hair's rants against NASA began to wane. Much to her disgust, the Prayer League's prayerful support for the Apollo missions didn't end with the biblical landing of Apollo 14. Another mission was following on its heels.

On July 26, 1971, three Air Force officers, David R. Scott, James B. Irwin and Alfred M. Worden, were scheduled to launch on Apollo 15. Aerospace Ministries distributed pamphlets containing material suitable for a special Apollo 15 mission church service, complete with hymns, prayers, and an invocation. In it Reverend Stout recounted a story of hope:

> Ten years ago a young officer just out of flight school was with a student pilot in a light plane that crashed on a desert of the southwestern part of the United States. Fortunately, the plane did not explode. When they were finally pulled from the totally demolished plane and taken to the hospital, it was discovered

that the young officer had two broken legs, a brain concussion, and many other injuries, including a broken jaw. And when he finally regained consciousness he had amnesia. His friends prayed for him in his recovery and in four months he was on his feet. He overcame that difficulty and in doing so set new goals for himself. His name is Lt. Col James Irwin. And he is going to the moon.

The invocation closed with a prayer written for the Apollo 15 crew by APL member Hallie Mills Dozier: "Our heavenly Father…open their eyes, we pray, to recognize Thy greatness in all they behold."

She could not have imagined how this prayer was about to be answered on Apollo 15.

"Jesus Christ walking on this earth is more important than man walking on the moon."

- James Irwin, Lunar Module Pilot, Apollo 15

33
APOLLO 15: ONE RED BIBLE

It was an iconic moment in the history of scientific inquiry. Dave Scott, the commander of Apollo 15, stood on the moon before a television camera, poised to replicate a famous experiment conducted more than four hundred years earlier.

"In my left hand I have a feather," Scott said, "in my right hand, a hammer. I guess one of the reasons we got here today was because of a gentleman by the name of Galileo a long time ago who made a discovery about falling objects and gravity fields, and we thought where would be a better place to confirm his findings than on the moon. And so we thought we'd try it here for ya."

Scott dropped the feather and the hammer at the same time, and 1.3 seconds later they simultaneously hit the lunar surface.

"Well, how about that," Scott went on. "Mr. Galileo was correct in his findings."

That summer of 1971 the Apollo 15 mission was thought to be the very essence of science. And it was. But there was another mission taking place beyond the view of the public. Three Russian cosmonauts had just set a new endurance record for time spent in space. Vladislav Volkov, Georgi Dobrovolski, and Viktor Patsayev had been living and working for the preceding twenty-four days orbiting earth on the Salyut 1 Soviet space station when they finally boarded their spacecraft for reentry.

It was a moment of triumph for the Soviet space agency, which had looked on as one Apollo mission after another landed on the moon and safely returned. Now, as the Soyuz 11 spacecraft tore back through the atmosphere and the parachutes blossomed over the steppes of Khazakstan, nothing seemed amiss—except that radio contact with the crew hadn't been reestablished since the reentry blackout. The spacecraft remained eerily quiet as a helicopter landed alongside. When recovery crews opened the hatch, they found the cosmonauts in their seats, harnessed in place. All three were dead.

An investigation revealed that Soyuz 11 had sprung a leak during reentry, depressurizing the cabin and killing the crew within seconds. The tragedy left the Soviet space program reeling.

"The price they paid was not fair," said a grieving Yevgeny Yevtushenko, the revered Russian poet. President Nixon acknowledged that the cosmonauts had contributed greatly to "the widening of man's horizons." The tragedy also gave rise to a growing debate in America as to why the U.S. was still going to the moon at all. The missions were expensive, and the Soyuz 11 tragedy demonstrated once again that manned space exploration was dangerous. To underscore their point, NASA critics pointed to another Soviet development.

In September of 1970, an unmanned Soviet probe had traveled to and from the moon, bringing back a lunar soil sample at a fraction of the cost of a manned mission. War was still raging in the jungles of Vietnam and there was a growing consensus that the U.S. couldn't afford to carry on both a war and a manned space program at the same time.

But NASA had a bold answer to that question. Apollo 15, it said, would be one of the great scientific expeditions in human history, destined to touch the soul of mankind. Apollo 15 Lunar Module Pilot Jim Irwin would call it a "Mission to the Heart of Creation."

He was a world-class field geologist and a true legend in his time. But Lee Silver's greatest gift was as a teacher, and that was why NASA officials called upon him to teach field geology to the astronauts. In 1969 Silver began by taking Lunar Module Pilot Jim Irwin and Commander Dave Scott into the interior of the Sonoran Desert in a mud-splattered four-wheel-drive truck. After days studying rocks under a blistering sun, Silver and his astronauts emerged in the lobby of a remote hotel, unshaven, clothes embedded with dust, their skin burned by the desert sun. The next day, in 100-degree heat, Silver once again led them through the barren and hostile landscapes, pointing out geologic features, faults, and the striations, shades, textures and general makeup of the variegated layers of rock. He instructed them how to interpret what they saw. He taught them to see the world around them as he saw it— through the eyes of a geological scientist. But geology, more than a scientific study, was an art that typically took years for garden variety geology students to fully "see" and comprehend. Scott and Irwin had only a few months to develop an intuitive sense of what to actually "see".

The Apollo 15 mission would carry with it a whole new level of ambition for NASA. Vicariously following the astronauts as they tromped around on the lunar surface would be a team of geologists, geochemists, and geophysicists in the Science Operations Room across the hallway from Mission Control. While the astronauts broadcast live pictures back to earth, the scientists could direct their movements, ask them to sample particular rocks, and help answer specific questions in real time. In the past, television had been merely a means of communicating an adventure to the people on earth. Now it would be harnessed in the service of science.

Without a doubt, however, the most distinguishing feature of the mission was the new lunar rover. While the Apollo 14 astronauts hauled what appeared to be a cumbersome lunar shopping cart on the moon, Apollo 15 had in essence a 500-pound electric dune-buggy capable of carrying two astronauts up to thirty-six linear miles over the lunar surface. Boeing had built only three. Its design was a clever piece of engineering. The two-seated vehicle, nicknamed "Chitty Chitty Boeing Boeing," had four-wheel drive and was carried to the moon collapsed and affixed to the outside of the LM. There was independent steering for both the front and rear wheels. Although a miracle of design, it looked homemade, something designed by a *Popular Mechanics* contributor and crafted inside a hobbyist's suburban garage. For instance, the tires were made of woven piano wire and the radiator was literally packed in beeswax, a substance known to withstand extreme heat. The price tag, however, was anything but ordinary—each rover cost $8 million.

While the rover afforded the ability to explore a vast area near the landing site, it also presented new dangers. What if the rover broke down several miles from the lunar module, well beyond walking distance? What if there was an emergency that required the crew to leave the surface right away? Everyone knew the answers, but as Apollo 15 Command Module Pilot Al Worden said, "Each astronaut resolves before each flight that they're not coming back. It's just something you get right with."

Like his crewmates, Worden had come into the Astronaut Corps by way of the Air Force. A graduate of West Point with two master's degrees in Engineering from the University of Michigan, he had served as backup command module pilot for Apollo 12 and in the fall of 1969 was selected as the command module pilot for Apollo 15. His commander, Dave Scott, was six feet tall, athletic, and handsome, and presented America's image of the astronaut. "We all hated him because he was so darn good looking," said test crewman Charlie Dry. "Everybody wanted to be Dave Scott."

Scott, an Air Force fighter pilot, joined the Astronaut Corp in 1963 with the third group. Three years later he and Neil Armstrong barely escaped death when Gemini 8 began to spin out of control due to a malfunctioning thruster and Armstrong aborted the mission. Scott made up for the lost opportunity with the success of Apollo 9.

Commander Scott was known as something of an over-achiever, even in the company of a famously over-achieving group. He stunned those around him with his ability to keep going and going, without rest and without sleep—like a robot. Before he and his crew began working with Lee Silver, Scott told them that he wanted the Apollo 15 mission to bring back the absolute maximum amount of scientific data. As one Apollo mission followed another, he wanted his mission to stand out.

In a world where marriages were endlessly tested, Scott's wife, Lurton, decided to meet her husband on his own turf. Shortly after learning the nature of her husband's mission, she signed up for an introductory class in geology at the University of Houston and learned the theory and vocabulary her husband was learning in the desert. After long absences, she could talk with her husband about what he and Silver, "the old cowboy," had been doing all day in the desert of California. During the turbulent days of Apollo, the Scott marriage seemed to thrive while others followed a rockier path.

Lunar Module Pilot Jim Irwin's life was one of those on a rockier terrain.

"When I was a lad," Irwin said, "the moon [had] a strange fascination for me…when I looked at the moon, a feeling overcame me and I just thought that someday I'd be able to go there. And I told our neighbors that I was going to go to the moon, I told my mother and father, and they all laughed at me. My mother was a little more direct. She said, 'Son, that's foolishness. Man will never be able to go to the moon.'"

At forty, Irwin was one of the oldest members of the Original 19. He had a quiet self-confidence about him, a maturity that set him apart. Perhaps almost losing his life and livelihood in the 1961 plane crash had reordered his sense of what mattered. The trauma fostered a devout religious faith, something Irwin carried with him long after his recovery. Few of his fellow astronauts knew of his faith, as he wasn't especially close to any of them. He reserved close friendship almost exclusively for his wife and children. Yet he was close to Dave Scott. The two complemented one another, and each knew it. Scott was naturally authoritative, while Irwin felt comfortable supporting his commander.

Although its purpose was expressly scientific, the upcoming Apollo 15 mission would prove to be an intensely spiritual journey for Irwin, one he found difficult to share with his wife. Soon the couple began growing apart, the two of them attending separate churches—Irwin the Nassau Bay Baptist

Church, and Mary the local Seventh Day Adventists. Mary Irwin's faith had been tested when her brother died in a hiking accident. "Somehow the death of her brother caused her to feel she ought to get closer to the Lord," Irwin said, "and the only way she felt she could do that was to go back to her own church." Their separate religious paths became a wedge in their marriage. When Irwin went to Scott for advice, his commander counseled patience.

"Everybody goes through it," he said. "It'll change."

The Irwins resolved to take a vacation in Acapulco—their first without their four children. But at dinner their first night at a restaurant overlooking Acapulco Bay, a waiter approached with a message from someone at NASA. Irwin immediately went to the phone to call Dr. Jim McGee, one of the agency doctors.

"Jim," he said, "there's been an accident at home. Three of your children are in the hospital."

The couple was on the next plane home, where they learned that three of their four children had been injured playing in a construction zone near their home. The injuries were serious but not life-threatening. Although the Irwins were relieved, their marriage continued to struggle.

"[Mary] distrusted the glory and fame and adulation that are heaped on astronauts," Irwin said. "It wasn't part of her life." Being the wife of an Air Force officer, as Mary Irwin had been, was strange enough. Being the wife of a test pilot was even stranger. "And being the wife of an astronaut is strangest of all," Irwin said. "Mary would be quite content to be the wife of a farmer."

When Jim learned in late 1969 that he would be the lunar module pilot for Apollo 15, life at home became more difficult. The emotional turmoil became so great that he considered quitting the astronaut corps. He even rehearsed what he would say to Deke Slayton: "Just take me off the crew. I don't think I'm going to be ready for the flight." When he told Mary of his plans, she shrugged and replied, "Do what you want."

To Jim's and Mary's surprise, it was at this point—precisely at their darkest hour—that their marriage began to show signs of life. As Jim put it, "Things slowly began to straighten themselves out for both of us." As the marriage slowly mended, Irwin found that his ability to focus on the mission improved dramatically. He began to develop a strange new awareness. "It was a feeling that God was very close to us and to the entire Apollo 15 mission. What I did not know at the time was that this feeling was destined for growth."

At first, God's proximity was beyond definition for Irwin. A month before liftoff, he shared it with the congregation at Nassau Bay Baptist Church by way of a poem by James Dillet Freeman cut out of a book and sent to him before the flight. The poem *"I am There"* expressed that God is with man when he is engaging in everyday business, as well as when he is on a great adventure.

"It was my feeling," Irwin said about the poem, "that God was with mankind as man felt his way out into space." The poem began:

> *"I am there when you pray and when you do not pray.*
> *I am in you, and you are in Me.*
> *Only in your mind can you feel separate from Me, for*
> *only in your mind are the mists of 'yours' and 'mine.'*
> *Yet only with your mind can you know Me and experience Me."*

Irwin's relationship with his church family steadily matured during the final weeks before the mission as he came to see a connection between his church and the moon shot. The mission, they decided, was an event they would share together by way of a "prayer pact." The document prepared by the church was simple enough. On it was a photograph of Nassau Bay Baptist Church and beneath it a verse from Mark 16:15:

> "Go unto all the world and preach the gospel to every creature. Our prayers go with James B. Irwin, David R. Scott and Alfred M. Worden—Apollo 15 Crew."

"When we left the earth, we carried a microfilmed copy of the prayer pact," Irwin said. "Two hundred and fifty church members signed in very small script."

As the mission fast approached, Jim and Mary Irwin were again a committed couple. The last few days, however, they spent apart. Dave Scott's wife, Lurton, had advised Mary to keep her physical distance from Jim, saying, "What if you're harboring a bug, and suddenly it erupted during the flight?"

"Don't ask me to come out [to see you]," she told her husband over the phone in Florida. "I want to come out so badly," she added.

Deep down, Irwin agreed that keeping their distance was for the best. This meant that he may have hugged his wife and children for the last time, and the thought broke his heart. "The last time I saw Mary and the family before the flight they were behind glass."

It turned out to have been the right decision. The day before the flight, Mary fell ill. She was grateful she hadn't "given in to my inner feelings" and gone to see Jim off.

When Apollo 15 launched on July 26, 1971, Irwin was singularly focused on the mission and its "scientific aspect." As the spacecraft thrust into space, he felt an overwhelming sense of release from all earthly cares. The violent ride was accompanied by an upwelling of emotion that nearly brought him to tears.

"I didn't have any notion of the spiritual voyage whatsoever," he said. "It would have been beyond my wildest imagination to guess that this flight would not satisfy me—that I would come back to earth a different person, bound for a higher flight."

Onboard Apollo 15 was the heaviest and largest payload ever carried into space. First there was the ALSEP, the Apollo Lunar Surface Experiments Package, a collection of scientific instruments that would measure the moon's geological and physical characteristics. Also aboard was the rover, increasing the weight of the lunar module by several thousand pounds. There were also other things that NASA officials had no idea were aboard.

During the run-up to launch, Scott had quietly gone about making a plaque with the names of the fourteen astronauts and cosmonauts who had died since the beginning of the Space Race, including the three cosmonauts aboard Soyuz 11. He also had a small figurine made, representing "The Fallen Astronaut" to leave lying on the moon. Irwin helped in the project by making sure the names of the Russians were spelled correctly, while Al Worden came up with the figurine.

There were other things onboard that, in light of the O'Hair lawsuit, NASA officials would rather not know about, including Irwin's prayer pact. And although few knew it, Irwin, a devout Baptist, and Scott, a member of the vestry in the local Episcopal Church, had brought along a small red leather Bible. When Irwin told friend Charlie Dry on the technical support crew about the Bible, Charlie asked, "Where are you going to put it? You're only allowed so much space and weight."

"That's okay," Jim said, "I'll take everything else out if I have to."

"And, by golly, he did it," said Charlie.

On the third day of the mission, as Apollo 15 neared the moon, gravity was nowhere to be found inside the command module *Endeavour*, creating a situation that, while not totally unpredictable, had the potential for creating a serious problem. There it was, a blob of tomato soup floating around the hatch.

"We take gravity for granted," said Worden. "Remove gravity and you can't keep your head in place—and soup doesn't stay in the spoon. We had a ball of tomato soup float around that might get into the instrument panel."

Food had to be mixed with various components to eat, and if the soup was too hot, they would let a spoonful float around the cabin to cool and catch it as it came back around. The crew would let each member's respective rations float in rotation around the cabin and grab theirs as it passed by. Rations were color-coded by crew member so they knew which to grab. One of the soups floated away from them. As the blob waltzed around the cabin, the bizarre mess threatened the spacecraft's electronics. Finally, they managed to throw a towel around it, which netted the blob and absorbed the liquid.

The moon grew larger and larger in the window, and on July 29, Apollo 15 slipped into the darkness of the backside of the moon. As he viewed this strange new world through the prism of his faith, Irwin knew his life was changed forever. He was overwhelmed by the sheer beauty of the moon, which he saw as evidence of "the nearness of God." This sense of awe and faith would dominate his thoughts in the coming days.

The jagged ridge of the moon's 12,000-foot Apennine Mountains towered over an immense canyon called the Hadley Rille, and both posed unique dangers as the *Falcon* attempted to land. As Scott and Irwin guided the lunar module over the mountains, the pitchover failed to reveal the craters they had studied on the pre-flight models.

"We had four craters lined up on the maps for our landing site: Matthew, Mark, Luke, and Index," laughed Scott. "We couldn't use 'John' because of Madalyn Murray O'Hair. [So] we called the last one Index, because Index was where we were supposed to land." But as they pitched over and peered out the window, they saw nothing resembling the craters. What came into view was a row of craters with only a vague resemblance—but it was indeed the landing site.

The next surprise came during descent when *Falcon*'s engine prematurely shut down. Even in one-sixth of earth's gravity, this was a problem, as the *Falcon* was nearly 2,500 pounds heavier than previous lunar modules.

"Man, we hit hard," Irwin said. "Then we started pitching and rolling to the side...I was sure something was broken. If the lunar module turns over on its side, you can't get back from the moon." For a good ten seconds, Scott and Irwin held their breaths.

From their vantage, it was impossible to tell what condition the craft was in. In any case, Mission Control broke the good news a few long seconds later: "You have a Stay." The two astronauts started pounding each other's backs "like we had just made a winning touchdown in a football game," said Irwin.

After preparing the lunar module for an emergency liftoff, they climbed to the top of the hatch to gaze at their surroundings, mesmerized by the soft, golden brown mountains and the ridge of the Hadley Gorge just a mile away.

The mountains reminded Irwin of Colorado mountains above the timberline in summertime.

"My first reaction was: this isn't a strange world at all," Irwin said. "I feel right at home."

"Okay, Houston," Dave Scott announced to Mission Control as he became the seventh man to set foot on the moon. "As I journey out here into the wonders of Hadley, I come to realize that there is a fundamental truth to our nature: man *must* explore. And this is exploration at its greatest."

Behind him was Jim Irwin, whose thoughts were somewhat less grandiose. "Oh, my golly," he said to himself, "I'm going to fall on my backside in front of all those millions of television viewers."

The first order of business was to deploy the rover, folded up like a lawn chair and affixed to the side of *Falcon*. Once they unfolded it, the astronauts climbed aboard and drove off at a whopping six miles per hour, galloping over the desolate landscape, bouncing into the black sky with each bump. The rover was fun to drive and had a navigation system to eliminate the difficulties that bedeviled Shepard and Mitchell on Apollo 14.

The first excursion was up Hadley Rille. The entire foray was documented by a panoramic lens of a television camera mounted to the rover, which allowed the Surface Geology Team in the Science Operations Room at Mission Control to guide them in their work. The astronauts' greatest challenge would be to find a primordial rock sample. Many of the geologists were doubtful Irwin and Scott would recognize, much less locate, one in the lunar environment.

Rare primordial rocks have very specific characteristics. Such a rock would be pure white and contain a mineral called *plagioclase*, matter that was formed when the moon cooled billions of years ago. Perhaps the greatest preflight concern by the geologists had been the question of moon dust. The astronauts would have to grab, with tong-like tools, as many different samples of rock as they could find, hoping that one would be the pure white *plagioclase* that scientists back home had told them to be on the lookout for. This mineral, the scientists suspected, was going to be genesis rock, a sample of the original lunar crust from around the time the moon was formed.

All that Scott and Irwin had to do was locate one under the blanket of powdery dust. Irwin was mindful that their chances were not great. He felt it would take divine direction, and prayer. "Perhaps the most important prayer of all was to petition God to allow us to find a sample of genesis rock in spite of the dust," he said.

On the first day of exploration, Scott and Irwin didn't find what they were looking for.

"Through all of this, I never lost the feeling of the closeness of God," Irwin said. That night, before falling asleep in *Falcon*, he felt the same anticipation he had experienced when he made the prayer pact at his church back home. "Please Lord," he prayed, "help us have a rewarding exploration tomorrow."

―――

The next morning Irwin and Scott emerged from *Falcon* and climbed aboard the rover, this time to explore Mount Hadley. Bouncing along at max speed, Scott drove 300 feet up the side of the mountain and then headed to Spur Crater. As they drove down the other side, the rover began to slide.

"The back wheel's off the ground," Irwin shouted.

Scott checked his speed. If they got stuck it would be a long walk back to the lunar module and the mission would be scrapped.

By now Irwin was in the throes of an experience he could only describe as mystical, enraptured by the sense that God himself was guiding him and his crewmate. As they came to a crater, old and badly mauled by millions of years' bombardment, they dismounted the rover and began their exploration across the dust on foot.

And there it was.

Irwin and Scott saw it at almost the same time; they could not have missed it. Incredibly, one peculiar-looking rock sat in the center of the crater perched on a sort of pedestal, almost as if a hand were holding it up. Even before they reached to grab it they realized it was not like any of the rocks they had seen before. Slowly Dave lifted the rock from atop its pedestal and turned it over. The underside was pure white.

"Oh man, look at that," he said. "Look at the glint. Guess what we just found! I think we found what we came for, a crystalline rock… Bag it up. I'll make this bag 196, a special bag." With a sense of awe and wonderment they carried the rock back to the lunar rover.

"How was it possible, in an environment where no wind blew, no rain washed?" Irwin thought. "How was it possible for Genesis Rock to have been shown to us with no disguising cover of dust?" To Irwin the answer was clear.

"I feel now that the power of God was working in me the whole time I was on the flight," he said. "I felt His presence on the moon in the most immediate and overwhelming way. There I was, a test pilot, a nuts-and-bolts type who had gotten rather skeptical about God, and suddenly I was asking God to solve my problems on the moon. I was relying on God rather than on Houston. Then there was my powerful desire to have a service on the moon, to witness. All this time God was taking over my life and I didn't even realize it."

The exuberant astronauts returned to the lunar module. Irwin was overcome. That night, he prayed in his hammock, floating in the faint gravity. In spite of the O'Hair lawsuit and the squeamishness of NASA, Irwin decided that he would say something the following day. As he drifted into sleep, he prayed that he would say it well.

As the astronauts attempted their final night of sleep, reports of the mission's success swept the world. A rock nearly as old as the solar system itself had been found on the moon. Newscasters were reporting the discovery of the "Genesis Rock."

Dr. Robin Brett, a scientist at the Mission Control Center, could hardly contain his excitement. "The knowledge we get from this mission," he announced, "will at least equal and perhaps exceed what we've learned from all the past missions combined."

His statement was bold, but by no means an exaggeration. The Genesis Rock would soon be analyzed by scientists, who would age-date the rock at 4.15 billion years—the oldest rock ever found by man. The discovery would confirm scientific theories around the world that the earth and moon were most likely created at the same time.

Their final night on the moon was brief, and both astronauts awoke deeply fatigued. Even Scott, famous for his superhuman reserves of energy, was exhausted.

Eight hours later, for the last time, they went out on the rover to explore. Soon Houston was calling for them to return to the lunar module and begin preparing for liftoff. Although exhausted, neither wanted their adventure to end. Each man had matters to attend to before leaving. As they headed back to the rover, Irwin took a moment to gather his thoughts.

"Oh, look at the mountains when they are all sunlit," Scott said. "Isn't that beautiful?"

Irwin saw his opening.

"Dave, that reminds me of a favorite biblical passage in the Psalms [121]," he replied. "I look unto the hills, from whence cometh my help. My help cometh from the Lord." Irwin paused for a moment, then immediately added, "Well, we do get quite a bit of help from Mission Control and Houston, too." He confided later, "I was afraid they might turn the radio off if I didn't give them equal credit."

Minutes later, and without much ado, Scott conspicuously announced that he was going to "clean up behind the rover." He had carried with him to the moon a few items "that had been left off the checklist," he said later, and neither astronaut let Houston in on the activity.

The Apostles of Apollo

While Irwin prepared the *Falcon* for liftoff, Scott drove the rover to a small rise about 300 feet from the lunar module. From that vantage, they could aim the TV camera at the lunar module for liftoff. Scott then produced the small red Bible. He made the final switch settings on the camera and then propped the Bible atop the rover's control panel. He left it there, he would later say, because "it was the most important document ever written—the instruction book for life." He didn't mention the Bible in his communications with Mission Control. It wasn't until years later that it surfaced in a parting photo of the lunar rover, a tiny red speck tilted against the center control panel, "so that those who came after would know that we came from a world of believers," said Irwin.

Scott then walked over to a small crater where he placed the figurine face down alongside the plaque commemorating the fallen astronauts.

During their return trip, Al Worden performed the first and longest deep space walk ever recorded from a command module. "One thing becomes clear when floating 240,000 miles from home," he reflected later. "God did it all."

As they approached the home planet, Worden directed *Endeavour's* entry into the earth's atmosphere. Blazing through a reentry fireball, upside down and backwards, the roll maneuver plastered them against the couch with a force seven times their body weight. It was physically impossible to raise their hands. Only two of the three parachutes deployed, and they hit the Pacific Ocean in yet another hard landing.

"Two out of three ain't bad!" the crew announced, confirming to Mission Control that they were fine. After splashdown, they heard a Navy SEAL tapping on the hatch window asking if they were "all right." All three of the Apollo 15 crew members were Air Force and had attended West Point. Asked later what the scariest part of the trip was for him, Worden replied: "Being rescued by the Navy."

Walking on the moon and seeing the earth from such a distance altered Irwin's life. As he put it, "I came back changed."

"Life is short," he said. "It's only what I do for God that will last for eternity." Upon his return, he was baptized at Nassau Bay Baptist Church and became a minister. "The Lord wanted me to go to the moon," Irwin said, "so I could come back and do something more important with my life than fly airplanes."

In 1972, after completing his post-flight activities, Irwin left NASA and formed High Flight Ministries, an interdenominational evangelical organization in Colorado devoted to telling his story of Christian rebirth and witnessing to others. Asked if he felt let down now that he was back on earth, he replied, "I've never had the letdown. The Lord has kept me on one great big high."

Later, his passion turned to the search for Noah's ark, which was believed to have come to rest in the mountain peaks of Ararat in northern Turkey. He had been among the mountains of Apennine on the moon. Now he would search for Noah's ark in the mountains of Ararat on planet earth.

He was sure it was there.

> "My walk on the moon was exciting, but my walk with the Son and my walk with Jesus is even more exciting. And it doesn't last three days, it lasts forever."
>
> *- Charlie Duke, Lunar Module Pilot, Apollo 16*

34
APOLLO 16: MOONWALKER MOMENT

"It seemed that drinking beer was in the flight controller's blood," said Sy Liebergot. "That made the 'Wheel' vital to our well-being."

The "Singin' Wheel," or "The Wheel," or "The Barn," was one of those unique institutions that rarely make it into the books. Nevertheless, it remained dear to the hearts of the flight controllers who made history during the Apollo era. "For those of us at Mission Control," Liebergot said, "the Singin' Wheel was what Pancho Barnes' Happy Bottom Riding Club was to the Edwards Air Force Base test pilots. It was where we went to celebrate and to blow off steam."

Situated just down the road from Mission Control, the Wheel was nothing fancy. From the outside, the red building resembled the 1930s barn it had been before the owners turned it into a bar. Inside, the floor sloped, tables and chairs were at a premium, and patrons waited their turn for the badly worn shuffleboard. At the bar, gallon jars held pickled pigs' knuckles, purple hardboiled eggs, and polish sausage. Streaking was popular at the time, which inspired a few flight controllers to strip and run through the night. Pitchers of beer were ordered and drunk, dollar bills heaped upon the center of the table, and the juke box played on. At the height of the Apollo missions, several flight controllers formed a band and called themselves "Backroom Swingers." The group was made up of two flight controllers on acoustic guitars; another on a tub bass made up of a wash tub, broom handle, and piece of rope; and Liebergot tapping away on a snare drum. The upstairs was a bar with a little dance floor surrounded by antique furniture and a standup piano. Only a few of the astronauts ever found their way to the Wheel.

"Most astronauts just considered us flight controllers part of the vast infrastructure on the ground to support them," Liebergot said. "And most of them didn't get to know us at all. Charlie Duke was an exception to that."

―⁂―

Charlie Duke was unique, even among the colorful personalities that made up the astronaut corps. Perhaps one of the most endearing traits of this South Carolinian was his down-to-earth nature. Raised a Baptist and baptized at age twelve, the 1957 graduate of MIT liked to be around smart but regular folks. He admits to being a people-pleaser. "I'd swing with the crowd," he said. "If there were some folks that were talking dirty, I'd fit in with them too. The next day I might be in church and would conform there as well." He loved country music, the sort played at the Grand Ole Opry. And he was fiercely dedicated to his work. Like many of his fellow astronauts, the commitment to his career took its toll at home. But Duke's wife, Dotty, wasn't one to give up on their marriage so easily.

"All my life," she said, "my dream had been the fairy tale 'Cinderella.' The dream of one day finding my Prince Charming, who would sweep me off my feet and, professing undying love, fulfill all my needs." When she met Charlie Duke, she thought he was that man. They married, went on their honeymoon, and then it happened. "When we returned ... I realized things were not okay. I was not first in his life, his job was first."

Even Duke recognized the problem.

"I did believe there was a God," he said. "But if there was really a God in my life it was my career. The Bible says seek first the kingdom of God. Love the Lord your God with all your heart and all your mind and all your soul. Well, God wasn't first in my life. I respected God, I honored God on Sundays, but my career was the most important love of my life."

After graduate school Duke served as a test pilot instructor for the Air Force. Following a stint at Edwards Air Force Base, he was invited to join the astronaut corps. To Dotty's lasting despair, the marriage didn't improve after the move to Houston. Charlie only became more intensely focused on his career. She'd hoped things would change when their son Charles was born. They didn't. When Charlie was named to the Apollo crew, she hoped again. But their relationship only grew worse.

"My job was to see that everything went well with the details of entertaining friends and relatives who came to view the launch," Dotty said, "taking care of the boys, and running a semi-organized house for the duration of the flight. There were press interviews, overseeing my wonderful family who had joined me in Houston, welcoming a constant stream of friends and neighbors, plus intently following the space adventures of Apollo 16."

For Charlie, Dotty was part of the larger team that would help him get to the moon. And she was. The problem was that the moon was more important to him than his wife.

The Apollo 16 crew was aware of how uncertain the life of a moon-bound astronaut really was. Even in the latter phases of the Apollo program, problems of every stripe continued to spring up out of nowhere, which inevitably meant delay.

"A delay usually resulted in at least one month's postponement," Charlie said, "and during that one month anything could happen." He had sharp memories of the run-up to Apollo 13 when he himself had caused the last-minute replacement of Ken Mattingly due to measles exposure.

"I felt terrible about it," Duke remarked, "but Ken was a good sport and never blamed me for missing that flight. He took it stoically, even though he was very disappointed."

As "punishment" for exposing Mattingly, Duke was given the label "Typhoid Mary" by his fellow astronauts. But history would show the illness to be a blessing. A few months later—shortly after the Apollo 13 ordeal—Slayton had asked Mattingly a question he would never forget: "Would you rather be the lunar module pilot on Apollo 18 or the command module pilot for Apollo 16?"

It was a seemingly innocuous question, but Mattingly sensed it was anything but.

"Well, I'd sure like to go down to the surface," he said.

"Well, I'll give you your choice," Slayton said cryptically, "but I would always take a bird in the hand."

Mattingly didn't respond right away.

"It's your call," Slayton concluded. "Just think about a bird in the hand."

At the time, Apollo 18 was in danger of being cut, but nobody knew for sure what would happen. Mattingly could have a chance at walking on the moon or the certainty of orbiting it. Mattingly decided it was "probably better to get near it than not go at all."

John Young would be the commander for Apollo 16, and the lunar module pilot would be none other than Charlie Duke. Just as Swigert had taken the center seat position for Mattingly on Apollo 13, Mattingly took Swigert's seat position as command module pilot on Apollo 16.

Still, as Mattingly had learned on Apollo 13, a bird in the hand was no guarantee. A few weeks before the launch of Apollo 16, Mattingly submitted a blood test to NASA doctors which came back with an "irregular" reading. In a moment of déjà vu, he was informed that he had elevated levels of bilirubin, which suggested that he might develop hepatitis. More blood tests were ordered, and Mattingly worried that another phantom medical problem would ground him. After an agonizing wait, NASA doctors decided that the problem posed no risk to the mission and he was cleared.

The unsettling dilemma gave the Apollo 16 crew a heightened sense that their mission might be scrubbed at any moment for any reason—maybe one they hadn't even thought of. And if they didn't go now, they weren't going to go at all. The Apollo era was drawing to a close. The launch date, everyone agreed, could not come soon enough.

"All I wanted to hear were the words, 'Liftoff!'" Duke said.

On one particular night before the launch, Charlie climbed into bed and fell asleep beside his wife. They were in Hawaii for a geological field trip and a little rest and relaxation. Sometime late that night, Duke was taken by a dream in which he found himself and Apollo 16 Commander John Young in the rover, headed north over the lunar Descartes highlands. On one of their planned stops they came to a ridge where he saw something unexplainable. There, extending before them, was a set of tire tracks. They seemed to be the tracks of a rover—but not *their* rover. Someone else, it seemed, was on the moon.

"We asked Houston if we could follow the tracks," Duke said, "and they said yes. So we turned and followed the tracks. We went about an hour or so and found this vehicle that looked just like the rover, and it had two people in it."

Duke pulled up one astronaut's visor. When he did, "I was looking at myself." Then he took a look at the other astronaut.

"Houston," Duke reported, "the other one looks like John."

"We can't explain all this," was Houston's reply.

In the midst of his dream Duke remembered being unafraid. The dream wasn't so much frightening as it was real.

"I didn't feel this was a premonition that we were gonna die up there on the moon," he said. "It didn't upset me. I felt comfortable that these two other astronauts looked just like us, and I took off one of the parts from this other vehicle so that we could show the people in Houston. And I knew that we made it home okay, because in my dream I can remember handing the parts to one of the lab technicians in Houston after we got back." The technicians determined the alien rover part to be over 100,000 years old.

Although the dream didn't upset him, Charlie decided not to share it with Dotty, as she "would be fearful that it was a premonition of our dying on the moon." At that moment there were simply too many other things to worry about, too many real things.

At dawn on April 16, 1972, the mission was still a Go, though the sense of uncertainty prevailed. That morning the crew took the elevator up to the spacecraft where the suit technicians assisted them. As Duke climbed aboard, he found a sign on his seat: "TYPHOID MARY'S SEAT." Then, opening his flight plan, he found a card that read: "STAY UNWELL GREETINGS." It was signed "Freddo, Stu and Ed" and had a picture of an evil-looking bug holding a dripping syringe. The backup crew—Fred Haise, Edgar Mitchell and Stuart Roosa—let Duke know they were standing by and more than ready to fill a slot should one open up at the last minute, a satirical punch at Duke's earlier measles exposure.

The moment served to dissolve a bit of tension as the crew prepared for launch. Young climbed into the spacecraft, followed by Mattingly. There was no fear, no reluctance. The countdown began, and Duke thought only one thing: *"Let's go, keep counting, no abort. Let's launch this beauty."* Even as he felt the massive Saturn engines roar to life, he was waiting for Launch Control to terminate the countdown. In the final seconds, he was still reviewing emergency exit procedures.

The fifth lunar landing mission had the typical smooth but agonizing liftoff. The command module was named *Casper* and the lunar module *Orion*. One minute into the flight they broke the sound barrier. Duke looked over at Young and Mattingly. They seemed "cool as cucumbers," especially Young, who was "the perennial Mr. Coolstone."

By 1972 the road to the moon was a well-worn path. The crew arrived in lunar orbit on April 19. At first Duke focused on the vast, grand scale of his view. Then his thinking unexpectedly changed. As the spacecraft wheeled about the moon, it turned intensely personal:

> As we came around the backside of the moon we had earthrise. And this blue and white jewel hung in the blackness of space. It was not a spiritual experience for me. I didn't feel close to God. I didn't think I *needed* God in my life. I had about all of God I wanted, which was about one hour every Sunday morning. But it was a philosophical experience as I held up my hand, and underneath my hand was the earth. And the thought occurred to me that there were four billion people under my hand. And you don't see the United States, or Europe, or Africa or South America—you just see earth. And the thought occurred to me there that we're all one on Spaceship Earth. And if we can all

learn to love one another, then man can solve his problems. But I had a basic problem in my life, a real problem—a real basic problem. If I didn't even love my wife, how could I love the rest of the world?

The moment would prove to be a tipping point. Many of the men who went to the moon would say the experience did not change them, and in many ways, Duke was among them. But he would carry this realization forward long after his return to earth.

First, however, they had to get to the moon. No two flights were alike, as each step of every mission brought forth its own set of problems to be solved before the prize of tramping about another world was realized.

Shortly after the lunar module undocked, Mattingly, flying solo in *Casper*, noticed a strange vibration in the command module's SPS engine as he prepared to fire it for an orbit change. Incredibly, the problem surfaced just as Duke and Young began to drift away. At first he had no idea what it was. But then he noticed the shaking occurred when he activated the secondary control system for the engine.

"You can feel that spacecraft shaking," he reported to Houston. "It's just oscillating like a wild man."

With that, Mission Control put the entire mission on hold. Young and Duke, who were happily floating toward the lunar surface, were ordered to delay their descent until the problem was resolved. Meanwhile, Mattingly continued to test the system, trying to discover precisely what was wrong. As he floated alone in the empty command module, he lovingly mumbled to his beloved *Casper*, "I be a sorry bird."

Houston had to remedy the situation quickly if Duke and Young were to walk on the moon. If need be, the lunar module could re-dock with the command module and use its engines to power them out of lunar orbit, as had been done on Apollo 13.

"If your heart could sink to the bottom of your boots in zero gravity," Duke said, "ours sank to the bottom of our boots. I mean, we knew this was real serious, that's our ride home, if that engine doesn't work you can't land because you've got to use the lunar module to get out of orbit… [It] was a bitter pill to swallow because you've trained for two years, you've come 240,000 miles, you were orbiting the moon one hour before you land, you look down and there's your landing site eight miles beneath you—and they're about to say come home."

"I can imagine the gloom and doom in the [lunar module] must have been terrible," Mattingly said.

Every potential fix was painstakingly explored. Three hours later, Houston directed Mattingly to initiate *Casper's* backup guidance system. When he did, the shaking suddenly stopped. The astronauts remained suspended, awaiting clearance. Houston wouldn't respond for several more agonizing hours. Then, out of the blue, it came:

"After having attended a meeting by management people in the back room," the CapCom announced, "the situation is Go for landing."

Suddenly the gloom lifted, and, said Duke, "our spirits lifted by ten thousand percent."

The Descartes region, the landing site for Apollo 16, was largely unknown, even though less than three years before, Apollo 11 had landed just 250 miles away. For the scientists in the Scientific Operations Room at Mission Control, it remained mysterious, believed only to be a region formed by volcanic activity. All any of them could say for sure was that Descartes held a catalogue of surprises, and no one expected that their theories on the moon's origins would soon be proven false.

By the time Young and Duke descended the final feet to the surface, they were six hours behind schedule. At 10:57 a.m. Houston time, John Young became the ninth man to set foot on the moon. As he did so, he raised two fists into the lunar sky in a gesture of triumph. A few minutes later, Charlie Duke became the tenth. *"Yahoo, Houston!"* he shouted in a fit of euphoria. "Here we are — we're ready to go."

Because of the late landing, there was talk in Houston of canceling the third moonwalk. The geology team was disappointed. New scientific data had emerged and advanced with each Apollo mission, and now a full third of the scientific payload of the current mission might be forfeited due to an operational glitch. William Muehlberger, the lead geologist for the mission, set his men to work through the night, preparing a case against canceling the moonwalk. The effort seemed to pay off. Mission managers agreed to postpone a decision on the third moonwalk for the time being. But ultimately it would depend upon how much water, oxygen, and power Duke and Young had at the time.

Meanwhile, after a series of glitches, the astronauts deployed the ALSEP. Once all the sensitive scientific instruments were in working order, they initiated an experiment that had been a long time in the making. It had been devised by Mark Langseth, a brilliant young geologist working for NASA. The purpose was to measure how much heat was flowing from the moon's interior, which required that a ten-foot hole be drilled into the lunar surface. The heat flow experiment, as it was known, flew first aboard Apollo 13, then

Apollo 15, wherein the drill got stuck after five of the required ten feet. But the readings Dave Scott received on Apollo 15 were astonishing, with the measured temperatures coming in at roughly twice what had been expected. Now Langseth's experiment was again poised to take place. All he needed was a single good reading. And that's when Langseth saw what was happening on the television monitor.

The camera showed Young standing in the middle of the field of scientific instruments, with spaghetti lines of cables extending out in every direction. The cable to the hapless heat flow experiment was wrapped around his foot. Before Langseth could relay the problem to the astronauts through Mission Control, Young took a few steps and the cable broke loose.

"Charlie," Young said.

"What?" Duke replied.

"Something happened here."

"What happened?"

"I don't know. Here's a line that pulled loose."

"Uh-oh."

This simple snag meant that much of the moon's interior would remain a mystery—or other tasks would have to be cancelled to make time to repair the cable, if it could be fixed at all. Young felt terrible.

"God, I'm sorry," he said. "I didn't—I didn't even know it."

"Tell Mark [Langseth] we're sorry," he said. "Is there no way we can recover from that?" he asked Houston.

The answer from Mission Control was no—not after the delayed landing.

In spite of the setback, the mission yielded extraordinary information. The rocks the astronauts found were not those geologists had expected. Instead of volcanic rocks, they were finding breccias, rocks welded together by the impact of meteors over the eons. According to Young and Duke, the geologists in the Scientific Operations Room were not only stunned, but in a state of denial.

"They thought we were going to find [volcanic] rocks," Young said, "and all we found were totally different kinds of rocks, very old rocks that were carved 3.29 billion years ago…and they thought they hadn't trained us right on the rocks."

Duke got the same impression. "I kinda think they thought maybe 'we wasted six years of geology training on these two dunces because they obviously don't know what they're lookin' at.' They argued with us all the way back from the moon that the rocks we had weren't what we'd really found," Young said. "So I told 'em, 'Why don't you just wait till you get these back on the ground and take a look at them yourself.' And they kept arguing over this. A bunch of science guys, of course."

In spite of this hang-up, the mission was going remarkably well. Problems of an unforeseeable nature, however, were brewing.

The news coverage for the mission provided by CBS was sponsored by Tang, the makers of the new orange drink. The publicity for the product was priceless, with heroic astronauts shown consuming the "Space Age" drink on an historic lunar mission before an audience of hundreds of millions. What Tang's makers didn't anticipate was the earthy nature of American astronauts.

Seeing the dust-covered astronauts on the surface, ground control quipped: "My kids don't get as dirty as you are."

"Yeah," Young said, "but I bet they're not having as much fun."

Tang executives were not amused. And it was about to get worse.

"I got the farts again," Young remarked. "I got 'em again. Charlie, I don't know what the hell gives them to me… I think it's acid in the stomach."

"It probably is," Duke responded.

"I mean, I haven't eaten this much citrus fruit in twenty years," Young continued. "And I'll tell you one thing, in another thirteen (expletive) days I ain't never eating any more. And if they offer to serve me potassium with my breakfast, I'm going to throw up. I like an occasional orange, I really do, but I'll be damned if I'm going to be buried in orange."

The exchange highlighted how much the astronaut's gourmet menu was loaded with fruit and potassium, but it didn't help Tang's case one bit.

Reverend Stout was in the press room with CBS anchorman Walter Cronkite, witnessing the Tang episode. Cronkite saw the incident as an opportunity for comic relief for the overworked flight controllers and composed a specially taped "breaking news release" meant for the ears of Mission Control only.

Stout managed a stone face as the broadcast began.

"Breaking News!" Cronkite's voice announced with all the flourish of a national news release.

Everyone rushed to the news desk to hear what came next.

"We've got a problem on Apollo 16," he announced soberly. "But don't worry about it, it's being taken care of. We have the backup crew drinking massive doses of Tang and farting all over the place and we're working out a solution right now."

After a moment of stunned silence, a burst of laughter filled the dead air in Mission Control.

On the moon, the Descartes highlands were not turning out to be what the scientists thought they were. Everywhere, for as far as the eye could see, were craters layered in breccias—and no evidence of volcanic rock.

Sixty miles overhead, Ken Mattingly remained in lunar orbit, filming the solar corona and the lunar surface, methodically collecting data. On the outbound journey the crew had played country music, which Duke and Young loved. Now that Mattingly was by himself, he played his favorite classical tapes, including Holst's *The Planets* and Berlioz's *Symphonie Fantastique*. He couldn't quite get over just how much fun he was having by himself.

There was, however, one completely unanticipated problem, a first for the space program: at some point during the outbound flight, Mattingly lost his wedding ring. When small items were lost in the cabin, they usually turned up in the air filters, but his wedding ring had not. As he carried out his busy schedule, he kept an eye out for the ring, but it never turned up.

"Well, I guess I lost it," he said, resigned.

Meanwhile, on April 23, Young and Duke prepared to lift off from the moon. An hour and a half later they docked with *Casper*. Duke and Mattingly were scheduled to conduct space walks on the return trip home. Duke was the first one out the hatch, dangling in space primarily as a safety observer as Mattingly retrieved experiments and film canisters from inside *Casper*. The two then exchanged places, and Duke climbed back inside. Mattingly was floating ten feet outside the hatch conducting an experiment when Duke saw a glittering object inside the capsule floating toward the open hatch. It was Mattingly's wedding ring. He reached for it—and missed.

"Well, lost in space…there it goes," he lamented, as he watched it drift out the hatch.

Then, five minutes later as he watched Mattingly working outside the hatch, he saw the floating ring again, this time headed straight for the back of Mattingly's helmet. It bounced off his helmet and headed back toward the spacecraft again. Mattingly, intent on completing his space experiments, had no idea. As the ring floated back through the hatch, Duke grabbed it. When Mattingly returned from his space walk, Duke was ready.

"Look at that," he said, holding up the ring.

"That's pretty cool stuff, Charlie."

Mattingly retrieved the errant wedding ring and the crew was soon on its way home, sailing through the velvety blackness.

"Coming back, I didn't see the wonders of the creation," Duke said. "The Psalmist says that 'The heavens declare the glory of God and the skies proclaim the works of his hands,' and I didn't see that then. I was more focused on how a big bang came about—not a guiding hand of God, if you will."

Apollo 16 was now in the books, as was the entire Apollo program for Duke.

"When Apollo was over," Duke said, "I had the thought: 'Well, it's all over. Now what are you going to do with the rest of your life?'"

Charlie prepared to go into business. But first he had some unfinished work to complete at home with his wife.

"Although our relationship was shallow," Duke said, "I rather liked it that way. I didn't have time to get involved in her troubles. I didn't want to talk out any problems because if we did, it usually ended up with an argument. I felt if she'd just change, everything would be okay, so I listened with only one ear as she poured out her deepest feelings."

Dotty came to realize as well that it would take years of work to mend their marriage. While Charlie received acclaim from a world of admirers, Dotty struggled to direct his focus back to their union. In his speeches Charlie would say, "The best thing that ever happened to me was my flight to the moon. It was the greatest experience of my life." The statement, repeated at nearly every public speaking engagement, hurt Dotty. "I wanted marrying me to be the best thing in his life," she said.

Desperate for a solution, she got down on her knees one night and asked God to help her heal her marriage. According to Dotty, she got her answer: *"Love Charlie,"* was the reply. *"Don't try to change him. Don't try to save him… Just love him."*

"With those words," she said, "I knew the Lord meant for me to love Charlie the way Jesus loves me—unconditionally."

Charlie's public speeches evolved as his marriage slowly healed. On one particular weekend, on the way home from a three-day retreat sponsored by a church at a nearby ranch, it suddenly all came together for Charlie Duke. All his astronaut training—all the search for fulfillment, all the years—were caught up in a single moment of truth. There in the car with Dotty at his side, he dedicated his life to Christ.

"The Apollo mission was a great accomplishment in my life," Duke said, "but my service to God through my faith is even a greater one. As Ecclesiastes says, the important thing is 'to know God, to love God, and to walk humbly before God.' And I want to be remembered as a man who did that.

"You don't need to go to the moon to find God," he said. "I didn't find God in space. I found Him in the front seat of my car on Highway 46 in New Braunfels, Texas, when I opened my heart to Jesus. And my life hasn't been the same since."

"In December 1972 I was the last man to walk on the moon. I stood in the blue darkness of the lunar surface and looked in awe at the earth. What I saw was almost too beautiful to have happened by accident. It doesn't matter how you choose to worship God... He has to exist to have created what I was privileged to see."

- Gene Cernan, Commander, Apollo 17

35
APOLLO 17: THE LONG LAST LOOK

The moon, scientists have said, was born in violence and searing heat, lived a brief life of boiling lava and massive collisions with chunks of space debris, then died geologically, an infant planet in a primitive state. Since that time, more than three billion years ago, the moon has been quiet, except for a few belches and an occasional fresh meteorite impact. The moon captured this evolutionary moment in time and preserved it virtually unchanged until touched by the bold curiosity of exploring man.

By December 1972, ten men had walked on the moon, fulfilling a centuries-old desire to explore its surface. Two landmarks left by these men wrote a new page in history for the desolate sphere. One rests on the southwestern edge of the Sea of Tranquility—it is a plaque attached to the ladder of the Apollo 11 descent stage, the site where men first set foot on the moon in July 1969. Another plaque lies just five hundred miles to the northeast of a region known as the Taurus-Littrow Valley. This one would close the story:

> Here Man Completed His First
> Explorations of the Moon
> December 1972, A.D.
> May the Spirit of Peace in Which We Came
> Be Reflected in the Lives of all Mankind

The plaque displays a map of Earth's two hemispheres and the near side of the moon. Beneath this are the signatures of Gene Cernan, Ron Evans, Jack Schmitt, and President Nixon. It was placed there by the crew of Apollo 17 in 1972. The mission's commander, Gene Cernan, was the last Apollo astronaut to walk on the moon.

There was the unmistakable sense that something for the ages—something that would be remembered for all time—was nearing an end. A preliminary sign came in July of 1972 with the news that Wernher von Braun was leaving NASA. The man who designed and supervised the building of the rockets now realized that his life's work was complete. Then, *Life* magazine, the iconic American publication, severed its longstanding relationship with the astronauts—a rich source of publicity and political capital for NASA and extra earnings for the astronauts themselves.

During the months preceding this, more than 13,000 workers at the Cape had lost their jobs, and another 900 were headed for the chopping block. Privately, NASA officials were concerned that employees would spend more time and energy preparing for their post-Apollo life and less on this final flight.

"Many of the Grumman troops literally worked themselves into unemployment when our lunar module went out the door at Bethpage," Gene Cernan said, "and more would be gone at the moment of liftoff."

Shoddy workmanship by departing employees was a possibility, and as the mission commander, Cernan took it upon himself to be vigilant for the slightest sign of neglect or complacency. During a visit to the Grumman facility on Long Island he spoke with the supervisor overseeing the last lunar module. "We're giving you our heart and soul on this one, Geno," he said. "This is the best LM that's ever gonna fly." Cernan was glad to see "a dedication to excellence bordering on ferocity," as he put it.

As one threat faded, however, another emerged—this one more sinister.

During the 1972 Summer Olympic Games in Munich, Germany, a group of heavily-armed Palestinian terrorists calling themselves Black September broke into the Olympic Village and took several Israeli athletes hostage. After an eighteen-hour gun battle, eleven of the athletes and coaches were dead. The scene was played out on the international stage and broadcast on live TV around the world. The tragic event, known as the Munich Massacre, weighed heavily on the upcoming flight of Apollo 17, as security experts feared that terrorist organizations might target a moonshot.

Security at Cape Kennedy was dramatically heightened at the direction of Charley Buckley, head of launch site security. But the Cape was not the most likely target. When security experts assessed the situation, they arrived at an alarming conclusion: the most vulnerable targets were the astronauts' children.

Heads of various security departments quietly suggested that armed guards park outside the astronauts' homes, with backup officers on standby should any threat materialize during the mission. Further, they proposed that security officers in unmarked cars pick up the children and drive them to and from school. According to Cernan, the wives were "almost wide-eyed with surprise and fright." So far, the children had been allowed normal upbringings and their families were determined to keep it that way.

As a compromise, armed security guards were allowed to park within sight a short distance from the house. Unmarked police cars would follow the school bus, and the children's classes would be watched over by well-dressed, polite, and very capable federal agents. "Amazingly," Cernan said, "the press never figured it out."

Meanwhile, planning and preparation for the Apollo missions moved forward at breakneck pace. While the final mission looked to be routine, no two lunar missions were alike. Indeed, the final moon shot would be different from its predecessors in a way that made it especially complicated: the Apollo 17 crew would include the first scientist with virtually no flight history, even in an airplane.

His name was Harrison Schmitt, but everyone called him Jack. The astronaut office called him "Dr. Rock." Schmitt had a degree in geology from the California Institute of Technology and a doctorate in geology from Harvard. After his graduation from Harvard in 1964, he worked for the U.S. Geological Survey in Flagstaff, Arizona, developing field geology techniques for the Apollo crews. It was there he learned how far his career could take him.

"When NASA and the National Academy of Sciences decided jointly that they would ask for volunteers from the science community to become astronauts," Schmitt said, "I thought about ten seconds and decided, yeah, that's a good idea."

It was arguably one of the shrewdest decisions a geologist had made in the history of the profession. In June of 1965, Schmitt was selected to join the first group of "astronaut-scientists" and began training to become a lunar module pilot, while simultaneously training his fellow astronauts to become geologists. It was an exchange of expertise that worked exquisitely well for the space program—until it came time to assign flights to individual astronauts. As the later Apollo missions were cancelled, seats on available flights were eliminated accordingly. Moreover, it became common knowledge that the scientific community wanted a scientist on the moon before the end of the moon missions.

Gene Cernan remembered a telling moment when the backup crew for Apollo 16 was announced. "At the Manned Spacecraft Center, groans were heard in many astronaut offices when Deke announced that the backup lunar module pilot [for Apollo 15] would be Jack Schmitt," he said. "The geologist! Jack was everybody's favorite scientist, but some aviator had just lost his seat on Apollo."

Although Schmitt had yet to get on with a prime crew, he was in line to become lunar module pilot for Apollo 18. Now, with Apollo 18 cancelled, everyone in the astronaut office could be reasonably sure that the lunar module pilot for Apollo 17 was about to lose his seat in the name of science.

Slayton had fought the change since it meant breaking up the crew of Apollo 17, which had proven itself a cohesive team over the years. Commander Gene Cernan, Lunar Module Pilot Joe Engle, and Command Module Pilot Ron Evans had trained together as backup for Apollo 14, only to have their lunar module pilot abruptly replaced on Apollo 17 at the bidding of Washington.

Cernan was vacationing with his wife Barbara at Las Brisas in Acapulco with fellow astronaut Ron Evans and his wife when he received the news from Slayton.

"Congratulations, Geno," Slayton said. "Apollo 17 is yours!"

Cernan was exultant—until he realized that Slayton had yet to tell him who his crew was going to be.

"Does this include Ron and Joe?" he asked.

"Well, not exactly. Ron's your command module pilot."

"What about Joe?"

"I need to talk to you about the rest of your crew."

"Why? What's happened?"

"Look," Slayton said, "get back here and we'll discuss it."

Thus far, the only men to have walked on the moon were *astronauts trained in geology*. The scientific community wanted a switch—they wanted a *geologist who was trained in astronautics*. In the end, that's what they got. In August 1971, Dr. Rock learned he would be the second-to-last man to walk on the moon.

"How I came to have that opportunity is something I will never be able to explain," Schmitt said. "When I received the call from Deke Slayton that his original recommendation for Joe Engle to fly on Apollo 17 had been overturned by headquarters, and headquarters had asked him to find some way to put me on a crew, he decided to put me on the crew for Apollo 17. I was excited about the whole thing, but tried not to show it too much."

Back in Houston, Slayton confirmed what the crew suspected. Cernan and Evans argued on Engle's behalf, but it was of no use. But this didn't stop Cernan from trying to make his point, and he didn't mince his words.

"I was being asked to fly a lunar lander down to the moon's surface with a scientist as my co-pilot!" he argued. "Good God, Jack had never flown anything before joining the program. He had done a good job learning how to fly little T-38 trainers, but for Christ's sake, Engle was an X-15 driver!"

So the crew of Apollo 17 was faced with an altogether new challenge. Jack Schmitt had little aviation experience, nor did he have Engle's convivial personality. Moreover, Joe Engle's wife Mary had become extremely close friends with Barbara Cernan and Jan Evans. Schmitt, on the other hand, was a bachelor with intellectual interests that made him seem all the more remote, especially to the wives. Even to their husbands, Schmitt came off as brash and hard to get to know. As Cernan said, "Jack just wasn't my kind of guy."

On first introduction, Schmitt usually came across as unlikable, and his arrogant nature made it hard for people to get close to him. But he didn't seem to care, and he wasn't opposed to getting into almost anyone's business.

"Jack Schmitt came into my office one day and started telling me how to run my office," Reverend Stout said. "He said he had a Ph.D. and to get a Ph.D. you had to do a lot of research, and, you know, he was there to 'help me.'" Stout immediately recognized who he was dealing with. "I appreciate your trying to help me," he responded politely, "but I too have a Ph.D. And I'm here ready to help you any time you need my help." Schmitt left his office.

Then there were professional matters the crew found downright preposterous. In one instance Schmitt put forth the idea of landing on the far side of the moon. The moon had the approximate land mass of Africa, and so far only a few acres had been explored. All of that had occurred on the near side, he argued. Why not venture farther afield?

The proposal was met with ridicule and disbelief, as there were obvious obstacles that made such a landing difficult and much too dangerous to execute. Everyone knew what happened to communications when a spacecraft slipped behind the moon—the crew lost contact with Mission Control. Schmitt was aware of this, but he had a solution: deploy a satellite in stationary orbit around the moon which would remain overhead and allow for constant communication with Houston. Undeterred, Schmitt went so far as to take his idea to NASA management—whereupon it was instantly shot down.

This episode revealed the growing rift between Schmitt and his crewmates. Cernan's and Evans' loyalties remained with Joe Engle, even though neither they nor anyone else could change the circumstances. So Schmitt found himself in an uncomfortable situation that wasn't entirely of his own making.

Meanwhile, NASA scientists were trying to pack every experiment they possibly could on board Apollo 17, aware that this was their last opportunity for live lunar research.

This left Commander Cernan in the position of having to constantly remind NASA scientists of "the definite limits of what we could carry and what could be done." One of the experiments involved five brown and white pocket mice, which would hitch a ride in the command module's cargo bay. They were part of an eminently important experiment, and they needed to be brought back alive.

That the Apollo program was coming to an end put everyone on edge. One day during the Apollo 16 mission, Schmitt was visiting with flight controller Sy Liebergot in the Spacecraft Analysis (SPAN) room and spilled a cup of coffee on Liebergot's console. Smoke began to pour from the console, and when Schmitt tried to open a panel to survey the damage, he knocked over a shelf of documents, setting off another chain reaction. More documents came crashing down, knocking over three more coffee cups. Flight controllers leaped out of the way to avoid being scalded and bumped into more people, knocking over still more documents. The chaos carried over onto a second row of consoles, which left more controllers dodging the cascade of coffee and documents.

"In about thirty seconds," Liebergot said, "Dr. Rock had started a chain of events that just about wiped out the SPAN room. We spent the next couple of hours cleaning up."

The incident was a symptom of the ongoing tension, but Gene Cernan had more important things to consider during the run-up.

"I had no time to worry about social discomfort," he said. "My paramount job was to mold the three of us into a team that could fly to the moon, not to make sure everybody liked Jack." With that, they plunged into training.

The landing site, chosen in February of that year, had been spotted by Al Worden during his lunar orbit on Apollo 15. The Taurus-Littrow Valley was something of a box canyon and an all-around rugged region of the moon where scientists hoped to find another geological treasure trove. The optimum time to land here was when the sun was shining at an angle to the lunar surface, so as to give the astronauts a better outline of the topography and to avoid landing atop a boulder or on the precipice of a chasm. To achieve this precise lunar arrival time, NASA scheduled the first night launch of a Saturn V rocket.

On Wednesday evening, December 6, 1972, nearly half a million people descended upon the Cape to witness the final launch of man's first missions to the moon. Among the crowd was the oldest living American, 130-year-old Charlie Smith. Born in Liberia in 1842, Smith had been tricked as a child into boarding a slave ship bound for the United States. He was in his early sixties when the Wright brothers flew the first airplane in December 1903 and a centenarian when the first atom bomb was detonated in New Mexico. After several countdown delays, he announced to a gathering of the press, "Ain't nobody going to the moon—me, you, or anybody else." But as the clock ticked past midnight, Smith was proven mistaken.

At 12:33:00 a.m. EST, Apollo 17 tore into the night sky in an incandescent blaze. At liftoff the reflected blaze trailing from the five Saturn engines bounced off the clouds, painting the instrument panel a violent crimson. The belching rocket heaved skyward for twelve minutes in an explosive burst of orange fire that seemed to Cernan like an eternity.

Old Charlie Smith appeared unimpressed. "I see them goin' somewhere," he said, "but it still don't mean nothin'."

When they were well on their way, Cernan took a break from the checklist and on the private radio loop asked Deke Slayton an oblique question to be sure the security guards back home were on alert.

"Don't worry, Geno," Deke said. "Everything is fine on the home front." Cernan was then told that his daughter, Tracy, was listening, and he reminded her from a hundred thousand miles away not to forget to feed the horses.

As the crew flew toward the moon, Cernan reported to the world, "Houston, I know we're not the first to observe this, but Apollo 17 would like to confirm the earth is round."

On December 11, Cernan and Schmitt descended to the surface in lunar module *Challenger* toward Taurus-Littrow. Four hours after touching down, the commander was wiggling his way out of the hatch and hopping down the ladder. The lunar soil glittered beneath his boots in the bright sun as if studded with millions of tiny diamonds.

"As I stood in sunshine on this barren sphere somewhere in the universe, looking up at the cobalt earth immersed in infinite blackness," said Cernan, "I knew science had met its match."

A few minutes later he was followed by Dr. Rock. No one was more keenly aware than Schmitt that the eyes of the world were watching him. From the moment he learned of his Apollo 17 assignment, he had immersed himself in the minute details of the mission. According to Cernan, "Jack led the life of an ascetic monk and had spent virtually all of his private time cuddled with schematic diagrams and tech manuals."

"I was focused on what needed to be done," Schmitt said. "We're talking in sheer dollars, something like a million [per] man minute on the moon's surface." That pure economic incentive was significant to Schmitt. Dr. Rock understood his mission: to make sure that each experiment was undertaken, even to the exclusion of absorbing the beauty around him.

Cernan's reaction was more emotional. He found the view of the earth from the lunar surface to be totally arresting. Although he had viewed the earth from space many times on Gemini 9 and Apollo 10, there was something different about seeing it from the lunar surface. So Cernan found Schmitt to be strangely unmoved.

"Hey, Jack," Cernan said during their first EVA, "Just stop. You owe yourself thirty seconds to look up over the South Massif at the earth."

"What?" Schmitt replied. "The earth?"

"Just look up there."

Schmitt took a quick glance. "You seen one earth, you've seen them all," he said, turning to study the lunar soil.

The response was true to Schmitt's dry sense of humor, but Cernan was not amused.

"It was so awe-inspiring," Cernan said, "you had to sneak a glance at it every chance you got… To me, it was like sitting on God's front porch, looking back home."

In Houston, the December weather was cold and wet. Barbara Cernan and 9-year-old Tracy had gone to early morning mass at Ellington. When they returned, the Cernan house was under siege by reporters, photographers, and well-wishers.

Barbara had done her best to hold the crush of press at bay while entertaining family and friends at the Cernan home. "Listening to them land on the moon today was just fantastic," she told a newspaper reporter, as visitors milled about the house behind her. "It's the happiest day of my life."

She was asked whether waiting was easier with her husband's third space flight. "If you think going to the moon is hard," she said with nervous excitement, "try staying home."

Over the years, Barbara had proven herself to be the perfect astronaut wife—always supportive of her husband's demanding schedule, always able to keep the household humming on her own. Now, with her husband's historic flight, she had to be ready with a thoughtful quote. But as her husband took his first steps on the moon, Barbara's easy, breezy togetherness began to crack.

Out of the blue, she turned to a friend and said, "I've just got to be alone for awhile."

But there was nowhere to go, no place to which she could retreat. Outside, a company of reporters had set up camp; inside was a maze of family and friends. Nevertheless, she retreated to the bedroom where it was quieter, but not quiet enough. So she went into the bathroom and took a hot shower. She needed some peace, and the pounding water would muffle the ambient noise. Stepping inside the cocoon of hot steam, she curled up on the shower floor and allowed herself to come undone.

Thirty minutes later, she was her old self again, her hair done, lipstick precisely applied. She emerged from the bathroom, smiling, confident, and ready for the show.

With each passing hour, the misunderstandings between the two moonwalkers seemed to dissolve as levity and elation eroded the tension.

During their first EVA, Houston informed the astronauts that they were already behind schedule, which meant they would have to pare down the day's list of geological work. "Jack wasn't a happy camper," Cernan said. Nevertheless, as the two astronauts bounded through a boulder field, Schmitt suddenly broke into song:

"I was strolling on the moon one day…"

"In the merry, merry month of December…" Cernan added, then feigned to correct himself: "…No, May…"

"May…" Schmitt carried on.

"May's the month."

"When much to my surprise, a pair of bonny eyes… be-dooppy-doo-doo…"

But as they unloaded gear from the "moonmobile," Cernan's rock hammer caught one of the rover's fenders, breaking it off. Several hours and one roll of duct tape later, they thought they had it fixed. They didn't. Once underway, a huge rooster tail of moondust shot up, spraying everything and reducing their speed to a slow crawl. "I needed to get that damned fender fixed," Cernan recalled, "and there wasn't a repair shop within 250,000 miles."

The crew continued to fall further behind schedule, which meant compressing the remaining time for geological exploration at Crater Emory. Meanwhile, orders from the scientists in Houston flooded in at such a pace that they frustrated even Schmitt, who was beginning to see just how exasperating his science counterparts in Houston could be. As the hours on the surface passed, Schmitt and Cernan came to develop a deep appreciation for each other.

"Just as Jack had become a pretty fair pilot," Cernan said, "I had become a pretty damned good lunar geologist... the result was that we were a doggone good team... Jack Schmitt belonged up there."

On the second EVA, the astronauts headed off in the rover to a crater called Shorty. It was here that Jack Schmitt proved his stuff as a geologist. As they bounded about the crater, the scientist made a surprising discovery. Where his feet had disturbed the surface, he noticed something orange.

"Hey! There is orange soil," Schmitt shouted. "It's all over!"

Cernan wasn't buying it. "My first feeling was that Jack Schmitt had been on the moon too long." Buzz Aldrin thought he'd seen a purple rock on the Apollo 11 moon landing and Cernan knew the tinted visor could be deceiving. But there it was.

"Hey, it is!" Cernan confirmed. "I can see it from here!"

The discovery would provide a view into the moon's geological makeup. Soil analysis would determine that the orange soil was really tiny beads of glass containing titanium, bromine, silver, zinc and cadmium. The moon was revealing itself to be far more diverse in character than previously thought. Instead of shades of black and white, it was a range of colors and composition. Although it held no forms of life, it contained a rich geological history.

With the completion of the final EVA, it was time close down and pack up for the trip home. Cernan and Schmitt had spent over seventy hours on the surface, twenty-two of which they had spent hiking through and driving over the mountains of the moon. They stowed the rock samples and equipment, and Schmitt climbed the ladder first.

Cernan paused to look back at the serene, mystic view before boarding the lunar module, knowing that whatever he said would be the last words spoken on the moon for years to come, maybe forever. For that brief moment, he experienced a surge of emotion. Attached to the ladder of the descent stage was the Apollo 17 plaque, which he read to the worldwide audience. Then he added a comment of his own:

"This is our commemoration that will be here until someone like us, until some of you who are out there, who are the promise of the future, come back to read it again and to further the exploration and meaning of Apollo."

The past three days had changed Gene Cernan.

"As I take man's last step from the surface," he continued, "I'd like to just say what I believe history will record. That America's challenge of today has forged man's destiny of tomorrow. And, as we leave the moon at Taurus Littrow, we leave as we came and, God willing, as we shall return, with peace and hope for all mankind."

As he turned to the ladder, he noticed a small sign apparently pasted there by an unknown worker, now likely unemployed. It read: "Godspeed the crew of *Apollo Seventeen.*" Cernan hesitated with his foot on the ladder, not wanting to end the moment.

"When I looked over my shoulder, there was the earth. And it's been looking at us for seventy-two hours. And that's home, and that's love, and that's life, and that's everything you understand is the past and the future."

He wanted to push a freeze button. One, two, maybe three minutes he stood there.

"And here I am standing on the moon ready to return to go back to the real world, to my real identity. I really wanted to understand more deeply what we had done—not just technologically, not just scientifically, but the meaning of us being there."

Cernan had caught a glimpse of infinity and the unlimited possibilities of mankind in a universe without boundaries. Overflowing with emotion, he lifted his foot from the soft moon dust, leaving behind a final boot print, and ascended the ladder. Once aboard, he and his crewmate prepared to leave. When the final preparations were complete and *Challenger* was ready to launch, Cernan silently thanked God for giving him the opportunity to go to the moon. He said a Hail Mary and made the sign of the cross. He then poised his left index finger over the yellow ignition button. The last words spoken from the moon were casual but decisive:

"Okay, Jack," Cernan said to the scientist at his side. "Let's get this mutha outta here."

> "We must abandon the view of ourselves as provincial limited victims of life, accidents of evolution, and start to view ourselves as limitless creatures of God, capable of exploring the vast reaches of space."
>
> *- Edgar Mitchell, Lunar Module Pilot, Apollo 14*

36
MOONDUST

If a traveler were to journey to the moon today, he or she would find all of the instruments and equipment left behind at the six Apollo landing sites—all of it sitting in static silence amidst a field of boot prints and tire tracks. Those tracks would appear almost exactly as they were left on this airless world several decades ago. Traversing the surface, the traveler would also find the priceless relics and messages that reveal the identity of those who had ventured here and why they had come at all. Among the most interesting artifacts to be found would be a figurine lying next to a plaque engraved with fourteen names. If the traveler looked carefully enough he might find the remnants of a small red Bible resting atop a rover parked near Hadley Rille.

Other things had been intentionally left behind on the rocky surface of the moon. There, six spindly lunar modules, six American flags, and three electric lunar rovers remain. Next to one of the rovers, written in the soft sand, are the initials "TDC," for Tracy Dawn Cernan, left there by her father. A future moon traveler might wonder why this particular rover is missing a fender. Gene Cernan brought the damaged fender 250,000 miles back to earth and returned it to Boeing, the original manufacturer, with a note: "This thing only went twenty miles and it's still under warranty. I want you to go back there [to the moon] and fix it."

Boeing engineers had a field day with it. They took the fender from the backup lunar rover and returned it to Cernan with a plaque that said, in essence: "Gene, here's the fender. If you want to use it, it's still back on the moon, so go ahead if you can get there."

Reverend Stout had watched Apollo unfold from its fiery beginning to the final slam of the hatch on Apollo 17. He witnessed first-hand the dedication each crew mustered to prepare for their mission. So entirely consumed were the astronauts with their missions that life outside of the Apollo space program was utterly ignored. "The mission was more important than ourselves, our lives, or our families," Frank Borman admitted.

There would be a price to pay for this singular focus. "NASA had taught them how to be astronauts," said Stout. "But they didn't teach them how to be ex-astronauts."

Now with Apollo ending and public interest waning, humanity's first space explorers faced "life after space." "I no longer have the luxury of being ordinary," said Cernan. "Trying to exist within the paradox of being in this world after visiting another may be why some moon voyagers tend to be reclusive."

When he returned to his family after his Apollo 10 mission, Cernan was brimming with euphoria as he hugged his six-year-old daughter, Tracy. "Punk," he said affectionately, "your daddy's gone closer to the moon than anyone ever has before. You know, it's real far away in the sky, up where God lives."

The little girl considered his words for a moment, perhaps remembering the times they had missed together, and replied, "Daddy, now that you've gone to the moon, when are you going to take me camping…like you promised?"

"Here I was trying to impress my little girl," Cernan recalled, "and she wanted to remind me of something that was much more important than listening to her father talk about going to the moon. My young daughter had just brought me down to earth and a world of hurrahs faded to almost meaningless insignificance. She wanted a daddy, not an astronaut."

Clearly the astronauts were aware of their place in human history. Their lives were forever changed. And a world that had long considered itself the center of its own universe was most certainly changed. The vast cosmic endeavor, however, had come at a price, not only in hard dollars, but in human life. Eight U.S. astronauts had perished during the space race. One of those names in particular begged for recognition.

In 1970 Ed White's sacrifice was recognized in a most public way. Bob Hope, the most famous entertainer on the planet at the time, decided to roll out *The Bob Hope Extra Special,* a televised four-and-a-half-hour benefit to raise money for the Edward White Youth Memorial Center Fund. Tom Stafford and Gordon Cooper agreed to serve as trustees. The show would honor all eight fallen astronauts and raise $360,000 to build a memorial center slated to go up next to White's church, Seabrook United Methodist in Houston. White himself had made the first contribution, donating his $500 prize from the Freedom Walk Award received for his historic Gemini 4 spacewalk in June of 1965. From that moment on, the fates of Seabrook Methodist and the space agency were inextricably intertwined.

Bob Hope unabashedly admired the astronauts and promised the special would be a "Texas-sized" event in the new Houston Astrodome. White's fellow astronauts immediately fell in line. But the prospect of filling the 50,000 seats in the nation's largest indoor sports arena concerned even the venerable Bob Hope. His biggest resource would be his army of celebrity friends, who volunteered to appear at no cost for what proved to be an unforgettable performance.

For music, Hope insisted on the Les Brown Band. He convinced Gregory Peck, who had just logged five months at Cape Kennedy working on the space movie "Marooned," to join the cause. Hope used all the charm he could muster to enlist a reluctant Cary Grant, who finally relented, saying, "Only Hope could get me to do a show like this one...I melt in front of a microphone."

The list of celebrities was a Who's Who of the time. David Janssen, Dorothy Lamour, Joey Heatherton, Glen Campbell and Robert Goulet all answered the call. Also on the bill were Bobby Sherman, Nancy Ames, Frankie Valli and the Four Seasons, the Friends of Distinction, Trini Lopez, John Rowles, the Step Brothers, and recent Heisman Trophy winner, O. J. Simpson.

But what captured the audience's attention most were Hope, Peck, Janssen, and Grant in a clever song sketch called "Showmanship" and a sentimental tribute to the astronauts themselves, entitled "We Love All Those Wonderful Guys."

Just before the close, Hope's name flashed in lights on the Astrodome scoreboard and the comedian came back onstage to announce that 46,857 people had paid to see the show, a record for the Astrodome and for any previous indoor variety show produced in the United States. Finally, Hope was presented with a life-size portrait of himself which would hang in the foyer of the Ed White Memorial Center. As the show closed, the audience was on its feet, applauding for as far as the eye could see in the half-light of that huge domed auditorium.

The extravaganza was so successful that Hope agreed to return to Seabrook Methodist Church for the dedication of the Ed White Memorial Center. In September 1971, he made good on that promise. But in the rush to finalize arrangements, the program organizers at the suburban church realized they had nothing to present to Hope as a token of appreciation. Someone suggested a lunar Bible.

This offered up a dilemma, as Stout had stored the lunar Bibles in a bank vault and the banks were now closed. He knew of one particular lunar Bible, however, that might be available—one presented to APL member Karen Shaw. Karen, married to a Seventh Day Adventist, confided in Stout that she had hidden the Bible in a closet for fear her husband would object to her owning a ceremonial artifact.

"So I rushed to her," Stout said, "and told her we needed it—and she gave it up right away." Stout turned the lunar Bible over to program organizers, and Karen Shaw's Bible became known as the "Hope Bible."

Once again, Ed White's fellow astronauts came forward. Rusty Schweickart, Gordon Cooper, Scott Carpenter and Tom Stafford were on hand, along with Seabrook Methodist pastor Reverend Bob Parrott and Bishop Kenneth Copeland, when the Hope Bible was presented to its namesake.

―᠊᠊᠊ᛝ ᚼ᠊᠊᠊―

As director of the Apollo Prayer League, Reverend Stout was responsible for maintaining the provenance of the lunar Bibles, and the task could not have fallen to a more conscientious steward. As a scientist, he possessed a natural penchant for detail. Throughout his career he had calculated and documented the nuances of complex events. Above all, he understood the importance of assiduously documenting the historical provenance of the first lunar Bibles, a duty he resolutely undertook immediately after the splashdown of Apollo 14. This was the third time he had watched and prayed for the biblical landfall, and he realized that documenting the Bibles was as important as the actual landing itself. These were not only Bibles, but historical artifacts.

Once the media drone had subsided, Stout took the lunar-landed Bible packet to a local newspaper office in La Porte, Texas, where a reporter and photographer were waiting to cover the opening of the first lunar Bibles. Taking a seat at a newspaper desk, he carefully opened the packet of biblical tablets that only weeks before had been to the surface of the moon. As he slipped them from their fireproof wrapping, the photographer's camera lens was hardly wide enough to capture Stout's radiant smile.

Later, he returned home with the packet and sat quietly at his table in a moment of solitude. He then removed a small engraving instrument and an NCR microscopic eyepiece from their case and arranged the items on the table in front of him. Then, one Bible at a time, he painstakingly engraved a unique microscopic five-digit serial number on each microfilm, designating its sequence and placement of that particular Bible on its historic journey. When finished, he entered the serial numbers in the Apollo Prayer League Lunar

Bible Registry in blocks of twenty, according to their original index group. To make certain his records were accurate, he asked APL secretary Byron Price to count them again. All were accounted for and properly serialized.

In the days and weeks that followed, Stout distributed many of the lunar Bibles to individuals closest to the project. Fifty-seven were designated for various individuals, politicians, dignitaries, and entertainers in recognition of their support. A few others were "segmented" into tiny fifty-page portions and given to APL family members and churches involved in the effort.

"Yul Brynner received a part of the Bible," said Stout. "Helen and I caught Dale Evans and Roy Rogers at a restaurant and presented his [Bible] there. The conversation drifted to Madalyn O'Hair, at which point Roy Rogers stuck out his tongue and gave her a 'raspberry,' saying that is what he thought of her."

Another was given to President Nixon, "since he's the only president whose name is on the surface of the moon," Stout said. George H.W. Bush, then U.S. Ambassador to the United Nations, also received a copy, along with Vice President Spiro Agnew, who responded in a letter, "Of my many mementos of the space program, your gift to me is one of the most prized."

Johnny Cash and June Carter asked for two segments of the lunar Bible as an expression of their lasting love for each other. When they met, Stout asked Cash for his autograph and then told the country singer that the name John Stout had appeared in an English folk song: "Who put the pussy in the well? Who put the pussy in the well? Little Johnny Stout put the pussy in the well!"

"Someday, John," Cash replied, "you will become very important." He then chimed in with his own country version of the tune: "Who put the Bible on the moon? Who put the Bible on the moon? *Little Johnny Stout put the Bible on the moon!*"

Alabama Governor George Wallace was so taken with the entire endeavor that when Stout gave him a segment of the lunar Bible, he granted him an honorary appointment. "All I had out of the military was a piece of paper," Stout said, "and Wallace sent me a gold-edged certificate of appointment as a colonel in the Alabama State Militia."

―――

Not long after the return of Apollo 14, Stout made the call he hoped one day to make. He phoned Ed White's widow, Pat, and asked if they could meet. Then, on an otherwise uneventful morning in the modest suburb of Seabrook, Texas, he presented her with a copy of the lunar Bible whose journey to the moon was inspired by her husband. Stout now somehow felt at peace, knowing that he had come full circle for the young astronaut.

The White family would always be close to the Seabrook Methodist congregation. During a visit to the church in 1971, Ed's father addressed its parishioners. Standing at the podium, Edward White Sr. spoke of the space program, the war in Vietnam, and the youth center dedicated to his son. At the time, his only other son, Air Force Capt. James B. White, had been missing in action in Laos for over a year.

"This program is the future of the United States," he said. "It's the new frontier and the last frontier. Cutting down on the space program is ridiculous and one of the silliest economic moves this country can make. It's like killing the goose that lays the golden egg."

Ed White was gone, but his presence, like that of the Apollo program, would linger in the hearts and minds of the congregation for years to come. Mary Brandt was a parishioner at Seabrook Methodist at the time of the Apollo 1 tragedy and memories of the lost astronaut still linger. "Ed always sat in the same place in church," Mary recalled. "After he was killed, it was so hard not to look and see if he was sitting there at the end of the pew on the aisle half-way down in the sanctuary."

From the outset, it looked as though the decade of the 1970s would hold economic challenges. The Nixon administration surmised that the downturn started in the autumn of 1969 and hit its low point in the fall of 1970. Key decision makers saw a recession looming and began cutting discretionary spending programs. NASA fell victim to the cuts. It would have cost another $20 billion to continue the program all the way through Apollo 20 and by now many in the American public couldn't remember why the country had gone there in the first place.

> "Man is free to shape his world. He is free to adventure into space for as far as he is capable. The sun, the moon, and the stars, the animals, and all of the goodness of the earth are given to sustain his life. The heavens and the earth which God brought into being are here for man to use. He must make good the trust given to him by God."
>
> *- Reverend John Stout,*
> *Director, The Apollo Prayer League*

37

END OF AN ERA

What began as a space race, in the end succeeded in giving all of humanity an expanded view of our universe. But with the lunar missions nearing an end, there was a very earthly reality to confront: America wasn't going back to the moon or anywhere else in space anytime soon. By the fall of 1972, the trickle of layoffs within the ranks of NASA and its contractors had grown into a tidal wave of bad news and pink slips. No one, it seemed, was immune.

The reality was emotionally difficult, not only for the astronauts, but for the scores of employees who had invested their time, labor, and so much of themselves into the grand endeavor of Apollo. It was a change that would have an enormous impact on millions of lives. John Stout was no exception.

Shortly before the final Apollo mission, Stout's supervisor invited him to lunch. "A must in get-a-longings with the boss," Stout said. John anticipated the usual shop-talk, but shortly after the food arrived, the conversation took an unexpected turn. His boss laid out what he wanted to say in a few direct words.

"John, we have terminated your position," he said matter-of-factly. "You are to terminate all those working for you and all programs that you were working on by 2:00 p.m. I'm sorry, I can't wait. I have to catch a plane." With that, he fished a dinner roll out of the basket, glanced at his wristwatch, and left.

At the time, Stout had twenty-seven programs running and so many people working for him that he had lost count. Although Apollo was winding down, other projects such as Skylab and the "reusable" Space Shuttle orbiter were on the drawing board and coming online within the next few years. In a major geopolitical turnaround, NASA had begun to collaborate with the Russians on a joint Apollo-Soyuz test project, the predecessor to the *Mir* program, whose purpose was to establish an inhabitable space station for use by cosmonauts and astronauts from thirteen countries. While these programs paled in comparison to the scale and majesty of Apollo, Stout believed he would still have a home at NASA for some time to come.

Information in those days was stored on IBM punch cards and Stout had saved all of them in six steel filing cabinets lent to him by Lockheed. He was now told that the filing cabinets, filling an entire wall of his office, would have to be returned. That night he and Helen went back to the Manned Spacecraft Center and dumped all the cards on the floor, filling the center of the office nearly a foot deep. They shoved the empty filing cabinets down the hall to the elevator and pasted a sign on them: "RETURN TO LOCKHEED."

Stout soon realized he was only one of many in this predicament. Five hundred NASA employees were laid off during that round of cut backs alone. The scope and pace of layoffs was relentless. Shortly after Stout was terminated, his supervisor was terminated.

What started as a challenge to the nation became a triumph for mankind. More than that, it became a philosophical and spiritual experience that united a deeply divided world, if only for a short time. Falling hard on the heels of the triumph, however, was the sad reality that Apollo was coming to an end.

If you had told NASA employees in the early 1970s that in forty years, human presence in space would amount to an earth-orbiting space station and Shuttle missions, they would not have believed it. When Apollo 17 splashed down in December 1972, conventional wisdom assumed that humanity was on the cusp of exploring the solar system. Some thought a permanent base on the moon to be the logical next step. Others felt going back to the moon would only drain precious resources needed for a mission to Mars—the ultimate goal of this generation of explorers. In either case, popular perception was that earthlings were on their way to the Red Planet.

In fact, plans for a Mars mission had already been drawn up. On August 9, 1969, just three weeks after Neil Armstrong and Buzz Aldrin walked on the moon, Wernher von Braun stood before the Senate Space Task Group to present a remarkably detailed proposal for a Mars mission. In retrospect it appears either majestically audacious or dreamily fantastic. In any case, von Braun was attempting to strike the funding iron while it was still hot.

The plan proposed a two-year journey to Mars using one or two nuclear-powered vehicles that would be assembled in earth orbit. With two vehicles, a sister ship could be used as a life raft in the event of "a major failure," as von Braun put it. The ship, or ships, would require 270 days to get to Mars and, according to the proposal, slip into Mars orbit "in the same fashion that the Apollo moon ship was placed into lunar orbit." Six to twelve astronauts would be in Martian orbit or on Martian soil for some eighty days before returning to earth, a journey that would take another 290 days. According to von Braun, the highlight of the return voyage would be a close fly-by of Venus. The target date for the journey was 1982, with work beginning immediately.

As extraordinary as this proposal appeared at that time, it was very real to von Braun and to the technicians and astronauts at NASA in 1969. President Kennedy had initiated the race to the moon only eight years earlier; hence, sending a team to Mars in thirteen seemed within the realm of possibility. But in the face of the U.S. "victory" in putting a man on the moon, the Soviet Union abruptly abandoned the idea of heroic space travel. The race was over, and so, presumably, was any reason to continue.

"A more ambitious day will inevitably come," Edgar Mitchell said. "Perhaps it will be our children or grandchildren, or even their children, but one day a craft from this shimmering blue dot will lower into a pale red Martian horizon. Then, gradually, imperceptibly, but inevitably, the shimmering blue dot will slowly recede in the view of the spacecraft that will carry our children's children throughout the ghostly white of the Milky Way. Still others will follow, and with them the ancient stories of their predecessors. Then they will leave the galaxy in order to make themselves in the image of God."

Mitchell's generation embodied a grander way of thinking, a way of viewing our world as only a few have seen it—from the vast perspective of another world. In a letter to Apollo 15 Lunar Module Pilot Jim Irwin, Mitchell summarized the impact of his experience. His language was neither technical nor heroic. He wrote as a brilliant astronaut who had come to see the lines between science and spirituality as nonexistent.

> As a result of my experience in space, any doubts that I had about the Universe being a Divine Creation evaporated, to be replaced with the certainty that the physical Universe and its creatures are the result of divine thought and purpose. I view the spiritual aspect of life as the most important part of human experience and believe that growth in the spiritual dimension is limited only by our individual unwillingness to see beyond our fears and selfish interests. To seek tranquility and joy by realizing one's true spiritual nature is the ultimate goal of all life.

Mitchell was not alone in this way of thinking. Apollo 12 astronaut Alan Bean's pastor, Reverend John Fellers, noted the melding of spirituality and exploration in a passage in the March 1970 Apollo Prayer League Newsletter:

"We humbly beseech Thee," he wrote in an open prayer, "that Thou will continue to guide and direct the exploration of space, that all may be done in this endeavor for the benefit of people throughout the earth to the end that we may have a clearer knowledge of Thee."

Here was the very thread that connected religion and science. Man's quest for knowledge was at its heart a quest to better understand the universe and its Creator.

The most renowned of the lunar Bibles, the Hope Bible, eventually made its way back to Reverend Stout after a fire destroyed much of Bob Hope's private collection of memorabilia. When Stout learned of the fire, he inquired about the Hope Bible and was told that, while some of the items had been salvaged, the Hope Bible was not among them. He had all but given up on recovering it when he received word that the Bible had been found. It was graciously returned to him and assumed its rightful place in the APL lunar Bible archives.

Madalyn Murray O'Hair, in the meantime, had all but disappeared from the headlines. As she slipped from public view in America, her interests shifted elsewhere. The First World Atheist Conference was scheduled for December 1972 in India, and O'Hair was busy expediting her visa.

―⁂―

It was an era of unprecedented change. In the 1950s only dreamers, lovers, and astronomers looked at the moon. Sixty thousand feet above the surface of the earth was a region that only experimental aircraft, weather rockets, and balloons could attain. Few scientists and even fewer average citizens could name all the planets of the solar system, much less discuss them with any degree of authority. Space was not a subject of common or daily interest except to astronomers and cosmologists.

The country was far different then. Many Americans had barely learned where Russia was on the map, families hand-cranked home-made ice cream in the back yards of their homes, and freshly-laundered clothes hung flapping on the clothesline. Television programs were broadcast in black and white, telephones had cords, and radios received only AM stations. Cell phones and panty hose had not yet been invented.

Twelve years later we were planning to explore the moon. Twenty years later we had been there and back.

In the end, it was the blending of scientific ideas, individual convictions, and unconventional courage that accounted for the success of Apollo. America's best from all ranks faced unparalleled challenges and succeeded, accomplishing what was thought impossible, and became American legends in the process. Among those lesser legends was a humble, relatively unknown pastor by the name of John Maxwell Stout.

Reverend Stout had lived a modest life, yet he would continue to show immense relevance to the world around him. During his life he brushed against the edges of greatness, always finding himself at the nexus of science and religion, ever pushing the boundaries of possibility.

In the years since Apollo, lunar Bibles have become treasured artifacts sought by Bible enthusiasts, collectors, and museums across the country, although they are sometimes confused with their unmarked counterparts flown on Apollo 13 or those distributed by the NCR at the 1964 New York World's Fair. But if one looks closely with the aid of a microscope, the serial numbers in John Stout's carefully engraved script come into view; and the viewer will know with some certainty that the Holy Scripture he is viewing is indeed a treasured fragment of space and biblical history.

As Reverend Stout prepared to leave NASA, he wondered openly if the end of Apollo would signal an end to reliance on God for those who worked there. His final newsletter echoed this emotion:

> When we look up to the heavens, are we reminded of God? If we have put Him out of our hearts, if He no longer belongs in our lives, if there is no place for Him in our society or in our schools, if we have become a nation "under God" in name only, or to put it another way, if we have put Him in a box and buried Him in some distant cosmic grave, then we may not see Him when we look up into the heavens. Russian astronauts were quick to report to us that they did not see our God up there. One of our own astronauts carrying an Apollo Prayer League Bible told us after his flight, "If you don't take Him with you, you are not going to find Him up there."

With the shadow of Apollo fast falling upon him, he made his last trip to the office to gather his personal belongings: the APL records; the microscopic eye piece and engraving tool; the picture of his beautiful wife, Helen, the love of his life; and the small framed photo of his son, Jonathan.

John Maxwell Stout walked out of his office at the Manned Spacecraft Center for the last time that evening in 1972, leaving behind a job that helped take twenty-four men to the moon, twelve of whom walked among its hills and craters. One landed the first Bible on it. Before closing his door, he turned for one last look. The view from his office window spanned a vista of a once-bustling epicenter of an ambitious new space program empowered to put a man on the moon—made all the more brazen, in retrospect, by the fact that nobody knew it was virtually impossible.

He closed and locked the door behind him.

"Fifty years after we are gone," he thought, "we may only be a footnote to what we have left behind. After another 100 or 500 years, we may be remembered only by our being associated with the fact that the Bible went to the moon. And while we were accomplishing this, we found something even greater—we found our own will, our own strength, our own future. We found ourselves."

Say to me, no more Apollo,
Say to me the job is done—
Then I say your words are hollow,
The work has just begun.

Alfred M. Worden, Command Module Pilot, Apollo 15
Excerpt from his poem "Apollo Lost" in Hello Earth

Photos

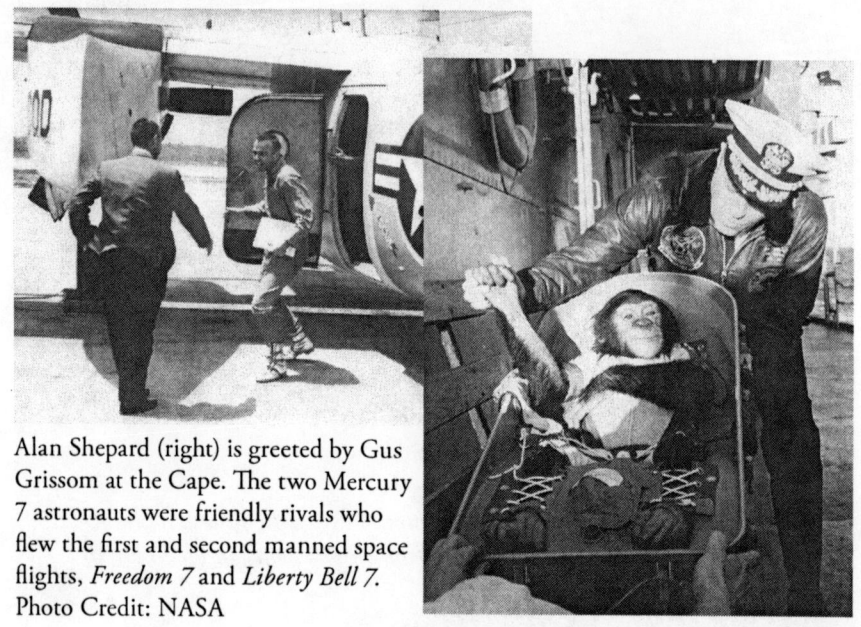

Alan Shepard (right) is greeted by Gus Grissom at the Cape. The two Mercury 7 astronauts were friendly rivals who flew the first and second manned space flights, *Freedom 7* and *Liberty Bell 7*.
Photo Credit: NASA

Chimpanzee Ham was rescued in good shape from his Mercury-Redstone 2 flight on January 31, 1961, after the rocket over-accelerated and splashed down in rough seas hundreds of miles off course.
Photo Credit: NASA

The Mercury 7 astronauts during survival training in Nevada. Portions of their clothing were fashioned from parachute material. From left: Gordon Cooper, Scott Carpenter, John Glenn, Alan Shepard, Gus Grissom, Wally Schirra, and Deke Slayton.
Photo Credit: NASA

Ed White space walk on Gemini 4
Astronaut Ed White uses the nitrogen-pressurized "zip gun" to propel
himself during his Gemini 4 space walk on June 3, 1965.
Photo Credit: NASA

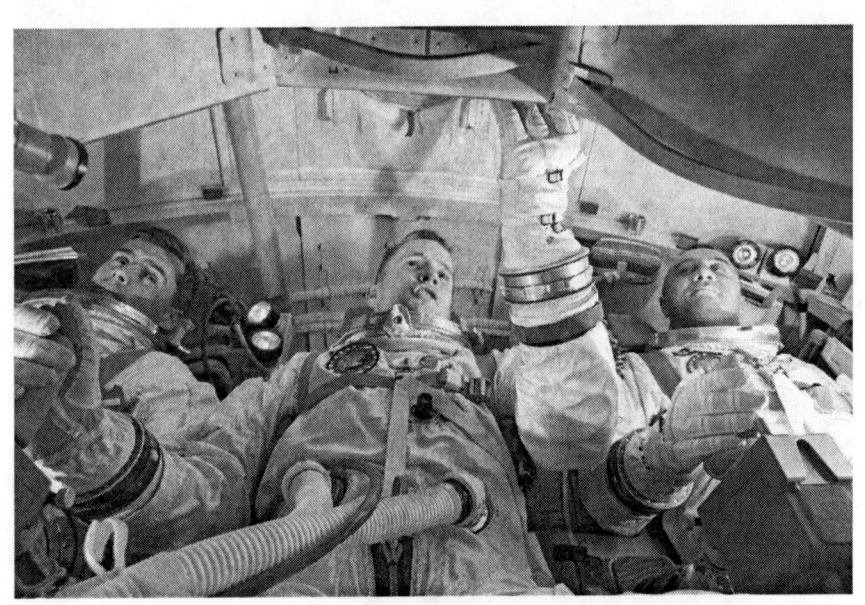

Apollo 1 crew
The Apollo 1 crew during training in the Saturn-1B AS-204 capsule.
From left: Roger Chaffee, Ed White, and Gus Grissom. The crew
died in a flash fire on the launch pad January 27, 1967.
Photo Credit: NASA

Reverend John Stout at Cape Canaveral in 1964 where he worked for NASA contractor, Pan American Airways, as manager of the Air Force Eastern Test Range information dissemination system and served as an industrial chaplain.
Photo Credit: Apollo Prayer League archive photo courtesy John and Helen Stout.

Madalyn Murray O'Hair, an avowed atheist, became infamous for her 1963 federal lawsuit banning prayer in U.S. public schools. She later filed suit against NASA in an unsuccessful attempt to ban religious acts in space.
Photo Credit: Courtesy www.fliker.com, photo by Alan White.

Apollo 12 astronaut Alan Bean (left) presents a printed leather Bible he carried on his flight to Reverend John Stout.
Photo Credit: Apollo Prayer League archive photo courtesy John and Helen Stout

"Earthrise" photo taken by the Apollo 8 crew Christmas 1968, showing earth for the first time as it appears over the lunar horizon. In an historic live broadcast that night, the crew took turns reading from the Book of Genesis. Photo Credit: NASA

Apollo 11 Commander Neil Armstrong prepares for a training exercise in the Lunar Landing Research Vehicle (LLRV). Armstrong later bailed out of the machine seconds before it crashed. Photo Credit: NASA

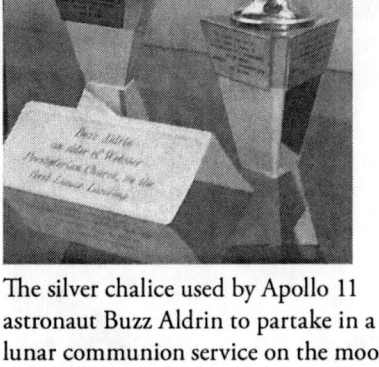

The silver chalice used by Apollo 11 astronaut Buzz Aldrin to partake in a lunar communion service on the moon. The 3 ½" chalice is displayed alongside the Presbyterian Church seal at Webster Presbyterian Church, Houston, TX. Photo Credit: C. L. Mersch courtesy Webster Presbyterian Church.

The silicon disc left on the moon by Apollo 11 crew, inscribed "From Planet Earth July 1969," measured roughly 1 ½" in diameter and contained microscopic images of good will messages from heads of 74 nations. Photo Credit: Courtesy of Tahir Rahman, *We Came in Peace for All Mankind*, (Photo enhancement by James A. Rendina)

Sy Liebergot, Apollo 13 EECOM Systems Flight Controller at his console in Mission Control at the Manned Spacecraft Center in Houston. Liebergot later served as EGIL Systems Flight controller for *Sklylab*.
Photo Credit: NASA

Apollo 13 Flight Director Gene Kranz lights up a cigar in Mission Control as Commander Jim Lovell appears on the big screen safely aboard the rescue carrier *USS Iwo Jima*.
Photo Credit: NASA

The *Singin' Wheel*, a favorite hangout of NASA mission controllers, was known for its pickled pigs' knuckles, shuffleboard, and abundance of beer.
Photo Credit: Courtesy Sy Liebergot

The Apollo 14 crew from left: Stuart Roosa, Alan Shepard, and Edgar Mitchell. On February 5, 1971, Mitchell succeeded in landing the first lunar Bibles on the moon on behalf of The Apollo Prayer League.
Photo Credit: NASA

Apollo 14 lunar module *Antares* rests on the surface of
the moon carrying the first lunar Bibles.
Photo Credit: NASA (Photo enhancement by James A. Rendina)

"A Moment of Reflection" painting by Ed Hengeveld depicts Apollo 15 Commander Dave Scott placing a red leather Bible on the console of the lunar rover before departing the moon.
Photo Credit: Courtesy Ed Hengeveld

The "Genesis Rock" retrieved by Apollo 15 astronauts Dave Scott and Jim Irwin from the surface of the moon is believed to be part of the lunar crust formed over four billion years ago.
Photo Credit: NASA

Apollo 17 Commander Gene Cernan, the last man on the moon, with his wife Barbara and daughter Tracy. The astronauts' children were assigned security guards during the mission due to terrorist threats surrounding the 1972 Olympic Games attack known as *Black September*.
Photo Credit: Courtesy Jerome Bascom, Apollo Mission Photos

The Apollo 17 lunar rover fender, accidentally damaged by a rock hammer in Cernan's pocket, was repaired on the moon with a plastic-coated flight plan cover and duct tape.
Photo Credit: NASA

Apollo 14 astronaut Edgar Mitchell returns the microfilm lunar Bibles to Reverend John Stout after his release from quarantine in 1971.
Photo Credit: Apollo Prayer League archive photo courtesy John and Helen Stout

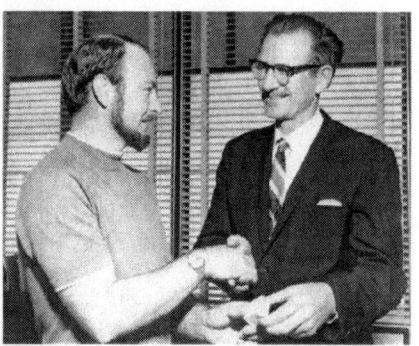

A smiling Reverend John Stout examines the first lunar Bibles from the Apollo 14 packet at a local newspaper shortly after their return in 1971.
Photo Credit: Courtesy *Bayshore Sun*, LaPorte, TX.

The Apollo Prayer League microfilm lunar Bibles produced by NCR and published by World Publishing Company contained the complete text of the King James Bible (actual size).
Photo Credit: Photo courtesy Cornelius Photography.

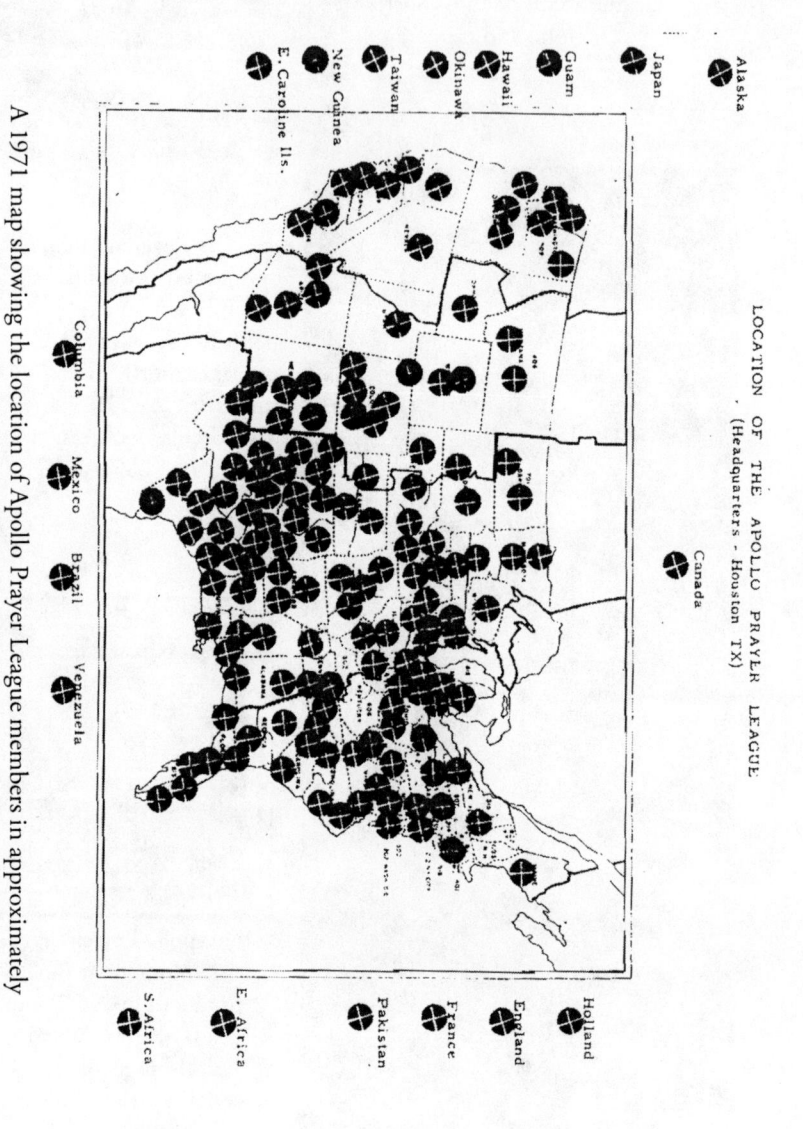

A 1971 map showing the location of Apollo Prayer League members in approximately 60 percent of all U.S. postal zones and 16 foreign countries.
Photo Credit: Apollo Prayer League archives courtesy John and Helen Stout.

I would be happy to forward any of your ideas or suggestions to the proper authorities here in Washington. Keep up the good work. I hope the next time you are in Texas we will be able to have another visit at the ranch.

Warmest regards,

Sincerely

Lyndon B. Johnson

Reverend John M. Stout
Caixa Postal 10
Lavras, Minas Gerais
Brazil

Excerpt from Senator Lyndon B. Johnson letter to Reverend Stout May 29, 1958.
Photo Credit: Apollo Prayer League archives courtesy John and Helen Stout.

Reverend John M. Stout, 2009
Photo Credit: Courtesy Lohman Photography, Pasadena, TX.

A rare encased lunar-landed Bible from the private collection of Reverend John and Helen Stout on display at the Dunham Bible Museum of America, Houston, TX.
Photo Credit: Courtesy Cornelius Photography.

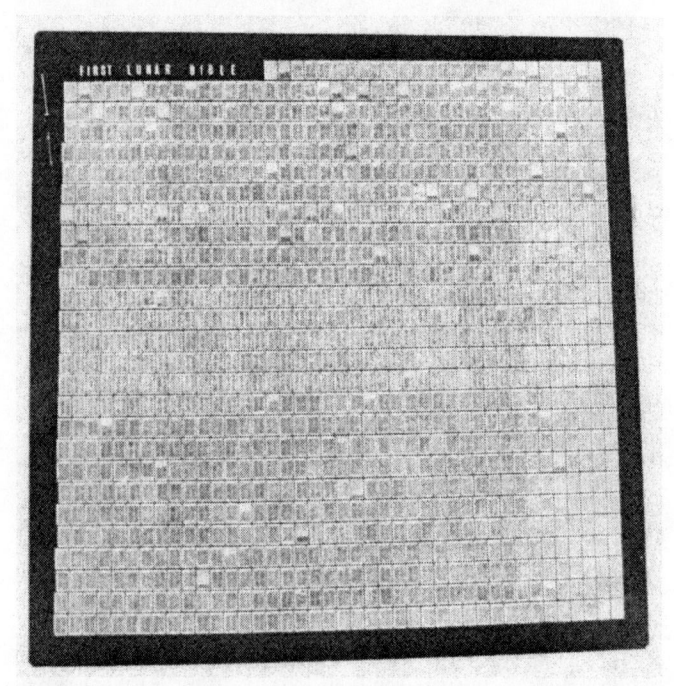

The original glass plate of the multi-focal "First Lunar Bible" made by APL-NASA employees at the Manned Spacecraft Center showing the Revised Standard Version (RSV). The King James Version was mounted on the opposite side.
Photo Credit: Apollo Prayer League archives courtesy John and Helen Stout.

BIBLIOGRAPHY

I am grateful for the many books that came before, rich with the research and recapitulation of events that comprise the foundation of this work. The kaleidoscope of stories woven together in these pages present a prism through which one can view the historical and spiritual fabric that made the Apollo program great—and took America along on the ride of its life.

Newspapers, magazine articles, and press communications of the time captured valuable insights into otherwise obscure stories and events. I am grateful to the many astronauts, their families, friends, parishioners, pastors, and co-workers whose corrections, suggestions, and additions added dimension and flavor to the stories.

Sy Liebergot and Jerry Bostick are among the many in NASA Mission Control who stepped beyond the bounds of documented events to provide a personal level of involvement rarely afforded an author.

I am grateful to Dr. Edgar D. Mitchell, who gave free and open access to his archives and through whose friendship I came to know the story of the First Lunar Bible. Were it not for him, the journey would never have begun.

I thank those who contributed and critiqued various parts of the manuscript during the many preliminary drafts, and whose encouragement and insights were invaluable: Virginia Dunn, Ed Hengeveld, Ann McCutchan, Sherry Jackson, Rick Killian, James Vance, Dennis Wharton, and Dwight Williams.

Lastly, I am grateful for the rare photographs and documentation preserved in the archives of The Apollo Prayer League by Helen and John Stout that offered the crucial threads of a story that was as incredible as it was unknown. Their trust and friendship in piecing together this page in history is beyond my ability to comprehend.

The following references and sources provided a mosaic of the personal journeys of those who took America to the moon and back.

Interviews and contacts

Buzz Aldrin (2010); Judy Alton, parishioner and historian, Webster Presbyterian Church, Webster Texas, Webster, TX (2008); Cathy Anello, niece of Buzz Aldrin (2009-2010); Jay Barbree, NBC Veteran Space Correspondent and author (2009); Ronald Berry, NASA Mission Control (2007-2008); Jerry Bostick, NASA Chief of Flight Dynamics Branch, Apollo, 1968-1973 (2009); Mary Brandt, parishioner and friend of late astronaut Ed White II, Seabrook Methodist Church, Seabrook, TX (2009); Anita Bryant, Christian singer and friend of Apollo 15 Lunar Module Pilot James Irwin (2008); Ed Buckbee, NASA Public Information Officer 1959-1968 (2007- 2008); Jerry Carr, Commander, Skylab 4 and Apollo Capsule Communicator, Apollo 8 and Apollo 12 (2006); Eugene Cernan, Apollo 10/Apollo 17 (2009-2010); Jeanette Chase, friend of the former Mrs. Neil (Jan) Armstrong (2008); Stephen Clemmons, NAA Senior Spacecraft Checkout Mechanic, Apollo (2009); Richard Crawford, former Mayor, Tulsa, Oklahoma, and friend of Neil Armstrong (2009); Madeline Aldrin Crowell, sister of Buzz Aldrin, Apollo 11 Lunar Module Pilot (2009- 2010); Sherry Darling, The Mary Baker Eddy Library (2006); Charles Dry, Technical Support Crew, Apollo (2008); Charlie Duke, Apollo 16 Lunar Module Pilot (2007-2009); Michael Esslinger, Apollo recovery historian (2008); David Frohman, Peachstate Historical Consulting (2005-2009); Lowell Grissom, brother of Apollo 1 astronaut Gus Grissom (2008); Fred Haise; Apollo 13 Lunar Module Pilot (2006); Paul Haney, NASA of Public Officer 1958-1969 (2006); Ed Hengeveld, artist and space historian (2010); Frank Hughes, NASA Mission Control (2006); Sylvia and Jack Kinzler, NASA Engineer and inventor of the lunar flag & module plaques (2007); Dick Koos, NASA Mission Control, (2007–2008); Sy Liebergot, NASA Apollo Environmental & Electrical Engineer (2007); Mott Linn, Curator, Clark University Robert Hutchins Goddard Collection (2006); Glynn Lunney, NASA Director of Mission Control (2008); Robin Manison, personal friend and fellow parishioner of Buzz Aldrin, Apollo 11 (2007); Larry McGlynn, space historian and artifact collector (2006-2009); T.K. (Ken) Mattingly, Apollo 16, Command Module Pilot (2006); Edgar Mitchell, Apollo 14 Lunar Module Pilot (2004 – 2010); James Paden, Clerk of Session, Webster Presbyterian Church (2008); Robert Parrott, retired pastor, Seabrook United Methodist Church, Seabrook, TX. (2009);.Robert Pierson, NASA Mission Control (2007–2008); Laura Shepard-Churchley, daughter of Alan B. Shepard (2009); Tom Stafford, Apollo 10 and Apollo–Soyuz Test

Project (2010); John M. Stout, Director, The Apollo Prayer League, NASA Chaplain, and ITT Senior Information Analyst (2009-2010); June Vernon, parishioner and friend of Jim Irwin Apollo 15, Nassau Bay Baptist Church (2008-2009); Guenter Wendt, NASA Pad Leader, (2009-2010); Edward White III, son of Gemini/Apollo Astronaut Edward White II (2009); Dean Woodruff, Pastor Emeritus, Webster Presbyterian Church (2007); Al Worden, Apollo 15 Command Module Pilot (2010).

Books and reference material

Aldrin, Buzz, and Malcolm McConnell, *Men from Earth*, 2nd Ed. New York: Bantam Falcon Books, 1993.

Aldrin, Buzz, and Ken Abraham, *Magnificent Desolation*, New York: Harmony Books, Random House Publishers, 2009.

Alton, Judith Haley, Patricia M. Brackett, and Donna Ray, *The Little White Church on NASA Road 1: From Rice Farmers to Astronauts*. Webster, TX: Webster Presbyterian Church, 1993.

American Academy of Achievement, *First American in Space*. Alan Shepard Interview, Houston, TX: February 1, 1991.

Apollo 8: Insider Stories. Arlington, VA: Public Broadcasting Company, 1997-2005 (Video).

Apollo 13: To The Edge and Back. Arlington, VA: Public Broadcasting Company, 1994 (Video).

Barbee, Jay, *Live From Cape Canaveral*. New York: HarperCollins, 2006.

Bean, Alan, *My Life As An Astronaut*. New York: Minstrel Paperback Original, Pocket Books, 1988.

Bizony, Piers, *The Man Who Ran the Moon*, New York: Thunder's Mouth Press (Avalon Publishing), 2006.

Boomhower, Ray E. *Gus Grissom: The Lost Astronaut*. Indianapolis: Indiana Historical Society Press (Indiana Biography Series), 2004.

Brooks, Courtney G., James M. Grimwood, and Loyd S. Swebson Jr, *Chariots for Apollo: A History of Manned Lunar Spacecraft*. Washington, D.C.: NASA Scientific and Technical Information Office, NASA SP-4205, 1979.

Buckbee, Ed, *The Real Space Cowboys*. Burlington, Ontario, Canada: Apogee Books, 2005.

Bryant, Anita and Bob Green, *Fishers of Men*, Old Tappa, NJ: Fleming H. Revel, 1973.

Burgess, Colin and Kate Doolan, *Fallen Astronauts*. Lincoln, NE: University of Nebraska Press, 2003.

Cass, Stephen, *Apollo 13, We Have a Solution*. New York: IEEE Spectrum, 2005.

Cernan, Eugene and Don Davis, *The Last Man on the Moon*. New York: St. Martin's Press, 1999.

Chaikin, Andrew, *A Man on the Moon*. New York: Penguin Books, 1994.

Collins, Michael and Charles Lindbergh, *Carrying the Fire: An Astronaut's Journey*. New York: Cooper Square Press, 2001.

Conrad, Nancy and Howard A. Klausner, *Rocketman: Pete Conrad's Incredible Ride to the Moon and Beyond*. New York: New American Library, 2005.

Cooper, Henry S.F., Jr., *13: The Flight That Failed*. New York: Dial Press, 1973.

Cunningham, Walter, *The All American Boys*. New York: Simon & Schuster Inc. 2004.

Duke, Charlie and Dottie, *Moonwalker*. Nashville, TN: Oliver Nelson, 1990.

Dickson, Paul, *Sputnik: The Shock of the Century*. New York: Walker Inc., 2001.

Godwin, Robert, *Apollo 13—The NASA Mission Reports*. Burlington, Ontario, Canada: Apogee Books, 2000.

Godwin, Robert. *Apollo 14—The NASA Mission Reports*. Burlington, Ontario, Canada: Apogee Books, 2000.

Gray, Mark, *Apollo 13: The Real Story*. Columbus, OH: Spacecraft Films, 2004 (Video).

Gray, Mark. *Apollo 1: The Apollo I Fire—Uncut*. Columbus, OH: Spacecraft Films, 2007 (Video).

Greene, Nick, *Apollo 13: The History of Space Exploration: The Successful Failure.* http://space.about.com.

Grissom, Betty, *Starfall.* New York: Thomas Y. Crowell, 1974.

Faith, William Robert, *Bob Hope: A Life in Comedy.* Cambridge, MA: De Capo Press, 1982, 2003.

Hansen, James R, *First Man: The Life of Neil A. Armstrong.* New York: Simon & Schuster, 2005.

Howard, Ron, *In the Shadow of the Moon.* Beverly Hills, CA: Imagine Entertainment, 2007 (Video).

Irwin, James B. and William A. Emerson Jr., *To Rule the Night.* Nashville, TN: Holman Bible Publishers, 1982.

Irwin, James B. and Charles Duke, *Man and His Universe.* The Second Baptist Family of Houston, Gary Moore Private Collection, undated (Video).

Jackson, Estelle Gifford, *The Eagle Has Landed.* Philadelphia: Dorrance, 1972.

Kraft, Christopher and James L. Shefter, *Flight: My Life in Mission Control.* New York: Penguin Group, 2001.

Kranz, Gene, *Failure is Not an Option.* New York: Berkley Books, 2000.

Lattimer, Dick, *All We Did Was Fly to the Moon.* Alachua, FL: The Whispering Eagle Press, 1985.

Lemle, Michael, *The Other Side of the Moon.* New York, NY, Lemle Pictures, 2004 (Video).

Lemle, Michael, *Our Planet Earth: The Global Perspective.* New York: Lemle Pictures, 1990 (Video).

Liebergot, Sy, *Apollo: A Race Against Time.* Multimedia Encyclopedia, 1995 (Video).

Liebergot, Sy, *Apollo EECOM.* Canada: CGP Inc, 2003 (Video).

Liebergot, Sy, with David Harland, *Apollo EECOM: Journey of a Lifetime.* Burlington, Ontario, Canada: Apogee Books, 2003.

Lindsay, Hamish, USB Honeysuckle Tracking Station, Australia.

Lovell, Jim, and Jeffrey Kluger, *Apollo 13* (previously *Lost Moon: The Perilous Voyage of Apollo 13*). Boston: Houghton Mifflin, 1994.

Mitchell, Dr. Edgar, *The Way of the Explorer*. New York: G. F. Putnam, 1996.

Murray, Charles and Catherine Bly Cox, *Apollo: The Race to the Moon*. New York: Simon & Schuster, 1989.

Murray, William J., *My Life Without God*. Eugene: OR, Harvest House Publishers, 1992.

NASA: *Gemini 4 Mission Commentary Transcript*. June 3, 1965, Tape 11, EVA-5 to EVA-12, and "Composite Air-to-Ground and Onboard Voice Tape Transcription of the GT-4 Mission," *NASA Program Gemini Working Paper No. 5035*, Houston: National Aeronautics and Space Administration Manned Spacecraft Center, Houston: TX, August 31, 1965.

NASA History: *Detailed Biographies of Apollo 1 Crew*. Mary C. White, National Aeronautics and Space Administration Manned Spacecraft Center, Houston, TX.

NASA Lyndon B. Johnson Space Center, Oral History Project: Interview with Jack Kinzler, Paul Rollins, and Jennifer Buchli, Seabrook. TX, January 16, 1998.

NASA Lyndon B. Johnson Space Center, Oral History Project: Interview with Thomas K. Mattingly III, Rebecca Wright, Costa Mesa, CA, November 2001.

NASA Lyndon B. Johnson Space Center, Oral History Project: Interview with Alan B. Shepard, Jr., Roy Neal, Costa Mesa, CA, November 1998.

NASA *Investigation into Apollo 204 Accident* Transcript, Oversight of the Committee on Science and Astronautics, U.S. House of Representatives, April-May 1967.

NASA *Report of the Apollo 204 Review Board* (RG-Z55), National Aeronautics Space Administration, Washington, D.C.: Government Printing Office, April 1967.

Pyle, Ron, *Destination Moon*. New York: Harper Collins, 2005.

Rahman, Tahir, *We Came in Peace for All Mankind*. Overland Park, KS, Leathers Publishing, 2008.

Reinhart, Al, *For All Mankind*. The Criterion Collection, New York: 1989 (Video).

Roth, Jeffrey, *The Wonder of It All*. Indican Pictures. Los Angeles, CA: 2006 (Video).

Schweickart, Russell, *No Frames, No Boundaries—Earth's Answer*. Explorations of Planetary Culture at the Lindisfarne Conferences. New York: Harper & Row, 1977.

Seaman, Ann Rowe, *America's Most Hated Woman*. New York: The Continuum International Publishing Group, 2003.

Shepler, John, *Alan B. Shepard: The Power of Not Giving Up Took Him to the Moon*, JohnShepler.com, 1998-2009.

Stafford, Thomas and Michael Cassutt, *We Have Capture*. Washington, D.C.: The Smithsonian Institution Press, 2002.

Thompson, Neal, *Light This Candle*. New York: Random House, 2004.

United States Court of Appeals, Fifth Circuit, Citation 432 F.2d 66, August 22, 1970, *Madalyn Murray O'Hair et al., Plaintiffs-Appellants vs. Thomas O. Paine*.

Watkins, Billy, *Apollo Moon Missions: The Unsung Heroes*. Westport, CT: Greenwood Publishing Group, 2005.

Webster Presbyterian Church 15th Lunar Communion Anniversary, Webster, TX: Webster Presbyterian Church, Pat Brackett Private Collection, 1984 (Video).

Wendt, Guenter, and Russell Still, *The Unbroken Chain*. Burlington, Ontario, Canada: Apogee Books, 2001.

Worden, Alfred M., *Hello Earth: Greetings from Endeavor*, Los Angeles, CA: Nash Publishing, 1974.

Zimmerman, Robert, *Genesis: The Story of Apollo 8*. New York: Dell Publishing, 1998.

Magazines, news articles, and print media

An abundance of anecdotal information was retrieved from news media reporting of events at the time and is too numerous to list here. While not exhaustive, the following articles gleaned from research, family and friends of the astronauts, and archival sources provided information not otherwise available through ordinary access.

"Aldrin Sought Minister's Aid for Right Words," *The Bridgeport Telegram*, July 19, 1969, Bridgeport, Texas.

"And it's been a long way, but we're here," Jerry Matulka, EarthToTheMoon.com.

"Apollo Crew Comes up with Homespun Names," *St. Petersburg Times*, May 24, 1969, St. Petersburg, FL.

"Apollo Eleven," *The Tulsa Tribune*, July 16, 1969, Tulsa, OK.

"Apollo Project Plagued by Trouble Even Before Fire," *New York Times*, January 29, 1967, New York, NY.

"Borman Tapes Wife's Yule Gift," *The Frederick Post*, December 26, 1968, Frederick, MD.

"The Brain and the Hillbilly Ride Apollo," *Pacific Stars & Stripes*, January 30, 1971, San Francisco, CA.

"Christmas Eve 1968 – Apollo 8 vs. Atheism," Martin J. Robbins, layscience.net, December 23, 2008.

Clemmons, Stephen, Wilmington, NC, online notes, July-August 2004.

"Climax Nearing In Race to Moon," *The Dallas Morning News*, May 18, 1969.

"Col. White Named Winner of '67 Astronautics Award," *New York Times*, February 8, 1967.

"Communist Pilot Catapulted from Crippled MIG," Life Magazine, June 8, 1953.

"Ed White 'Mr. Standee' of Space Team," The Valley Independent, June 2, 1965. "Edward White Sr. Believes in His Country," Houston Chronicle, May 30, 1971.

"First Lunar Bible Assembled in LP," *La Porte Baytown Sun*, April 8, 1971.

"First Man to Step on the Moon Will be Buzz Aldrin," *Chicago Daily News*, 1969.

"Four Days of Apollo 13's Return Filled With Fear, Faith," *Tulsa Daily World*, April 19, 1970.

"Future of Space Program: A Lot Depends on Apollo 14 Trio," *U.S. News & World Report*, Feb 1, 1971.

"Grief Expressed The World Over," *New York Times*, January 19, 1967.

"Inquest on Apollo," *Time* Magazine, February 10, 1967.

"LBJ Leads Nation in Tribute to Moonmen," *The Valley Independent*, January 28, 1967.

"Just One of Those Days," *Houston Chronicle*, March 2, 1971.

"Madalyn O'Hair and a College Girl," *Independent Press-Telegram*, July 3, 1971, Long Beach, CA.

"The Lives They Lived: Alan Shepard; 15 Minutes of Fame," Thomas Mallon, *New York Times,* January 3, 1999.

"Mankind's Voyage," *The Christian Science Monitor*, April 20, 1970.

"'Mentalgram Sent By Astro Mitchell," *Stars & Stripes*, Feb 15, 1971, San Francisco, CA.

"Microfilmed Bibles Traveled to the Moon," *The Houston Post*, March 28, 1971.

"Mitchell: 'He Grabbed The Experiment and Shook It Until It Broke,' " *Washington Star*, January 31, 1971.

"Mitchell Took Bible To Moon," *The Dallas Morning News*, April 2, 1971.

"The Remarkable Reverend Stout," *The Eastside News*, July 1, 1971.

"Roosa: 'I'm Gonna Take Time to Look At The Scene and Appreciate It,' " *Washington Star*, January 31, 1971.

"Of Nixon, Kennedy and Shooting the Moon," *New York Times*, July 17, 1989.

"Scientists See Valued Treasures from Apollo 15," *The Dallas Morning News*, August 3, 1971.

"See Big Delay in Apollo," *The Valley Independent*, January 28, 1967.

"Shepard: 'If a Person Wants Something Enough He's Just Got to Hang In,' *Washington Star*, January 31, 1971.

"Shepard's Méniére's," *San Francisco Examiner*, March 18, 1971.

"Sketches of the 3 Apollo Astronauts Killed at The Cape," *New York Times*, January 28, 1967.

"A Slight Hitch In Astro's Speech," *The Dallas Morning News*, (undated 1971).

"Soviet Military Charged With Space Race Surrender," *The Dallas Morning News*, July 21, 1969.

"Space Agency Gets Prayer Petitions," *The Dallas Morning News*, March 8, 1969.

"Space Tragedy," *New York Times*, January 29, 1967.

"Teacher Uses Ingenuity To Get Satellite Photos," *Fort Worth Star-Telegram* (undated).

"Three Astronauts as They Saw Each Other: 'Gus Had Doubts, Ed Was All Go, Roger Saw Mars,'" *Today Magazine*, January 28, 1967.

"Three Astronauts Clawed at Hatch", *New York Times*, January, 1967.

"Three Valiant Astronauts Yearning for the Moon Died Side by Side," *The Register*, Danville, VA, January 29, 1967.

"The 3 Astronauts," *Life* Magazine, February 3, 1967.

"Two to be Buried at Arlington, Third at West Point," *New York Times*, January 29, 1967.

"U.S. Pacific Chief Sees United States On Losing End in 'War On Communism," *Lowell Sunday Sun*, October 6, 1957.

"White, Edward H. III: "I Felt Red, White, and Blue All Over,' " *Life* Magazine, June 18, 1975.

"What on Earth Do the Mitchell's Do?" *Fort Lauderdale News and Sun-Sentinel*, November 16, 1980, Fort Lauderdale, FL.

"White Buried At West Point Military Base," *The Valley Independent*, January 31, 1967.

The Washington Post: William Hines interview with Ed White, Washington: D.C., December 1966.

"Yearning to Tread on the Moon Obsessed Apollo Astronauts," *Corpus Christi Caller-Times*, Sunday, January 29, 1967

Other sources

Numerous articles and documentation were obtained from archived files of newspapers and magazines across the country, the astronauts' personal archives, space collectors, universities, various national archives, NASA Dr. Martin Luther King Jr. Library, and from the Library of Congress and Nixon files under the Freedom of Information Act.

Original AP news wire releases by aerospace writers Howard Benediot, *Driving for Science*, November 23, 1972; Howard Benediot, *Ten On The Moon*, December 2, 1972; Paul Recer, *Lunar Lessons of Apollo*, November 27, 1972.

Apollo Prayer League Newsletters and Bulletins: Pre-Apollo 12, November 1969; Pre-Apollo 13, March 1970; Relief efforts and Madalyn O'Hair [undated]; Thanksgiving newsletter, November 1970; Pre-Apollo 14 Prayers, January 1971; Post-Apollo 14 The First Lunar Bible, est. May 1971; Pre-Apollo 15 church service program, July 1971; Post-Apollo 15 [undated]; Pre-Apollo 16 [undated].

Reverend John and Helen Stout's personal archived records, newsletters, and press releases; Apollo Prayer League Lunar Bible Registry, bylaws, and minutes; photographs; and correspondence to-from The Apollo Prayer League, the Apollo astronauts, NASA officials, U.S. government officials, and other outside organizations.

INDEX

Aaron, John
   and Apollo 1, 59
   and mission control, 62
   and Apollo 12, 170
   and Apollo 13, 203–206, 210
Abbey, George, 199
abort alarms, Apollo 11, 128-129
abort switch problems, Apollo 14 232-235
Aerospace Emergency Relief, 159–160
Aerospace Ministries, 47, 49, 159–160, 225, 248
Agnew, Spiro, 289
AGS (Abort Guidance System), 107
The Albuquerque News, 227
Aldrin, Andrew, 115
Aldrin, Edwin Eugene Jr., (Buzz), 64
   Apollo 11, 83, 108-110, 115-119, 121, 122-126, 128-130, 134-138,141-144, 146, 154,
   Gemini 12, 112
   background of, 110-112
   communion on the moon, 115-119, 131, 133–134
   Manison family, 144-148
   reading from Psalms, 140
Aldrin, Eugene, Sr., 110, 117, 130, 132
Aldrin, Fay Ann, 110
Aldrin, Janice, 115
Aldrin, Joan, 115, 119, 133, 143, 144,146,147
Aldrin, Maddy, 110
Aldrin, Marion, 110. 124
Aldrin, Michael (Mike), 115
Allen, Bill, 40

ALSEP (Apollo Lunar Surface Experiments), 256, 269
Amarillo Globe-Times, 155
American Atheists, 49, 93, 150, 153, 157, 162. *See also* atheists; O'Hair, Madalyn Murray
American Eagle (Gemini 4), 27
American Society of Professional Photographers, 5
Ames, Nancy, 287
Anders, Valerie, 78, 91
Anders, William (Bill), 72, 155, 216
   Apollo 8, 74, 78, 79, 80–83, 85
   Christmas Eve broadcast, 82, 85–87
Antares, xvi, 230–235, 241–242, photo of, 304
Apennine Mountains, 257, 262
Apollo 1, xvi, 27-29, 30-34, 37-43, 50-52
   fire, 34–36
   inquest into fire, 50–59
Apollo 4, 61–63
Apollo 5, 64
Apollo 6, 65
Apollo 7, 66, 67–71, 226
Apollo 8, 72-75, 77-91, 94
   Christmas eve reading, 82, 85–87
   Earthrise, *xv,* 83
Apollo 9, 94-98
Apollo 10, 98, 100-101, 102-108, 281
Apollo 11, 96, 105,106, 108, 109-143
   photo of silicon disc left on lunar surface, 301
   landing on moon, 125–126, 130, 171

communion 115-119, 131, 133–134, 140
Apollo 12, 169-170, 171-174, 175-176, 177
   lightning strike, 170–171
   Signal Condition Equipment (SCE), 170–171
Apollo 13, 110, 180, 190-191, 193, 194-197, 200, 202-214, 217-222
   explosion,197-198
   investigation of, 229
   oxygen tank problems, 101, 191–192, 196, 197–198, 199, 220
   reentry, 216–217
Apollo 13 Review Board, 229
Apollo 14, 229-236, 239-242, 243
   photo of crew, 303
   lunar Bibles on, xvi, 222, 224, 226, 237, 244–245
   paranormal experiment, 243–244
   problems in descent, 232-235
Apollo 15, 248-249, 250-262
   Genesis Rock, 258-260
   plaque for fallen astronauts, 256, 261
   red leather Bible on, 256, 261
Apollo 16, 263-273
Apollo 17, xiii, 229, 274-284, 292
Apollo 18, 176-177, 180, 229, 265, 277
Apollo 19, 180
Apollo Lunar Surface Experiments Package (ALSEP), 256, 269
Apollo Prayer League (APL), xvi, 46-49, 51, 93, 116, 132, 142,156, 157-159, 160, 162-164, 186-187, 189, 208, 222, 224-226, 237 248-249, 288-189, 294, 295
   Voice of America, 132
   formation of, xvi, 44-49
   map location of members, 308
   mission of, 160–161
   right of astronauts to pray, 157–159
   significance of lunar Bible, 162
Apollo Prayer League Lunar Bible Registry, 288–289
Aquarius, 198, 203–204, 206–207, 212, 217
Arlington National Cemetery, 41–42
Armstrong, Jan, 39, 124, 138–140, 226
Armstrong, Mark, 139
Armstrong, Neil, 40, 142, 149, 154, 158, 172, 252-253, photo in LLTV, 301.
   Apollo 11, 2, 82,108, 109–110, 112, 121, 122-123, 125–126, 128-129, 130, 131, 134, 135, 137-138, 141, 171
   first step on moon, 134
   background of, 113
   crashing the LLTV, 113-114
Armstrong, Rick, 139
AS (Apollo-Saturn)-204, 28, 30–31, photo of, 299
AS-204 Review Board, 52
Austin Presbyterian Theological Seminary, 5

Babbitt, Donald, 34, 35
Bales, Steve, 128–129
Barnes, James, 155–156
Baron, Robert, 54–55
Bassett, Charlie, 51, 99–100
Bay of Pigs, 12
Bean, Alan, 114, 166, 169
   photo of, 300
   Apollo 12 mission, 171–177
   background of, 167–168
   and lunar Bibles, 163–164, 177, 186
Bean, Clay, 166
Bean Amy, 166
Belo Horizonte (Brazilian newspaper), 6
Bergman, Jules, 52
Berle, Milton, 214
Berry, Charles, 191, 205, 207
Berry, Ron, 205, 207
Bibby, Cece, 13–14

The Bob Hope Extra Special, 286–287
Bonney, Walter, 2
Boot Hill, 105
Borman, Frank, 57, 58, 80, 85–88, 92,120, 137, 154, 155, 213, 286
   Apollo 8, 74, 76–81, 87–88, 90–91
   Gemini 6, 66
   and Apollo 1 fire, 44–45, 57
   Christmas eve broadcast, 85–89
Borman, Susan, 76, 77, 81, 91
Bostick, Jerry, 71, 195, 210, 219
Bourgin, Si, 86–87
Boyle, Edward, 243
Brand, Vance, 185
Brandt, Mary, 290
Brett, Robin, 260
Brooks, Dave, 191
Brooks Air Force Base, 26, 183
Brynner, Yul, 289
Buchanan, Pat, 120
Buckbee, Ed, 69, 92
Buckley, Charley, 275
Buckner, Jim, 84
Burton, Clint, 197, 202
Bush, George H. W., 177, 187–188, 222, 225, 289

Cadena, Marta. *See* Kranz, Marta Cadena
Campbell, Glen, 287
Cargill, Eugene, 194
Carpenter, Scott, 1, 40, 253, 288, photo of, 298
Carr, Jerry, 81, 143, 170
Carter, June, 289
Casey, Chuck, 65
Cash, Johnny, 289
Casper, 267, 268–269, 272
Censorinus, 107
Cernan, Barbara, 277, 278, 281–282, photo of, 306
Cernan, Eugene (Gene), xii, 88, 231, 234, 274-277, 279, 286, photo of, 306
   airplane crash at St. Louis, 99–100
   Apollo 10, 98–99, 102, 105–107, 113, 281
   Apollo 17 280–285
   Gemini 9, 99, 100, 102, 281,
Cernan, Tracy Dawn, 280,281, 285, 286, photo of, 306
Chaffee, Martha, 38, 39
Chaffee, Roger, 28, 40, 42, 60, 83, photo of, 299
   Apollo 1, 28-29, 31-32, 35, 44-46
   death of, xvi, 34-36, 38–39
   as U-2 pilot, 28
Chaffee, Sheryl, 38
Chaffee, Steve, 38
chalice, communion, 116-117, 119, 123, 131, photo of, 301
Challenger, 284
Chapin, Dwight, 212
Charlie Brown, 102, 104–105, 107–108
Chase, Bill, 139
Chase, Jeanette, 138–139, 140
Chicago Daily News, 112
Chimpanzees (chimps), 1
Christian Science Monitor, 221, 236
Christmas Eve broadcast, from Apollo 8, 85–88
CIA, Soviet space launches and, 8, 74, 94
Clemmons, Stephen, 55
Cole, Ed, 12
Collins, Michael (Mike), 79, 83-85, 216
   Apollo 11, 108, 123, 125–126, 138, 141, 142-143
Columbia, 123, 125, 138, 139, 143
communion on the moon, 115–121, 131, 133–134, 140
Communist Manifesto (Marx), 47
Cone Crater, 190, 236, 239–240, 245
Conrad, Jane, 201, 219
Conrad, Pete, 168-169, 201, 219
   Apollo 12 mission, 170–177
   background of, 166–167
Constantine, Emperor, 162

Cooper, Gordon, 1, 11, 40, 179–180, 286, 288, photo of, 298
Copeland, Kenneth, 288
Cotton Bowl, 225
Cronkite, Walter, 8, 62, 88, 124, 141, 271
Cuban Missile Crisis, 28
Cummings, E. E., 94
Cunningham, Walter (Walt), 1, 37, 66–71, 169

"The Day of the Fire," 44
DeMoss, Bob, 38
Descartes highlands, 266, 269, 272
Diaz, Gloria, 125
Dickison, Benny R., 127, 244
Dimaline, Ernest, 42
Distinguished Service Medal, 245
Dobrovolski, Georgi, 250
Dobrynin, Anotoly F., 40
Dole, Bob, 225
Dozier, Hallie Mills, 248
Dry, Charlie, 256
Dry Gulch, 105
Duke, Charlie, 106, 129, 130, 138, 196, 206, 214, 273
  Apollo 16, 265, 267–272
  background of, 264–265
  dream, 266
  German measles, 190–191, 265, 267
Duke, Charlie, Jr., 191
Duke, Dotty, 264, 265, 266–267, 273
Dunham Bible Museum of America (Houston), photo of lunar Bible, 310
Durst, Norman, 225

Eagle, 121, 123, 126, 129–131, 134–136, 137–138
Earthrise, xv, 83, photo of, 300
Ed White Youth Memorial Center, 286-288
Edmund Scientific, 187
Edward White Youth Memorial Center Fund, 286

Edwards Air Force Base, 81, 111, 113, 180, 182, 239, 263–264
Einstein, Albert, 225
Eisele, Donn, 66–69
Ellington Air Force Base, 24, 88, 91, 159, 224, 281
Endeavor, 256
Engle, Joe, 45, 87, 105, 208, 231, 239, 277–278
Engle, Mary, 278
Esquire magazine, 177
Evans, Dale, 277, 278, 289
Evans, Jan, 278
Evans, Ron, 231, 275, 277
Eyles, Don, 233, 234

Faget, Maxime, 121, 224
Falcon, 257–259, 261
Fallaci, Oriana, 172
Fellers, John, 294
Finegan, Pat. *See* White, Pat
first lunar Bibles (see lunar Bibles)
First Lunar Bible, 224-226, 244, photo of, 310.
First Lunar Bible Honor Roll, 225, 244–246
Fischer, Regina, 149
Flannigan, Peter, 229
Four Seasons, Frankie Valli and the, 287
Fra Mauro, 190, 194, 235, 239, 240, 242, 245
Freedom 7, 8, 10, 12–13, 181
Freedom Walk Award, 287
Freeman, James Dillet, 254–255
Freeman, Ted, 51
Fry, Loretta, 158

Gagarin, Yuri, 9, 10, 12, 136, 150
Galileo, 250
Gemini 3, 15
Gemini 4, 16–17, 19–22, 27, 43, 62, 287
Gemini 6, 66, 69–70
Gemini 7, 66, 79
Gemini 8, 252–253

Gemini 9, 99-100, 102, 281
Gemini 11, 169
Gemini 12, 112
General Motors, 167
Genesis Rock, 258–260, photo of, 305
Gilruth, Robert (Bob), 8, 109, 119–120
Givens, Ed, 51
Glenn, John, 1, 3, 7, 8, 13-14, 123,158, photo of, 298
Goddard, Robert, 110, 124, 146
golf on the moon, 241–242
Gordon, Dick, 40, 128, 166–167, 168–169, 171–172, 175–176
Goulet, Robert, 287
Graham, Billy, 225
Grant, Cary, 287
Griffin, Gerry, 170
Grissom, Betty, 33, 38, 39
Grissom, Dennis (father of Gus Grissom), 52
Grissom, Gus, xvi, 1, 8, 12, 16-17, 21, 28, 40, 44-46, 53, 56, 60, 83
   photos of, 298, 299
   Apollo 1, 28-29, 30–36, 37-38
   background of, 12–13
   Capcom for Gemini 4, 16-17, 21
   death of, xvi, 34-36, 38-39, 41-42
   Gemini 3, 15-16
   Liberty Bell 7, 14–15, 55
Grissom, Mark, 38
Grissom, Scott, 38
Grumman (Co.), 64, 72, 95, 137, 184,185, 206, 228, 275
Guidepost magazine, 47
Gumdrop, 96, 102
Gurney, Edward, 54–55

Hadley Rille, 257, 258, 285
Haise, Fred W. (Freddo), 142, 176, 182, 231, 234-236, 240-242, 267
   Apollo 13, 182, 190, 193, 197, 203, 206, 211–212, 217
   background of, 182–187
Haise, Mary, 199
Haldeman, H. R., 213–214

Ham (chimpanzee), 1–2, 9, photo of, 298
Hamblin, Dodie, 138, 139
Hammock, Jerry, 200
Haney, Paul, 3, 158, 166
Hannigan, Jim, 101, 203
Harrington, Bob, 247
Hasselblad camera, 106, 173–174
Heatherton, Joey, 287
Hedgeveld, Ed, painting of Dave Scott, 305
Hello Earth (Worden), 297
Henry, Jack, 133
Heselmeyer, Bob, 203
High Flight Ministries, 261–262
Holiday Inn, 2, 12, 13
Hope, Bob, 286–288, 294
Hope Bible, 288, 294
Hour of Power sermons, 224
House, Paul, 179
House, William, 179
Houston-Astros, 196
How Great Thou Art (James), 238–239
Humphrey, Hubert H., 41, 43

"I am There" (Freeman), 254–255
Index (lunar feature), 257
International Bible Society, 247
Intrepid, 164, 171, 174–175
Irwin, James (Jim), 248-249, 251, 253-254, 261-262, 293
   Apollo 15, 256–261
   background of, 253–256
Irwin, Mary, 254–255

Jackson, Bradford, 162, 163
Jackson, Estelle, 162, 163, 189
James, Sonny, 238–239
Janssen, David, 287
Jesus, 225
Joe's Crater, 105
Johnson, Harold, 20
Johnson, Lady Bird, 42–43
Johnson, Lyndon B., 5, 27, 40–41, 42, 43

excerpt of letter to Rev. Stout, 309
Jones, Don, 90
Journal of Brazil, 6

Kapryan, Walter, 223
Kattah, Pedro, 132
Kennedy, Jacqueline, 26, 27
Kennedy, John F., 7, 26, 27, 50, 60, 71, 130, 225,
   assassination of, 26
   moon landing proclamation, xiii, 12, 18–19
   PT-109 torpedo boat, 26
Kennedy, Robert F., 77
Kennedy, Ted, 125
Kent State University, 223
Kerwin, Joe, 195, 207–208, 216–218
Khrushchev, Nikita, 12
Kilmer, Alfred Joyce, 142
King, Martin Luther, Jr., 65, 77, 225
King James Bible, 105, 161–162, 187, 224, photo of NCR lunar, 307
Kinzler, Jack, 20, 117, 119, 120–121, 175
Kissinger, Henry, 216
Kittyhawk, 230, 232, 238, 242
Komarov, Vladimir, 57–58
Koos, Dick, 9–10, 89, 127–128, 129, 144
Kopechne, Mary Jo, 125
Kraft, Christopher (Chris), 70, 72–73, 75–76, 77, 83, 88, 89, 109, 200
Kranz, Gene, 58-60, 62, 72-74, 84, 88, 95-98, 127-128, 129, 140, 193, 194, 200, photo of, 302
Kranz, Marta Cadena, 95, 98, 127

Laitin, Christine, 86–87
Laitin, Joe, 86–87
Lamour, Dorothy, 287
Landwith, Henri, 2
Langley Research Center, 52
Langseth, Mark, 269–270
Lanzkron, Rolf, 51

Legler, Bob, 214
Leonov, Alexi, 20, 154
Les Brown Band, 287
Liberty Bell 7, 14–15, 55
Liebergot, Sy, photo of, 302
   on mission control, 62–63
   incident with Jack Schmidt, 279
   and Apollo 13, 195–200, 202–203, 217, 220
   on Singin' Wheel, 263–264
Life magazine, 2, 138, 139, 152, 275
Lincoln, Abraham, 225
Lindbergh, Charles, 109, 227
LLTV (Lunar Landing Training Vehicle), 113–114
LMs (Lunar Modules)
   on Apollo 10, 105
   description of, 95
   on Apollo 4, 62–64
   on Apollo 5, 64
   on Apollo 9, 96
Lopez, Trini, 287
Lousma, Jack, 194–195, 197, 199, 203, 211, 217–218
Lovell, James (Jim), 10, 40, 66, 81, 155, 181,190, 222
   photo of, 302
   and lunar Bibles, 186-187, 222
   Apollo 8, 74, 78–83, 89–90
   Apollo 13, 189, 193–194, 197–198, 206–207, 211, 217
   Gemini 7, 66
   Gemini 12, 112
   swapped from Apollo 14, 182, 186, 188
Lovell, Jeffrey, 91
Lovell, Marilyn, 74–75, 78,,81, 91, 186, 188, 199-201, 209, 219-220
Lovell, Susan, 209
Low, George, 51, 60, 72–74, 104, 109
Luke (lunar feature), 257
Luna 15 (Soviet spaceship), 134
lunar Bibles, xvi, 162, 163.177, 186-187, 222, 224–226, 244–245,

288-289, 294,-295, photo of NCR, 305.
Lunar Bible Honor Roll, 225, 244–246
Lunar rover, 252, 258, 282, photo of, 306
Lunney, Glynn, 51, 69, 70, 72, 101, 108, 137-138, 194, 202-203, 205, 209

MacArthur, Douglas, 110
Manison, Robin, 119, 144, 147–148
Manison, Tom, 115, 119, 144–145, 147–148
Manison, Tommy, 145–146, 147
Manned Orbiting Laboratory (MOL), 182
Mark (lunar feature), 257
Mars missions, 292–293
Marx, Karl, 47
Matthew (lunar feature), 257
Mattingly, Ken (T.K.), 45, 89,
    Apollo 13, 182, 190, 193, 199-202, 206, 210-219
    Apollo 16, 265–269, 272
    background of, 182–186
    exposed to German measles, 191, 192, 267
Maxey, Edward, 243
Mayer, John, 73
McCain, Adm. John, 75, 90
McCord, James, 225
McDivitt, Jim, 74, 95, 96-97
    Gemini 4 mission, 16–17, 20–22, 57
McDonnell, James (Mr. Mac), 33
McDonnell Aircraft facility (Cape Canaveral), 12, 15, 20, 33, 52, 54
McDonnell Aircraft facility (St. Louis), 99-100
McGee, Jim, 254
Méniére's Syndrome, 15, 178–179
Mercury 7, 1–7
Mercury 7 astronauts, photo of, 298
Mercury-Redstone rocket, 11–12, 14

MET (Mobile Equipment Transporter), 239–240
microfilm lunar Bibles. *See also* lunar Bibles, photo of NCR, 307
"Mission to the Heart of Creation," 251
Mitchell, Billy, 110
Mitchell, Edgar, 50, 66-67, 100-101, 180-181, 182, 196, 202, 206, 221, 242, 245, 246, 267, 285, 293, photos of, 303, 307
    Apollo 14, 223, 227, 230–237, 238–243
    background of, 228–229
    carrying lunar Bibles, 222, 224, 226, 244–245
    paranormal experiment, 243–244
Mitchell, Louise, 234, 236
Mobile Equipment Transporter (MET), 239–240
MOL (Manned Orbiting Laboratory), 182
Molly Brown (Gemini 3 capsule), 15–16
Mondale, Walter, 52–53
Mount Hadley, 259
Mount Marilyn (Sea of Tranquility), 83, 182, 207
Muehlberger, William, 269
Mueller, George, 53
Munich massacre, 275
Murray, John Garth, 150
Murray, Madalyn May (O'Hair), 150–152. *See also* O'Hair, Madalyn Murray
Murray, Robin, 152–153
Murray, William (son of Madalyn Murray O'Hair), 150–152
Murray, William J., Jr., 150
Murray vs. Curlett, 49, 151

NAA (North American Aviation), 31, 33, 34, 40, 51, 52, 54-55, 56, 57, 58, 60, 74, 184, 205
NACA (National Advisory Committee for Aeronautics), 20
Nassau Bay Baptist Church, 255

National Annual Convention of
American Atheists, 49
National Cash Register Company (See
NCR)
Navajo Indians, on moon people, 92
NCR (National Cash Register
Company) microfilm bibles, 161–
162, 177, 186–187, 224, 288, 295
New York Daily News, 139
New York Times, 67, 76
Nixon, Richard, xiv, 122, 135, 220,
229, 245, 251, 275, 289, 290
confides in Stuart Roosa, 223
contingencies for Apollo 13, 208,
212, 213
lunar surface plaque, 119, 120
Noah's Ark, 186, 262
North American Aviation (see NAA)
Northwestern University Traffic
Institute (Evanston, IL), 244
Norton, Homer, 225

Odyssey (command module), 190,
195–196, 198–199, 202–205,
217–219
O'Hair, Jon, 149
O'Hair, Madalyn Murray, 93, 134,
140, 142, 154, 155-158, 161,
188, 221, 246-248, 289, 294,
photo of, 300
background of, 49, 149–153
letter to Reverend Stout, 157
Murray vs. Curlett, prayer in schools,
49, 150
O'Hair vs. Paine, 116, 118, 149,
153–155, 157, 161, 248
public debates, 247-248
O'Hair, Robin, 149
O'Hair vs. Paine, 116, 118, 149, 153–
155, 157, 161
Oklahoma Hills (moon feature), 105,
106
Omega watch, 123
Original 7, 1–3, 7, 67, 179, 180
Original 19, 180, 183, 185, 253

Orion, 267
Osis, Karliss, 244
Oxygen Tanks on Apollo 13, 101, 191–
192, 196–199, 220

Paine, Thomas, 116, 118, 149, 153,
157, 192, 209, 220, 248
Parrott, Bob, 288
Patsayev, Viktor, 250
Patten, Mary Ellen, 103, 105
Paul VI, Pope, 40, 92, 136, 154, 187,
213
"Peace in Space" treaty, 40
Peale, Norman Vincent, 47, 186
Peck, Gregory, 287
Personal Preference Kit (PPK), 116
Peters, Bill, 203
Phillips, Maj. Gen. Samuel, 52–53
Phillips Report, 52–53
Pierre Marquette Discovery Medal,
141–142
The Planets (Holst), 272
Playmate, 173
The Power of Positive Thinking (Peale),
47
Prayer for Vision, Faith and Works
(Weld), 84
Price, Byron, 289
Prodan, John, 183
Profitt, Dick, 79
Propst, Gary, 34, 35
Puddy, Don, 101
Purser, Paul, 160
Pyle, Ray, 51

Rathman, Jim, 12, 167
Ream, Bud, 37
Red leather Bible
Apollo 12, 161-163, 177
Apollo 15, 256, 261
painting of Apollo 15, 305
Reedy, George, 5
Revised Standard Version (RSV), 224,
photo of First Lunar Bible, 310
Reynolds, Frank, 134

Rhine, J. B., 243–244
Roberts, Bernard J., 244
Roberts, Jack, 154, 248
Rogers, Roy, 289
Rogers, William, 216
Roosa, Joan, 239
Roosa, Stuart, 202, 222, 267, photo of, 303
   Apollo 1 fire, 34-36, 238
   Apollo 14, 180, 181, 227, 229–232, 238–239, 242, 245
   background of, 180–181, 239
Roosevelt, Franklin R., 225
Rose, Rod, 84, 89–90, 100
Roths, John Henry, 150
Rowles, John, 287

Safire, William, 119–120
Sagan, Carl, 141
Saturn 1B rocket, 30, 33, 64
Scheer, Julian, 85, 116, 154
Schirra, Walter (Wally), 1, 34, 39, 57, 66-67, 96, 124, photo of, 298
   Apollo 7, 66, 67–71
   Gemini 6, 66, 69–71
Schmitt, Harrison (Jack) "Dr. Rock," 107
   Apollo 17, 275, 276, 277, 278, 280–283
Schuller, Robert H., 224
Schulz, Charles, 102
Schweickart, Rusty, 74, 94, 96–97, 288
Scott, David R. (Dave), 253, 254, 255, 256, painting of Scott with Bible, 305
   Apollo 8, 74
   Apollo 9, 96–97
   Apollo 15, 248, 250, 251, 257–261, 270
   background of, 252–253
   Gemini 8, 252–253
   Genesis rock, 258-259
   plaque for fallen astronauts, 256, 261
   red leather Bible, 256.261

Scott, Lurton, 139, 253
Sea of Tranquility, 83, 98, 102, 105–106, 131, 134, 274
See, Elliott, 51, 99–100
See, Marilyn, 100
Senate Space Task Group, 292
Sevareid, Eric, 141
Shaw, Karen, 288
Shaw, W. J., 244
Sheaks, Larry, 189, 198
Shepard, Alan B., 1-2, 8, 13, 57, 178-179, 180, 181, 202, 222, 243, 245
   photos of, 298, 303
   Apollo 14, 226, 230–236, 239–242
   background of, 227–228
   Freedom 7, 8-12
   golf on the moon, 241
   Méniére's Syndrome, 15, 178–179
Shepard, Laura, 11
Shepard, Louise, 10, 11, 179, 227
Sherman, Bobby, 287
Silver, Lee, 251, 253
Simpson, O. J., 287
Singin' Wheel, 196, 199, 263–264, photo of, 303
Skylab program, 167, 169, 176, 292
Slayton, Deke, 1, 34, 35, 57, 63, 74, 109, 112, 113,116, 118,129, 132, 169,173,178-179, 180-181, 183, 185,186, 192-193, 202, 240, 254, 265, 277-278, 280, photo of, 298
Smith, Charlie, 280
Smithsonian Institution, 222
Smylie, Ed, 212
Snoopy, 102–103, 104–105, 106–108, 173
Socrates, 225
Soyuz 1, 57–58
Soyuz 11, 250–251, 256
Space Shuttle orbiter, 292
Spaulding golf club, 241
Spann, Pat, 139
Spider, 96, 102

Sports Illustrated, 62
Sputnik, xiii, 3-4, 5-6, 111
Sputnik 2, 4
St. Christopher Episcopal Church, 84-85, 89, 91
Stafford, Mary Ellen, 103, 105
Stafford, Thomas Sabert, 103
Stafford, Tom, 40, 41, 66, 179, 215, 286, 288,
    Apollo 10, 98, 102–108
    airplane crash in St. Louis, 99–100
    background of, 103
Starlite Inn, 2
Step Brothers, 287
Stout, Helen, 4-5, 23, 26, 47-48, 164, 289, 292
Stout, James, 163, 244
Stout, Joe F. II, 45, 225
Stout, John Maxwell, vii, ix, xvi, 4, 6, 7, 23-26, 44-49, 92-93, 116, 122, 132, 141, 157-165, 170, 177, 178, 186-188, 202, 208-209, 213, 221-222, 224-226, 233, 236, 244-245, 246-249, 271, 278, 286, 288-289, 291- 292, 294-296
    photos of, 300, 307, 309
    background of, 4–6
    formation of APL, xvi, 46, 47–49
    Aerospace Ministries, 47, 49, 149, 225, 248
    Apollo 1, 44-45
    Apollo 11, 116
    Apollo 12, 162-164, 177
    Apollo 13, 186-188, 208–209, 213, 222, 224, 226
    Apollo 14, 222, 224-226, 233, 234, 236-237, 244
    Apollo 15, 248-249
    effort to put Bible on the moon, 46, 161–163, 177 186-188
    on prayer, 157, 160–161, 178
    U.S. International Observatories of Satellites, 6
    Sputnik, 5-6

Voice of America, 132, 134, 208, 221, 222
    with George H.W. Bush, 177, 187–188, 222, 225, 289
    with Harrison Schmitt, 278
    with Johnny Cash, 289
    with Lyndon B. Johnson, 5, letter, 309
    with Madalyn O'Hair, 49, 93, 116, 157-158, 159, 161, 162, 165, 188, 221, 246-248
    with Norman Vincent Peale, 47, 186
    with Roy Rogers, 289
    with Walter Cronkite, 271

Stout, Jonathan, 26, 289, 295
Strauss, Richard, 133
Stump, Adm. Felix, 4
Surveyor 3, 171, 174
Swigert, Jack, 190-191, 192–194, 197–198, 203–204, 207, 211-213, 215–218, 265
Symphonie Fantastique (Berlioz), 272

Tang, 271
Taruntius (moon crater), 83
Taurus-Littrow Valley, 274, 279, 280
Texas A&M, 4, 225
Thompson, Floyd, 59
Time magazine, 223
TLI (Trans-Lunar Injection), 73, 79, 104
"Trees" (poem), 142
the Trench, 195, 198, 204–205, 207, 210
Truman, Harry, 225
TV Guide, 88
2001: A Space Odyssey (movie), 133

United Bible Societies, 187
U.S. International Observatories of Satellites, 6

Valli, Frankie, 287
Van Tassel, Roy, 41–42

Voice of America radio network, 132, 134, 208, 222
Volkov, Vladislav, 250
von Braun, Wernher, 9, 18, 40, 61, 65, 94–95, 110, 275, 292-293
Von Ehrenfriend, Dutch, 59
Voting Rights Act, 42

Wagon Road (moon feature), 105
Wallace, George, 225, 289
Washington, George, 225
Washington Star, 29
WCC (World Council of Churches), 225
Webb, James E., 52, 53, 56, 58, 73–74, 116
Webster Presbyterian Church, 117–118, 122–123, 132, 144–145, 147
Welch, Louie, 39
Weld, G. F., 84
Wendt, Guenter, 10, 71, 101–102, 123, 230
Whatley, Seagel, 156
White, Bonnie Lynn, 38
White, Edward, III, 38, 61, 138
White, Edward, Sr, 25, 39, 50, 61, 290
White, Edward (Ed), II, 17, 19-22, 24-28, 31, 34–35, 42-43, 50 52, 53, 56, 60, 83, 111, 136, 161, 224, 225, photos of, 299
  background of, 17–18, 19, 25-26, 27
  Apollo 1 fire, 34–36
  death of, xvi, 34-36, 38–43, 50
  Gemini 4, 16, 17-22
  The Bob Hope Extra Special, 286–287
White, James B., 39, 290

White, Pat, 25–26, 38–40, 224, 289
Wilkes, Marshall, 51
Wilkes, Mildred, 51
Williams, Clifton Curtis (C.C.), 169, 170
Winborn, Conrad, 26, 37, 41–42

Woodruff, Dean, 115–118, 119, 123, 132–133, 143, 149
Woodruff, Floy, 119, 147
Worden, Alfred M. (Al), 141, 248, 252, 255, 256, 261, 279, 297
World Council of Churches (WCC), 225
World Publishing, 161–162
Wright Brothers, 230, 280
Wydler, John, 55

Yankee Clipper, 174–175
Yevtushenko, Yevgeny, 251
Young, Jonathan (John), 190, 266-267
  Apollo 10, 98, 102, 105
  Apollo 16, 265, 266-270
  Gemini 3, 15

zip gun device, 20–21, photo of, 299
Zond 5, 94